REVIEW OF FISHERIES

IN OECD MEMBER COUNTRIES

1986

ORGANISATION FOR ECONOMIC CO-OPERATION AND DEVELOPMENT

Pursuant to article 1 of the Convention signed in Paris on 14th December, 1960, and which came into force on 30th September, 1961, the Organisation for Economic Co-operation and Development (OECD) shall promote policies designed:

- to achieve the highest sustainable economic growth and employment and a rising standard of living in Member countries, while maintaining financial stability, and thus to contribute to the development of the world economy;
- to contribute to sound economic expansion in Member as well as non-member countries in the process of economic development; and
- to contribute to the expansion of world trade on a multilateral, non-discriminatory basis in accordance with international obligations.

The original Member countries of the OECD are Austria, Belgium, Canada, Denmark, France, the Federal Republic of Germany, Greece, Iceland, Ireland, Italy, Luxembourg, the Netherlands, Norway, Portugal, Spain, Sweden, Switzerland, Turkey, the United Kingdom and the United States. The following countries became Members subsequently through accession at the dates indicated hereafter: Japan (28th April, 1964), Finland (28th January, 1969), Australia (7th June, 1971) and New Zealand (29th May, 1973).

The Socialist Federal Republic of Yugoslavia takes part in some of the work of the OECD (agreement of 28th October, 1961).

Publié en français sous le titre :

EXAMEN DES PÊCHERIES
DANS LES PAYS MEMBRES DE L'OCDE
1986

© OECD, 1987
Application for permission to reproduce or translate
all or part of this publication should be made to:
Head of Publications Service, OECD
2, rue André-Pascal, 75775 PARIS CEDEX 16, France.

The Committee for Fisheries approved this Review for the twentieth consecutive year at its 60th Session, on 29th June to 1st July 1987.

The Council agreed to its derestriction in November 1987. The statistical data are sometimes preliminary.

Also available

PROBLEMS OF TRADE IN FISHERY PRODUCTS (December 1985)
(53 85 02 1) ISBN 92-64-12775-5   384 pages   £14.00   US$28.00   F140.00   DM62.00

EXPERIENCES IN THE MANAGEMENT OF NATIONAL FISHING ZONES (April 1984)
(53 84 01 1) ISBN 92-64-12566-3   160 pages   £7.20   US$14.50   F72.00   DM32.00

INTERNATIONAL TRADE IN FISH PRODUCTS. Effects of the 200-mile Limit (August 1982)
(53 82 01 1) ISBN 92-64-12318-0   192 pages   £8.80   US$17.50   F88.00   DM44.00

* * *

NATIONAL POLICIES AND AGRICULTURAL TRADE (July 1987)
(51 87 04 1) ISBN 92-64-12976-6   334 pages   £12.00   US$25.00   F120.00   DM52.00

AGRICULTURE IN CHINA. Prospects for Production and Trade (July 1985)
(51 85 04 1) ISBN 92-64-12741-0   84 pages   £6.50   US$13.00   F65.00   DM29.00

BIOMASS FOR ENERGY. Economic and Policy Issues (October 1984)
(51 84 06 1) ISBN 92-64-12632-5   136 pages   £7.50   US$15.00   F75.00   DM33.00

AGRICULTURAL TRADE WITH DEVELOPING COUNTRIES (June 1984)
(51 84 02 1) ISBN 92-64-12579-5   114 pages   £5.60   US$11.00   F56.00   DM25.00

ISSUES AND CHALLENGES FOR OECD AGRICULTURE IN THE 1980s (April 1984)
(51 84 01 1) ISBN 92-64-12572-8   160 pages   £13.00   US$26.00   F130.00   DM58.00

* * *

MULTILINGUAL DICTIONARY OF FISH AND FISH PRODUCTS (February 1984) published by Fishing News Books Limited, 1 Garden Walk, Farnham, Surrey, England to whom orders should be sent.

Prices charged at the OECD Bookshop.

THE OECD CATALOGUE OF PUBLICATIONS and supplements will be sent free of charge on request addressed either to OECD Publications Service, Sales and Distribution Division, 2, rue André-Pascal, 75775 PARIS CEDEX 16, or to the OECD Sales Agent in your country.

TABLE OF CONTENTS

General Survey .... Pages 7 - 39
Tables to the General Survey .... Pages 40 - 52

COUNTRY NOTES

	Page		Page
Australia	53	Japan	131
Belgium	63	Netherlands	138
Canada	68	New Zealand	142
Denmark	79	Norway	148
Finland	84	Portugal	159
France	89	Spain	172
Germany	103	Sweden	183
Greece	108	Turkey	189
Iceland	116	United Kingdom	192
Ireland	121	United States	200
Italy	124	Yugoslavia	219
		EEC	222

Statistical Annex to Country Notes .... Page 228

GENERAL SURVEY

I. EXECUTIVE SUMMARY

In 1986 the world catch set a new record. Preliminary figures suggest that world landings were in the 89.2 to 89.3 million ton range and thus well above the 1985 level of 85.2 million tons. This increase was mainly due to an increase in Peruvian and Chilean catches mostly used for non-human purposes. In the OECD area, however, more countries faced diminished catches than countries experiencing improvements. Total OECD landings amounted to just less than 30 million tons, marginally above that of the previous year. The decrease in total OECD landings was partly due to lower harvests of fish for industrial purposes. In the meantime, the largest relative gain was for Denmark where catches were up by 6 per cent; however while Danish landings of industrial fish increased, catches for human consumption were down. The largest relative drop in catches is noted for Belgium, Germany and Sweden; all three countries experienced around a 13 per cent decrease.

Overall landings in the North-east Atlantic, one of the most productive fishing areas in the OECD, were down by around 3.5 per cent to 6.6 million tons; landings for human consumption in this area were up almost 4 per cent to 3.4 million tons. In the North-west Atlantic only Canada recorded increased landings, while United States landings showed a decrease. Combined east coast landings of these two countries at 3 million tons were some 3.5 per cent below the previous year's level. In the South Pacific, Australia and New Zealand reported slightly lower catches; in the meantime their values were well up. Japanese landings at 12.7 million tons were 506 000 tons above the 1985 harvest. Aquaculture continues to play an increasingly important role; estimates suggest that annual production stemming from this source in the OECD is in the 1.7 to 2.0 million ton range. Most of this production takes place in Japan where production amounts to around 1.2 million tons per year.

Excluding meal and oil production, trade in fish and fish products was above the 1985 level, both in quantity and value. Prices for the products most traded internationally (groundfish and shrimps) were also well up while tuna faced mixed results depending on end use. As regards presentation, it seems that trade in fresh products experienced the largest increase. For trade in non-human products, meal fared well but oil experienced difficulties. This was due to the fact that major importers substituted fish oil with other oils and fats and, as a consequence, demand fell and prices fell considerably.

1986 was, however, characterised by a number of events whose effects may extend further. The steep depreciation of the United States dollar had a major impact on trade flows and prices. European exporters have, to a certain degree, left the United States market to the benefit of Canadian exporters experiencing price increases. Thus, Canadian export earnings were well up in 1986.

Although international cooperation in fisheries management and trade took place within various bodies during 1986, recourse to solving fisheries problems bilaterally seem to be on the increase; fish stock management is a multi-disciplinary issue which, in many cases, will have to draw on the expertise of international agencies. Also management decisions with regard to one stock of fish have repercussions on other stocks and species. Thus there is a need for international cooperation. At best, administrations use international agencies for gathering the background data, however, more often than not the data and advice are disregarded at the negotiation level. Consequently, many fish stocks remain exploited at a too high level.

As reported in the country chapters, consumption of fish and fish products is increasing in most Member countries. A change from previous years is, however, that also consumption of fresh products is on the increase. In the meantime some markets continue to depend largely on the imports of low cost products, e.g. canned tuna, surimi products, etc. while other markets seem to have switched towards relatively expensive products such as fresh fish. The latter has especially been the case for the United States.

II. MANAGEMENT

a) <u>International cooperation</u>

This section reviews major management measures taken by international management bodies, to which OECD Member countries are partners, and in which these cooperate on the management of highly migratory species and stocks straddling the EEZ boundaries.

The International Commission on South-east Atlantic Fisheries (ICSEAF) opened for 1987 a total quota of hake of 411 000 tons. This was a decrease by 70 000 tons or 15 per cent compared with 1986. The ICSEAF hake quotas for 1986 and 1987 are given in Table 1.

Table 1

ICSEAF HAKE QUOTAS, BY COUNTRY, 1986 and 1987

(tons)

	1986	1987
Spain	124 656	106 437
Portugal	13 458	11 499
Germany	(	7 243
France	(25 428	7 243
Italy	(	7 243
Japan	9 445	8 700
Other	308 104	262 635
Total	481 000	411 000

The TACs adopted by the North Atlantic Fisheries Organisation (NAFO) for 1986 concerning the North-west Atlantic are given in Table 2 (figures in brackets refer to 1985).

Table 2

NAFO QUOTA ALLOCATION, 1985 AND 1986

(tons)

	Cod		Redfish		American plaice		Squid	
Canada	16 055	(16 055)	11 150	(11 150)	54 350	(48 440)	(a)	(a)
Iceland	–	(–)	–	(–)	–	(–)	–	(–)
Japan	–	(–)	400	(400)	–	(–)	2 250	(2 250)
Norway	1 200	(1 200)	–	(–)	–	(–)	–	(–)
Spain		(11 340)	–	(–)	–	(–)	2 250	(2 250)
Portugal		(4 815)		(1 900)		(350)		(500)
EEC(c)	18 810b)	(2 655)	3 100b	(1 200)	960b)	(625)	(a)	(a)
Other	9 900	(9 900)	30 350	(30 350)	1 600	(1 585)	145 000	(145 000)
Total	45 965	(45 965)	45 000	(45 000)	57 000	(51 000)	150 000	(150 000)

a) The sum of this allocation may not exceed 135 750 tons.

b) Includes Spain and Portugal.

c) The Community, Spain and Portugal objected to the quotas allocated by NAFO for 1986.

An access agreement for United States tuna fishing vessels was concluded between the United States Government and the sixteen Member countries of the South Pacific Forum Fisheries Agency. In return for unlimited access to rich tuna fishing grounds in the South Pacific, the five-year agreement provides for the payment of US$12 million by the United States to the Agency. Of this amount, US$2 million is payable by the United States tuna fishing industry while the remaining US$10 million is to be paid via public funds.

The 1986 quota of 158 000 tons of yellowfin tuna was set by the Inter-American Tropical Tuna Commission (IATTC) with possible increments of two extra quotas of 13 610 tons was overfished. A total of 234 100 tons of yellowfin was actually taken in the Commission Yellowfin Regulatory Area (CYRA). In addition to these catches, 57 000 tons of skipjack and 4 900 tons of bluefin tuna were taken by the contracting parties, however, these two species remain unregulated.

b) *Supranational measures*

1986 TACs and quotas for Member States of the EEC were adopted in December 1985. However, they were amended at various occasions with due regard to further advice from scientists and negotiations with third countries. Further, at the end of 1986 it was decided to introduce certain amended technical measures for the conservation of fishery resources in the light of the latest scientific advice. The most important changes were increases in mesh sizes when fishing in the North Sea and west of Scotland and Rockall and fishing within the 12-mile coastal zone.

c) *National measures*

New Zealand passed in 1986 the necessary legislation for introducing individual transferable quotas (ITQs). The system was further extended to cover shallower water fin fisheries. It is the intention of the New Zealand government to manage all fisheries were appropriate through this system. In an aim to reduce pressure on stressed fish stocks the Government has introduced a "buy-back fishing right scheme".

One of the aims in conservation and managing measures of the Icelandic Government has been to create greater flexibility, avoid rigidity and make a greater leeway for vessel-owners. The law governing management is now valid for a two-year period while some of the most important and controversial regulations have become law. Further, the system which allows vessels over 10 grt to choose between a catch quota and an effort quota has been made more flexible in the sense that a certain amount of an unused quota can be transferred to the following year.

In Norway the stock situation for cod continued to improve in 1986. Research indicates that the stock is increasing even if the situation in 1986 still was vulnerable. Meanwhile, the stock situation for saithe north of 62°N as well as in the North Sea is poor. North of 62°N some fishing grounds even had to be temporarily closed due to a large share of undersized fish. The situation for the capelin stock has changed dramatically in the Barents Sea. Fishing was allowed only during the winter season; the fishery was completely closed during the summer. In 1987 there will be no capelin quota since the spawning stock biomass is too low. Also the shrimp stock in the Barents Sea declined during 1986 and it seems that the stock will remain at a low level for the next few years.

In the United States three new fishery management plans (FMPs) were implemented in 1986 while five FMPs were amended. The most important change was the complete replacement of the Northeast Multispecies FMP, establishing i.a. minimum sizes for cod, haddock, yellowtail flounder, pollock, American plaice, winter flounder and white flounder.

III. POLICIES

a) **Financial support**

It is convenient to classify financial assistance to the fishing industry in two broad categories, i.e. structural adjustment programmes and income support programmes. Naturally, this classification could be more refined since income support programmes also have structural consequences, for instance in contributing to maintaining in activity firms which are less efficient, while industrial adjustment programmes also have an effect on income. In addition to these two broad categories of subsidies, governments implement special programmes to face exceptional situations such as ecological accidents, or sudden closures of markets. Apart from these particular circumstances which have not affected Member countries in 1986, financial aid programmes do not change much from year to year.

i) **Programmes for structural adjustment**

In 1986, the main event in the area of financial aid was the new programme of the European Economic Community. It is a ten-year framework, ECU 800 million is set for the first five years (1987-1991). This amount covers the following measures:

Restructuring and renewal of the fishing fleet
Modernisation of the fishing fleet
Development of aquaculture and structural works
 in coastal waters
Exploratory fishing
Joint ventures
Adjustment of capacity
Port facilities
Search for new markets
Specific measures and operational costs

This new programme brings together three series of measures which expired at the end of 1986. In addition, the entry of Spain and Portugal into the Community justified the elaboration of a new programme. The principal objectives of this regulation are to increase the competitiveness of the European fishing industry through modernisation, thus decreasing the costs of production, and to reorientate the activity of certain parts of the fleet and to adapt the capacity of this fleet to the fish stocks available. However, the pursuit of these objectives must not lead to developing excess fishing capacities, this explains the size of the item "adjustment of capacity" where ECU 64 million is set to finance permanent withdrawals of vessels from the fleet.

In OECD Member countries, as far as subsidies are concerned, stability is the rule. It should be noted, however, that Australia has modified the Fishing Industry Research Trust Account in order to direct more funds towards development. Sweden has used, for structural purposes, a programme which was first aimed at supporting fishermen's income. To be eligible for the price supplement, fishermen will now have to sell their catches to buyers cooperating in a special association. This rule is comparable to that of the withdrawal price programme of the EEC. Finally, in Canada, a programme to help finance the construction of small inshore vessels was discontinued.

For the rest, as much as it can be estimated through available information, only the levels of funds allocated to the different programmes have changed and it is not always possible to tell whether these changes reflect trends or only conjunctural variations. In fact, in most Member countries even the amount of subsidies, in constant prices, has changed very little. This is the case in Finland, Italy, the United States and the United Kingdom. However, a slight tendency to decrease can be detected in Sweden and Denmark. Finally, some countries continue to allocate little or no subsidies to their fishing industry, this is the case for New Zealand and Iceland.

ii) Programmes for income support

Concerning income support programmes, nothing special marked 1986. Besides the EEC Member countries, through the Common policy, the main OECD Member countries which have income support programmes are Norway, Sweden (as already mentioned) and to a lesser extent, Canada. Finally, several governments participate in the financing of vessel insurances. On all these points, there are few major changes. In 1986, only the levels of funding have changed according to the needs, but these fluctuations do not reflect any significant transformation in the industry.

b)   Measures for improving economic efficiency

It is clear that the vast majority of the measures of financial assistance aim at improving economic efficiency. But, the profitability of the industry also depends on a good management of the resource. Hence, it is in this area that most governmental initiatives can be found. In 1986, New Zealand extended its system of individual quotas to inshore fisheries and Australia introduced it for some fisheries. Iceland has also slightly modified its system to make it more flexible and more efficient. Finally, some aspects of the management mechanisms of the United States and of the EEC have been analysed and debated publically, although for now no major changes have been implemented in the resource allocation mechanism.

c)   Other government action (fishery agreements)

Fishery agreements may be divided into the following two categories:

-- Agreements on joint managed stocks;

-- Agreements on stocks exclusively under management of one coastal State.

In general, the first mentioned type of agreement involves some kind of reciprocal arrangement; for OECD countries they involve some 4 to 4.5 million tons annually. For agreements on stocks exclusively under management of one coastal State, it has been estimated that they involve around 2.5 million tons of fish annually, however, this excludes cases where the agreement specifies maximum capacity rather than catches. In return for fishing possibilities various benefits are given in exchange.

In the North Sea most stocks are under joint management between the USSR and Norway and Norway and the EEC. The USSR-Norwegian fisheries agreement established and allocated the most important TACs (as shown in Table 3) for 1987. As opposed to previous years, both parties agreed that the capelin stock is under such pressure that no fishery will be allowed. Compared with 1986, TACs were increased for cod (+160 000 tons) and haddock (+150 000 tons) while the TAC for blue whiting was kept unchanged and that of capelin decreased by 120 000 tons.

Table 3

NORWAY-USSR ALLOCATION, 1987

(tons)

	Total TAC	Norway	USSR	Others
Cod	600 000	342 000	202 000	56 000
Haddock	250 000	92 500	132 500	25 000
Blue whiting	385 000	-	385 000	-
Capelin	0	0	0	0

The Norway/EEC agreement on the main joint stocks covers a total allowable catch of around 1.6 million tons, of which more than one-third is herring. For 1986, Norway and the EEC agreed on a TAC for North Sea herring of 570 000 tons. The TACs for North Sea cod and for plaice for 1986 were cut by 80 000 tons and 30 000 tons respectively, whereas the TACs for haddock and for saithe were increased by 23 000 tons and 40 000 tons respectively. The agreement further comprises arrangements on other joint stocks totalling around 540 000 tons as well as mutual fishing rights in each other's exclusive zones, adding up to about 30 000 tons for the Community and 40 000 tons for Norway. A summary of the most important TACs for shared stocks is given in Table 4.

Table 4

TACs AND QUOTAS FOR EEC-NORWAY SHARED STOCKS

(tons)

Species	TAC	Zonal attachment Norw. Zone	Zonal attachment EEC Zone	Norway's total quotas[a]	EEC's total quotas[a]
Cod	170 000	28 900	141 100	8 550	161 450
Haddock	230 000	52 900	177 100	30 900	199 100
Saithe	240 000	124 800	115 200	119 800	120 200
Whiting	135 000	13 500	121 500	13 500	121 500
Plaice	180 000	12 600	167 400	2 600	177 400
Mackerel[b]	55 000	40 000	15 000	39 200	15 000
Herring	570 000	200 000	370 000	200 000	370 000

a) Including allocations to third countries as well as transfers.

b) Includes a Swedish fishery for mackerel in ICES IIIa at a traditional level.

The Japan-USSR fisheries agreements belong to those agreements where huge amounts of fish are involved. Under the reciprocal fishing agreement, however, the 1986 fish quota allocated to Japan within the USSR zone was drastically reduced to 150 000 tons compared with the 1985 quota of 600 000 tons (the same decrease in tonnage was allocated to the USSR within the Japanese zone). For 1987, the quota for Japan increased to 300 000 tons while the quota for the USSR increased to 200 000 tons. Under the Japan-USSR Fisheries Cooperative Agreement, which provides for Japanese catches of Soviet origin salmon, the 1986 quota for Japan decreased to 24 500 tons from the 1985 quota of 37 600 tons and the 1987 quota remained unchanged from the 1986 quota.

Agreements on access to stocks exclusively under management jurisdiction of one coastal State has been accentuated with the introduction of the new regime of the sea. For some flag States, they continue to be an important means of obtaining necessary supplies; their positive adjustments to the new fisheries regime have not been successful. For coastal States with a developing fishery sector these agreements may provide a unique opportunity to acquire know-how and technology so that later on they will be able to exploit the resources in their EEZs. For developed coastal fishing nations these agreements are the only means of opening marketing channels.

In 1986, the United States allocated a total allowable level of foreign fishing of 826 000 tons compared with 1 476 000 tons in 1985. This is a considerable decrease from previous years, however the amount of fish traded over-the-side has increased greatly and has thus substituted foreign fishing with domestic. The most important countries taking part in these fisheries during 1986 were Japan with an allocation of 475 000 tons, the Republic of Korea with 116 200 tons and Poland which received an allocation of 78 000 tons. Alaska pollack, yellowfin sole and mackerel are the three most important species.

Canada allocated a total of 264 200 tons to foreign fishing ventures off the Atlantic coast. This corresponds to 16 per cent of the 1.68 million tons allocated to both domestic and foreign fishermen in 1986. Countries benefiting from these allocations included the EEC, Poland, the USSR and Japan; main species involved were cod, redfish, Greenland halibut and squid. On the west coast of Canada a small allocation of 23 764 tons was opened for Polish and Soviet vessels.

In the OECD area, Japan and Member States of the EEC, in particular, Spain, France and Germany are those countries benefiting most from various allocations within coastal States EEZ. Most of these agreements, however, are with developing contries and although the species involved are varied, tuna and shrimps are of particular importance.

Joint ventures have been another solution emerging from the introduction of the new regime of the sea to overcome supply difficulties for the flag State's fleet. Benefits for the coastal States are transfer of technology and use of the resource. Among OECD countries providing the resource, Australia and New Zealand are the most important while Japan and Spain are the most important countries providing fishery technology. Outside the OECD, Latin American countries, in particular, have provided resources.

There should be no doubt that agreements on joint stock management are a necessary part of future management in those areas of the sea where fishable stocks are restricted to more than one coastal State's EEZ. For such stocks, collaboration on their rational use will continue. Concerning arrangements on access to stocks under the management jurisdiction of one coastal State, the future may be more bleak. In general, coastal States' objectives are to seek to develop their domestic fishing capacity with a view to fully exploit the resources under their management jurisdiction. In the meantime, arrangements with developing countries will be more important in the intermediate stage of this process.

IV. PRODUCTION

a) <u>Fishing fleet</u>

Fishing fleet statistics do not always allow for easy comparison between countries. In fact, there are as many definitions of fishing vessels as there are Member countries. Most countries set a lower length or power limit over which the vessels are counted (i.e. less than 5 grt in Denmark). Some countries include vessels fishing in inland waters (Belgium, Netherlands). In addition, in the context of these limits, some countries count all the registered fishing vessels, while others (i.e. France, Portugal, Sweden) only count active vessels. However, in this case, the definition of activity varies from country to country. Furthermore, all classification by size of fishing vessels must be interpreted with care, since, firstly, the definition of the criteria may have changed through time (i.e. grt) and secondly, the methods to measure them may not always be the same.

The principal event in 1986 was probably the decision by the European Community to legislate on the fishing fleet (Council Reg. No. 2930/86). This regulation establishes, for the EEC, the definitions as well as the ways to measure the length, breadth, tonnage, engine power as well as the date of entry into service of fishing vessels. The Council decision is to be put in parallel with the new European Community Programme of aid to the fishing industry mentioned above. Indeed, the programme wil also create an administrative register for all the European fishing vessels. Although the precise characteristics of the register have not yet been established, it is clear that the industry is moving towards a greater harmonisation of fishing fleet statistics.

In fact, statistical information on fishing fleets is somewhat sketchy. Nevertheless, it seems that, in 1986, the total OECD fishing fleet remained relatively stable. But this general stability reflects mouvements in different directions depending upon the countries. Thus, while some countries have increased both the number and the tonnage of their vessels (Norway, Finland, Sweden), others have reduced them (Germany, New Zealand) and for others the size of the fleet has remained stable (Belgium). However, it is probable that the main changes, i.e. modernisation, adjustment of vessel capacities, took place inside the fleets. As a general rule, these adjustments are difficult to estimate through aggregated data, however the case of the United Kingdom is interesting in this respect, since from 1985 to 1986, the number of vessels has increased while the total tonnage has decreased.

V. CATCH

Total OECD landings are estimated to have increased marginally to a level around 30 million tons. In the North-east Atlantic, landings decreased by around 9 per cent mainly due to a heavy decrease in Norwegian landings. Total landings in the North-west Atlantic were down by around 3.5 per cent due to a decrease in United States landings.

In the North Pacific, Japanese landings were up slightly as were United States domestic landings. However, Canadian landings were down by around 6 per cent. In the South Pacific, landings decreased in Australia as well as in New Zealand.

Total landings by major producers of tuna within the OECD, i.e. France, Japan and the United States, were up slightly. The Japanese catch of skipjack tuna increased by 93 000 tons to 408 000 tons, while its catch of other tunas decreased by 57 000 tons to 334 000 tons.

Preliminary figures indicate that returns on landings are well above those of 1985 in most Member countries, even though catches have gone down. The largest increases are noted for Canada and the United States, while Germany experienced another rather heavy decrease in the total value of landings. Japan expects a decrease in the total value of landings despite increased landings by volume.

a) North-east Atlantic

Preliminary figures on total Norwegian landings show a considerable drop for the third consecutive year. At 1.82 million tons landings were down 11 per cent or 229 000 tons, the lowest figure since 1964. However, while landings of fish for direct human consumption showed an increase by 4.5 per cent landings of fish for other purposes, decreased by 22 per cent. This was due to the considerable drop in catches of capelin, almost 365 300 tons or 60 per cent. Of demersal species total landings of cod increased by almost 8 per cent to 263 000 tons. Landings of haddock increased by 33 000 tons to 57 800 tons while landings of saithe decreased by as much as 73 700 tons to 127 900 tons, mainly due to weak year-classes. Landings of herring for human consumption increased by 70 300 tons to 157 400 tons while landings of herring for other purposes increased only slightly or by 15 600 tons to 167 600 tons. Increased landings of blue whiting could only slightly counterbalance the heavy decrease in catches of capelin. Landings of shrimps, which in 1985 reached a record of 91 000 tons, decreased by 38 per cent or 34 800 tons.

Landings in Iceland did not reach the record level from 1985, 1.67 million tons. However, at 1.66 million tons the decrease amounted to only 1 per cent or 20 000 tons. Also Iceland experienced a drop in catches of capelin, but by not more than 98 000 tons or 10 per cent. The catch of cod, which from an economic point of view, is the far most important species to the Icelandic fishing industry, increased by 42 700 tons to 365 000 tons. Of other demersal species landings of saithe and catfish increased while landings of redfish and haddock decreased. However, the changes were relatively small. Of the pelagic species landings of herring increased by 33 per cent to 65 800 tons while landings of shrimps increased by as much as 43 per cent to 35 000 tons.

Landings by Community vessels showed a very slight decrease in 1986. However, Danish landings, totalling 1.84 million tons, increased by almost 6 per cent. Meanwhile, landings for human consumption decreased by 3.8 per cent while landings for other purposes increased by almost 10 per cent. Landings of cod, which is by far the most important species also in Danish fisheries, decreased by 11 per cent to 153 000 tons. Also landings of herring were down, by 22 000 tons, while landings of mackerel increased somewhat. Overall landings in the United Kingdom decreased by almost 7 per cent or 51 600 tons. Haddock landings increased again in 1986 and haddock replaced mackerel as the most important species in terms of quantity. While cod landings decreased by 15 000 tons landings of saithe increased by almost 4 000 tons. Landings of herring increased by as much as 34.5 per cent to almost 100 000 tons. In France, the volume of landings increased in 1986 compared with 1985. German landings at 204 800 tons were down almost 11 per cent - the value also decreased by 6.4 per cent to DM 328.3 million. This decrease was due to a fall of 29 000 tons in landings of frozen fish caught by the long distance fleet primarily in fishing grounds in the North-west Atlantic. Meanwhile, landings by Irish vessels increased by a fair amount; a total of 208 500 tons was landed for human consumption compared with 160 300 tons a year earlier.

Both Swedish and Finnish landings were down in 1986 reflecting i.a. poor harvesting and stock conditions in the Baltic and Kattegat/Skagerrak areas. In Sweden a total of 173 300 tons (1985: 198 200 tons) of fish valued at SKr 646.2 million (1985: SKr 656.4 million) were landed. Landings of the two most important species i.e. herring and cod, were both down; herring at 75 500 tons was down 13 per cent while cod landings at 43 400 tons was 15.4 per cent lower than in 1985. A total of 117 400 tons valued at Mk 237.1 million was harvested by Finnish vessels. Of this some 99 000 tons would have been herring, two-thirds of which was used as fodder.

Landings of the major groundfish species in the North-east Atlantic (cod, saithe, haddock and plaice) showed mixed results. Cod landings showed an increase in 1986 after several years of declination and totalled 953 000 tons, an increase of 2.1 per cent. Also landings of haddock were up (+14.7 per cent to 265 000 tons) while landings of saithe decreased again, in 1986 by 19.6 per cent to a total of 240 000 tons. Landings of plaice showed only a small change.

Overall landings of mackerel increased again since landings in Norway increased considerably. Also landings of herring increased notably especially in Norway where total landings of herring amounted to 325 000 tons, an increase by almost 86 000 tons. However, prices on herring for human consumption seemed to have continued the downward trend observed in 1985.

Overall landings of fish for reduction in Iceland, Norway and Denmark, the major countries engaged in this type of fishery, decreased again. While landings increased in Denmark they were down in Norway as well as in Iceland due to the heavy decrease in landings of capelin.

b) <u>North-west Atlantic</u>

Reaching 1 211 400 tons, total Canadian landings on the Atlantic coast were up 42 100 tons or 3.6 per cent in 1986. Landings of groundfish, which account for around two-thirds of total landings by volume, were up by around 6 500 tons although landings of cod decreased by 8 600 tons. Total landings of pelagic and other fin fish increased by 5 per cent or 13 200 tons. However, landings of herring and mackerel decreased by 6 and 11 per cent respectively. Landings of crabs decreased by 3 000 tons or 6 per cent to 42 400 tons while landings of lobster increased by 9 per cent to 35 300 tons.

Canadian landings not used for human consumption at 78 800 tons were almost 10 000 tons lower than in 1985, a decrease which was caused mainly by lower landings of herring. Herring continued, however, to be the main species harvested for purposes other than human consumption. Main utilisation includes bait, roe and reduction to meal and oil.

United States landings on the east coast, including the Gulf, amounted to 1.79 million tons, 7.7 per cent below the 1985 catch of 1.94 million tons. Of the 1986 catch, 1 085 000 tons or 60 per cent were menhaden for reduction, the remainder being a wide variety of species. It is not expected that the menhaden fishery will develop in the next few years, as the fishing effort is presently too high. East coast landings correspond to 65 per cent of total United States landings, and their value at US$1.63 billion to 59 per cent of total landed value.

Landings of the major groundfish species on the United States east coast amounted to 140 400 tons which compares with 162 250 tons in 1985. The groundfish species include butterfish, Atlantic cod, cusk, flounder, haddock, hake, ocean perch, pollock (saithe) and whiting. Landings of flounder (blackback, fluke and yellowtail) at 47 800 tons and US$105.4 million, were down 10 000 tons and US$3 million by quantity and value respectively. For the third consecutive year landings of cod declined; they reached 27 770 tons in 1986 some 10 000 tons less than in the previous year. Fishing intensity on the cod stocks is still too high for any immediate improvement to be expected in landings; as a consequence, various management measures have been instituted in 1986, i.a., minimum landing size and mesh size in the trawl fisheries.

c) <u>South Pacific</u>

Landings in New Zealand by domestic vessels amounted in 1985 (latest available statistics) to 160 000 tons, a decrease by 3.6 per cent compared with 1984. Landings of finfish, including tuna, decreased by 3.4 per cent while landings of shellfish decreased by 4.8 per cent. Orange roughy (deep sea perch) continued to be the most important finfish and landings were up by 26.5 per cent to 26 700 tons. Landings of rock lobsters, the most important species among shellfish, continued to be stable at around 5 000 tons. Regarding joint ventures, landed quantities decreased by about 7 per cent to 145 000 tons, mostly due to diminished catches of squid.

Also total Australian landings fell in the 1985/86 fishing season, by 2.2 per cent to 152 000 tons. However, landings of finfish increased by 3 500 tons to 84 700 tons while landings of shellfish decreased by 7 000 tons to 62 400 tons. Decreased landings of especially scallops but also of rock lobsters accounted for this development.

d) <u>North Pacific</u>

Total Japanese landings reached 12 677 000 tons. This compares with 12 171 000 tons in 1985 and 12 820 000 tons in 1984, when the overall Japanese harvest reached a record high. By quantity the most important species landed were as follows: sardines 4 215 000 tons (+9 per cent from 3 866 000 tons in 1985), mackerel 955 000 tons (+24 per cent from 773 000 tons in 1985), tuna 742 000 tons (+5 per cent from 706 000 tons in 1985) and Alaska pollack 1 400 000 tons (-9 per cent from 1 532 000 tons in 1985). These four species account for 58 per cent of all landings of fish.

In the Pacific, total United States domestic landings amounted to 877 300 tons in 1986 and thus was well above the 1985 level of 831 800 tons. Of this total, the salmon fisheries accounted for just over one-third at 298 800 tons (-9.4 per cent compared with 1985) with a value of US$494 million (+12.3 per cent) or 46 per cent of overall landed value. Other important species landed in the Pacific fisheries included Alaska pollack at 59 200 tons, Pacific cod at 47 200 tons and Pacific flounders at 28 900 tons.

United States joint venture catches in the North Pacific, more commonly known as over-the-side sales, mostly for processing at sea on foreign factory ships, increased to 1.25 million tons compared with sales of 950 000 tons in 1985. Of the total harvest (including foreign allocations) joint ventures accounted for 55 per cent in 1986 compared with 38 per cent in the previous year. The main reason for this increase has been the considerable decrease in total allowable levels of foreign fishing. Main species involved are Alaska pollack, Pacific cod and squid. Foreign countries involved in the processing of the United States caught fish are Japan, the USSR and South Korea.

Canadian catches in the Pacific were down by 6 per cent. Groundfish species were up 2 600 tons while landings of pelagic and other finfish were down by 14 000 tons or 10 per cent. Total landings of salmon, which increased dramatically in 1985, were down by around 4 per cent. However, landings of pink salmon decreased by 22 per cent.

e) Tuna

A total of 742 000 tons of tuna was landed in 1986 in Japan and was up 5 per cent on the 706 000 tons landed in the previous year. With 408 000 tons corresponding to 50 per cent of the total, skipjack tuna continued to be the most important species by volume; in 1985 the corresponding figures were 315 000 tons and 45 per cent. 136 000 tons of bigeye tuna was next compared with 149 000 tons in 1985. The remaining was distributed as follows: yellowfin tuna 113 000 tons (1985: 134 000 tons); albacore 46 000 tons (1985: 58 000 tons) and bluefin tuna 22 000 tons (1985: 22 000 tons).

United States fishermen harvested a total of 251 800 tons of tuna (1985: 234 200 tons) which not only included landings in the United States but also landings to United States processors abroad, as well as transhipments. Yellowfin tuna was the most important species with 132 800 tons followed by skipjack with 107 700 tons. The value of United States tuna fisheries reached US$217 million, only some US$5 million more than in the previous year. An increasing part of these landings take place in Puerto Rico where most of the United States canneries have established factories. In 1986, 84 per cent of the total tuna landings took place in Puerto Rico compared with 30 per cent five years earlier.

French landings of tuna totalled 98 700 tons and were thus marginally up on the 96 900 tons landed in 1985. No change in species composition took place; 55 100 tons of skipjack, 42 100 tons of yellowfin and 1 100 tons of albacore tuna were landed.

f) Aquaculture

Attracted by high profits aquaculture production continues to increase in Member countries. During 1986, it was reported that trials with sea bass, bream and well as the more usual species such as cod, were undertaken with a view to using them in aquaculture. These trials were successful. It remains, however, that by volume seaweed, oysters and mussels are the most important species while, by value, salmon, catfish and yellowtail are in the forefront. Table 5 provides an overview of production in the OECD.

Table 5 shows that by quantity, Japan, the United States and Spain are by far the largest producers. However, there is a wide difference between these countries as to the species they produce. In Japan the production consists of around 46 per cent seaweed and laver (nori), 31 per cent oyster and shellfish, the remaining 23 per cent is fish (the most important being yellowtail, eel and sea bream). The United States production consists of around 50 per cent catfish, 23 per cent trout and 13 per cent crawfish, the remainder being spread over a wide variety of species. Finally Spanish production consists of 93 per cent mussel culture.

Table 5

AQUACULTURE PRODUCTION IN SELECTED MEMBER COUNTRIES, 1984-1986

(tons)

	1984	1985	1986
Canada			7 000
Denmark	26 000	27 330	27 800
Finland	9 300	10 000	9 500
France	110 000	129 000	
Germany	24 000		
Greece	1 600	2 100	2 320
Ireland	13 600	13 640	
Japan	1 208 000	1 184 000	1 284 000
Norway	29 900	33 800	49 700
Spain	227 820	239 820	259 150
Sweden	2 100	2 520	3 100
United Kingdom	16 300	19 331	22 850
United States	213 700	242 000	

Aquaculture production in Norway, the leading producer of salmon among Member countries continues to increase at a very rapid rate as shown in Table 6. Salmon production increased by 16 000 tons or by more than 50 per cent. Meanwhile, production of rainbow trout decreased by 17 per cent or almost 900 tons. The number of delivering sea farms remained almost the same as during the preceding year while the total number of licences for salmon and trout farming increased by 82 to 689.

Table 6

DEVELOPMENTS IN NORWEGIAN AQUACULTURE, 1983-1986

	1983	1984	1985	1986a)
Salmon (tons)	17 298	21 881	29 473	45 494
Rainbow trout (tons)	5 405	3 569	5 141	4 248
Total (tons)	22 703	25 806	33 796	49 742
First-hand value (NKr million)	776	960	1 434	1 717
Number of delivering sea farms	362	410	487	500
Total licences for salmon and trout farming per 31/12	483	586	607	689

a) Preliminary figures

As seen in Table 5, Japan is by far the largest aquaculture producer in the OECD with a total production in the 1.2 million ton area. In 1985 and 1986, Japanese production was broken down as shown in Table 7. A total of

41 800 establishments were engaged in marine culture in 1985 a figure which has been constantly decreasing over a number of years.

The prospects for aquaculture production, volume wise, are very positive. Indeed, improvements in hatchery techniques and development of feed stuff sufficiently fine for the very small mouths at the early stages of production could turn many more fish species suitable for aquaculture. However, as reported by Norway as well as other countries with considerable aquaculture interests, e.g. the United Kingdom and Ireland, prices received at the farm have recently shown a downward trend. Marketing strategies will have to be further developed with a view to increasing consumer demand. A possible solution may be product development combined with further market segmentation.

Table 7

JAPANESE AQUACULTURE PRODUCTION, 1985 AND 1986

(tons)

	1985	1986
Marine culture	1 088 000	1 190 000
Yellowtail	151 000	146 000
Oyster	251 000	252 000
Nori laver	352 000	458 000
Wakame seaweed	112 000	135 000
Inland water culture	96 000	94 000
Eel	40 000	37 000
Trout	19 000	20 000
Common carp	19 000	20 000
Total	1 184 000	1 284 000

VI. PROCESSING AND MARKETING

a) *Utilisation and processing*

Statistics on utilisation of catch are given in Table 3 of the Statistical Annex to the General Survey. It should be noted that changes in utilisation only take place slowly as changes in consumption are fairly slow and gradual and as changes in processing methods per se are a costly affair. In Finland the quantities of landings devoted to fresh/chilled use fell five percentage points corresponding to around 6 800 tons, a significant amount as overall landings were about 127 000 tons. These quantities were used instead for animal fodder in fur farming and fish culture. Due to unfavourable landings of fish for reduction (down 281 200 tons) combined with a notable increase in landings for human consumption, Norway experienced changes in utilisation.

Fuelled by favourable prices in the major export markets (Germany and the United Kingdom), Iceland reports a marked increase in the part of the catch utilised fresh or chilled. This increased from 141 200 tons in 1985 to 175 100 tons in 1986, of which 62 000 tons was capelin exported fresh for reduction. Also the amount of fish devoted to salting reached 190 300 tons, up 8 per cent on 1985. A similar development is noted for Norway; here the catch utilised for fresh, chilled purposes increased more than one-third from 134 200 tons in 1985 to 180 100 tons in 1986 due, inter alia, to landings of herring for human consumption.

The United States reports increased domestic processing both on land and factory vessels. This development follows i.a. federally funded research and development projects which provide grants for harvesting, processing and marketing projects. In 1986, 15 factory trawlers and 3 factory motherships were operating in the Alaska fishery processing Pacific cod and pollock fillets as well as headed and gutted cod, pollock, rockfish and sablefish. In addition, the first United States surimi processing ship came into operation. It is forecasted that in 1987 11 new factory trawlers will enter into service as well as several new surimi processing vessels. Another new development was the introduction of a "display auction" on the east coast, the first of its kind in the United States.

Mainly due to low world market prices for fish oil, alternative ways of utilising fish oil other than the traditional outlets, e.g. margarine and fats industry, was tried. At least three new areas in which fish oil may be used have been explored with success. At present there is a small market for fish oil for medical purposes in helping to alleviate cardio-vascular disease; this appears to be a very promising market with potential for further increases. As fish meal/oil production is a high energy consumer some plants, in light of the development of crude oil prices, have been using fish oil as fuel. Finally, fish oil may be used more in the food business, as fish oil properties are regarded as being more healthy than other fats; thus a major outlet in food processing may develop.

In the meantime, development of food technologies may give rise to competition for raw material as there may be more than one single end-use of a given species. A case in point would be blue whiting in both surimi processing and the fish meal/oil industry. It appears evident that fishing costs will increase and that the end use of the resource (e.g. surimi - fish oil for the pharmaceutical industry, fish meal for fish farming) will be a question of the prices obtainable for the final products - and hence profits. What may evolve is the necessity to regulate these fisheries even more than has been the case until now.

b) <u>Marketing - Demand/Supply/Prices</u>

Prices for groundfish in all major markets were up considerably during 1986. This was caused by the continuing high demand and shortage of supplies of most groundfish species. While prices on the United States market for standard 16 1/2 lb. blocks of cod were quoted at US$1.25 to US$1.30 per lb. at the beginning of 1986, they steadily increased and were quoted at US$1.50 to US$1.60 per lb. at year end. Even with this increase of 20 to 25 per cent, demand continued to be strong. In Europe, also, prices for whitefish were up; in the United Kingdom average import prices for fresh, chilled cod stood at £830 per ton compared with £622 per ton in 1985 (+33 per cent). Industrial blocks of cod averaged £2 052 per ton up 23 per cent on the previous year. As a result of the depreciating dollar vis-à-vis the European currencies, exporters successfully directed more of their products to the European markets towards the end of the year, but without major impact on price developments.

Tuna prices were generally lower in 1986 compared with 1985. The main reason for the lower prices of tuna for canning was improved landings, particularly in Japan but probably also in the major tuna fishing or canning countries. Demand for canned tuna in Europe and in the United States was well up on 1985 fueled by lower prices.

The increased supplies of cultivated salmon together with the fall in the United States dollar exchange rate, meant that consignments to the European markets, particularly from Norway and Scotland, increased considerably. As a consequence, prices fell throughout the year. On the French wholesale market at Rungis, salmon prices were quoted at around FF 66 per kg. in the first weeks of 1986, while towards the end of the year prices were between FF 35 per kg. to FF 40 per kg., the top prices being paid during the Christmas period - a period of traditionally high demand. Other markets experienced similar developments with regard to salmon. 1986 was probably the first year when the rapid development in finfish aquaculture, i.e. salmon, showed pessimistic signs for the future. It appears that with the expected further increases in production of species other than salmon, and the fact that investors continue to be attracted by the high returns obtained from aquaculture, pressure will be brought to lower prices.

Estimates on future production of species from aquaculture have varied widely and with the experience already gained it does not seem wise to provide an outlook for the too distant future. In the meantime, however, 1986 showed that salmon prices were rather flexible, an experience which in all likelihood could be repeated with other species. In this regard it should be taken into account that, at least in Western Europe, any increase in food consumption is unlikely while a shift in demand is more likely to reflect changes in relative prices in addition to what may be fueled by "la mode", i.e. consumer awareness of nutritional value. Should these suppositions hold it seems likely that consumption of fish will increase to the detriment of meat consumption. The United States, one of the major markets already shows signs of moving in this direction.

Prices, retail and wholesale, were considerably up in the United States in 1986. The 1986 index of ex-vessel prices was recalculated and set at 100; this compares with 89 for 1985. Hence ex-vessel prices, as an average for all fish, would have been up 12.4 per cent on 1985. Two reasons may explain this development. Tight supplies from all major sources, especially towards the end of the year and a continued strong demand well supported by the health drive as reported in the 1985 Review of Fisheries. The increase in prices was apparent in all presentations.

Arrivals on the Tokyo wholesale market reached 865 000 tons in 1986, up 1.5 per cent from 852 000 tons during 1985. The average price of these arrivals for 1986 was Y 852 per kilo, slightly lower than that of 1985 (Y 863 per kilo). However, for the three major products, price decreases were far larger than the average. For frozen bigeye tuna arrivals increased by 9 800 tons to 53 200 tons and the price fell 10 per cent to Y 1 016 per kilo. Arrivals of frozen salmon, at 28 900 tons, show little change and arrivals of salted salmon at 22 300 tons were down by 4 per cent. Meanwhile, the average price decreased by 24 per cent in the case of frozen salmon and 15 per cent in the case of salted salmon.

Total 1986 year-end cold storage holdings in the United States were 2.4 per cent lower than that of the previous year. This was due to a decrease in holdings of fillets and steaks, in particular cod, which were down 58 per cent to 4 656 tons and a decrease of 24 per cent of round and dressed salmon to 19 145 tons. In the meantime, holdings of blocks increased by almost 40 per cent to 20 651 tons; the most significant increase was for minced blocks, up from 1 009 tons in 1985 to 5 722 tons in 1986. In Japan total cold storage holdings were up 5.5 per cent to just above 1.2 million tons. The biggest changes were noted for mackerel, up 54 per cent to 126 233 tons and octopus, up 55 per cent to 30 692 tons. Overall tuna holdings were down 4.5 per cent to 60 583 tons, mainly caused by a 36.6 per cent reduction in holdings of yellowfin tuna at 10 949 tons.

On the French Rungis wholesale market, total deliveries of fish and crustaceans increased by 5.3 per cent to reach 108 907 tons; just above one quarter of which was imported. This total was made up of 64 552 tons of seafish (+1 per cent); 8 202 tons of freshwater fish (+38.7 per cent) of which fresh salmon accounted for 3 224 tons (+88.5 per cent); and 36 335 tons of shellfish and molluscs (+7.7 per cent). Market surveys show that French consumers bought more filleted fish in 1986 than in 1985 and that whole fish was slightly down. Regarding species, purchases of whole plaice and herring (relatively cheap) and cod fillets were up while anglerfish and dab, both expensive species, were down.

Based on data for imports and landings, the average value and thus, presumably prices, were well up on the United Kingdom market. This was particularly the case for fresh chilled fish. For the three most important items, cod, plaice and haddock, average import values were up 33 per cent, 18 per cent and 39 per cent, while the average landed values for the same species were up 16 per cent, 11 per cent and 11 per cent. For frozen products, similar price developments occurred, e.g. the average import value of frozen fillets of cod with skin was up 14 per cent. In spite of these price increases, the United Kingdom reports that consumption of fish in 1986 was steady compared with 1985.

National landings of fresh fish on the German market declined somewhat in 1986 compared with 1985, but was almost fully compensated by direct landings by foreign vessels as well as fresh fish imports. As a consequence, the part of quantities marketed in the three seafood auctions, i.e. Bremerhaven, Cuxhaven and Hamburg and stemming from imports, increased to 51 per cent from 48 per cent in the previous year; these three auctions handling 77 000 tons of fish in 1986. In addition, landings of frozen fish by the German long distance fleet again experienced a decrease; landings were 48 800 tons compared with 74 100 tons a year earlier. Germany became increasingly dependent on supplies from foreign sources. As regards prices these do not seem to have reflected this situation to any great extent, which may be explained by the fact that prices for substitute products, e.g. chicken and pork, followed a downward trend during 1986.

VII. INTERNATIONAL TRADE

In 1985 OECD imports amounted to US$14 878 million, up 10.3 per cent from 1984. (Some caution should be taken with these data as they are expressed in current United States dollar). Of this amount US$7 937 million was imported from other OECD countries. OECD exports in 1985 reached US$8 294 million, 11.1 per cent up on 1984, and of which US$7 285 million was intra-OECD exports. Hence, the global trade deficit of OECD countries in fish and fish products which was US$6 024 million in 1984 increased in 1985 to US$6 629 million. These data thus confirm the trend of a worsening trade position in fish products of the OECD vis-a-vis non-OECD countries.

Three tables giving an overview of trade flows are given in the Statistical Annex to the General Part. Table 4 shows that in 1985 the three major importers, the EEC, the United States and Japan, covered 88 per cent of total OECD imports. As regards origin, intra-OECD trade amounts to an average 53 per cent of the total but, as will be noted, there is a considerable difference between product groups. While the OECD can supply itself up to 87 per cent of dried, salted, smoked fish; self-sufficiency in crustaceans and molluscs is only 29 per cent.

Since 1980, OECD imports of fish and fish products have increased from US$12 157 million to US$14 878 million, i.e. an increase of 22.4 per cent (see Table 8). However, while intra-OECD imports increased by 16.4 per cent to reach US$7 937 million in 1985, OECD imports from outside sources were up 30.2 per cent to US$6 937 million, especially for two product categories, i.e. crustaceans and molluscs, and canned and prepared fish. For the former group, imports from outside sources increased 41.2 per cent over the five-year period, while from inside the OECD imports were up by only 18.2 per cent. For canned and prepared fish, the OECD countries increased their imports by 33.2 per cent from outside sources, while imports from other OECD countries decreased by 1.7 per cent, reflecting increased competitiveness by LDCs. In relation to total imports, non-OECD countries supplied 46.6 per cent in 1985 compared with 43.8 per cent in 1980.

At least four events occurring in 1986 merit attention, as they all would have had an implication for trade among OECD Member countries. The United States dollar depreciation, in particular vis-a-vis European currencies and the Yen made the United States market less profitable, at least for its traditional exporters of groundfish and crustaceans. Canada was, however, an exception, as its currency appreciated relatively less than others with respect to the United States dollar and as opposed to its major competitors Canada would have benefited from this situation. Meanwhile, the United States imposed a countervailing duty of 5.82 per cent on fresh groundfish, excluding fillets, when imported from Canada. A third major event was a devaluation by 12 per cent of the Norwegian kronor in May 1986, which at least in the short run would have made Norwegian exports more competitive. The fourth major event was the accession of Spain and Portugal to the EEC.

United States imports of fish and fish products set a new record in 1986. Amounting to US$7.6 billion, imports were up by 13 per cent on 1985. Of this amount, US$2.8 billion represented imports of non-edible products. In quantity, imports reached 1.3 million tons of edible products, up by 102 000 tons from 1985. Also United States exports increased; at

US$1.4 billion, they were up by US$272 million compared with 1985. A partial reason for this 24 per cent increase in exports is attributed to the decline in the value of the United States dollar vis-a-vis its major trading partners.

Table 8

OECD IMPORTS & EXPORTS BY PRODUCT GROUPS
1981-1985

(US$ million)

	1981	1982	1983	1984	1985
Total OECD imports	12 629	12 945	13 306	13 483	14 878
Fish, fresh, chilled (excluding fillets)	1 896	1 764	1 643	1 737	1 633
Fish, frozen (excluding fillets)	1 853	2 043	1 772	1 889	2 275
Fish, salted, dried	756	717	714	663	783
Crustaceans, molluscs, fresh, frozen, etc. (excluding prepared or preserved	4 335	4 720	5 203	5 288	5 568
Fish/Crustaceans, prepared and preserved	2 180	2 056	2 234	2 238	2 459
Total OECD exports	7 988	7 630	7 600	7 459	8 294
Fish, fresh, chilled (excluding fillets)	882	843	919	952	1 145
Fish, salted, dried	922	700	584	505	537
Crustaceans, molluscs, fresh, frozen, etc. (excluding prepared or preserved	1 458	1 519	1 605	1 539	1 682
Fish/Crustaceans, prepared and preserved	1 742	1 547	1 607	1 592	1 757

1986 overall Japanese imports amounted to 1.87 million tons, an 18.5 per cent increase over the 1985 volume of 1.58 million tons. However, by value it decreased compared with Y 1 176 billion in 1985, corresponding to a decrease of 3.4 per cent. Of this amount, imports of fresh, chilled or frozen fish comprised about 30 per cent or Y 411.1 billion (1985: Y 415.1 billion), crustaceans and molluscs 46 per cent or Y 527.8 billion (1985:531.5 billion) while the third largest imported products were salted, dried or smoked products with 7 per cent and Y 80 billion (1985: Y 104.3 billion). By species and compared with 1985 increases are noted for shrimps, total imports of which amounted to 222 054 tons (+15.9 per cent) valued at Y 325.2 billion (marginally down from 1985: Y 335.6 billion). In the meantime exports showed a considerable decline. Valued at Y 217.5 billion in 1986, they were 24.4 per cent lower than the Y 287.6 recorded in 1985. Exports of canned fish fared poorly at Y 36.4 billion, they were almost 36 per cent lower than in 1985 (1985: Y 56 763 million).

Of the six major EEC markets, i.e. United Kingdom, Germany, France, Denmark, Italy and Spain, only the Italian imports were down in quantity on 1985, excluding fish meal and oil. French imports at 536 000 tons were up 10 per cent, Danish imports were up 8.7 per cent to 493 600 tons. Germany imports were 474 000 tons up 10.7 per cent and United Kingdom imports up

8.6 per cent to 418 000 tons, while imports into Italy were down 16.8 per cent to 395 000 tons. On the export side, the three principal EEC countries (the Netherlands, Denmark and the United Kingdom) were faced with mixed results. United Kingdom exports at £329 million were up 25 per cent (+10 per cent by volume). Dutch exports at Gld 1.87 billion were up by 4 per cent (+9.4 per cent in volume). However, while Danish export earnings were up 10.5 per cent at DKr 11.2 billion, the quantities exported were marginally down.

Although the quantity exported by Norway fell almost 100 000 tons to 730 300 tons in 1986, total value of these exports increased by 10.3 per cent to reach NKr 8.9 billion. This development was i.a. caused by a considerable fall in exports of meal and oil products whereas exports of fish for human consumption and especially of fresh, chilled products were well up. Canadian exports increased considerably; at 590 900 tons valued at C$2 422 million in 1986 they were up 6 per cent and 30 per cent respectively, by volume and value. By value the largest relative gain is noted for canned fish, up to just above 60 per cent. In the southern hemisphere Australian exports, which consist mainly of shellfish, were well up; valued at A$512.5 million exports was 22.1 per cent above the 1985 figures. New Zealand exports at NZ$ 657.3 million were 20.6 per cent above 1985 figures.

The overall level of trade, when measured in value, during 1986 was above that of the previous year. Quantity-wise the major product groups traded internationally were also well up. However, as shown in the following sections, product groups were faced with mixed results with regard to price developments. Most Member countries report both increased imports and exports. There were, however, a few exceptions but the decreases were few and in most cases marginal. For 1987 reduced supplies of groundfish will be of major importance as TACs of the most important groundfish stocks in EEC waters have been reduced. On the other hand most pelagic species should either remain at the 1986 level or eventually increase, as will be the case for herring. For shrimps, the cold water species are expected to be in short supply which should leave room for an increase in trade in warm water species.

As the following section will show in detail, trade in fresh, chilled and frozen products was well up in 1986. This was particularly the case for groundfish species the reason being an increased demand in major markets, i.e. the United States and Europe. As supplies of groundfish throughout the year were tight, prices moved upwards. As regards presentation, trade in fresh chilled products, fillets or whole has been especially successful.

Demand for shrimp was well up in 1986, leading to an increased trade. This was the case for all three markets, i.e. the United States, the Japanese and the EEC markets. Prices for shrimps also moved upwards sustained by an increase in demand. Most of the increased trade came from warm water species purchased mainly in India and Indonesia. But perhaps more important was the large increase in exports of aquaculture species from Taiwan, Ecuador and China. By quantity the level of trade in cold water species was lower than in 1985 due to a drop in landings.

a) *Fresh, chilled*

1985 saw a minor decrease, over 1984, in total OECD imports of fresh, chilled whole fish or fillets (see Table 9). Thus, over the last five years, it would seem that total OECD imports of these two items have stabilised at around US$2 100 million. In the meantime, the OECD exports of fresh fish and fillets increased by 21 per cent from 1984 to 1985 when it reached a record of US$1 323 million. As such, the OECD trade deficit in fresh chilled products has been narrowed considerably.

1986 United States imports of fresh fish amounted to 137 700 tons valued at US$355.6 million up 9.8 per cent and 37.1 per cent respectively. However, while imports of the lesser valued species such as whole cod, cusk, haddock, hake and saithe were down, imports of fresh salmon, flounders and roundfish were up. Again in 1986 the imports of fresh salmon set a new record with 12 900 tons valued at US$77.8 million, most of which was transported by air from Norway. The majority of other supplies of fresh fish originated in Canada. In fact Canada exported 103 000 tons of fresh seafish, whole or in fillets to the United States during 1986, only marginally above the 1985 level. However, the value of these fresh fish exports at C$278 million was 38 per cent above the 1985 level. There was a change in composition of these exports towards more filleted fish and less whole fish. Recent developments in the consumer drive towards apprasing the nutritional value of fish seem to have borne fruit.

Table 9

OECD IMPORTS & EXPORTS OF FRESH, CHILLED WHOLE FISH OR FILLETS, 1981-1985

(US$ million)

	1981	1982	1983	1984	1985
OECD total imports	2 157	2 034	1 950	2 083	2 049
United States	726	657	595	721	538
Japan	272	229	298	312	314
EEC	849	827	850	855	945
Germany	201	196	201	195	207
France	186	185	188	191	206
OECD total exports a)	1 000	962	1 052	1 093	1 323
Canada	68	82	95	124	179
Japan	11	5	11	14	18
Norway	79	84	119	134	168
EEC	740	677	711	687	771
Denmark	272	243	220	217	223

a) Does not include United States exports.

For the European market, fresh fish is most commonly transported by train or lorry. Denmark, Sweden and the Netherlands are the three major exporters of fresh products; their combined exports amounting to 293 700 tons in 1986 marginally down from the 294 100 tons exported in 1985. The principal importers of fresh fish are Germany, France, Italy and the United Kingdom, where combined imports in 1986 amounted to 424 300 tons around the same level as 1985. In the meantime, Norway has developed an important export of fresh/chilled products which by value exceeds any other country's exports of these items. In 1986, Norway exported salmon worth NKr 1.458 million.

However, and in addition to those quantities traded "over-the-side", direct landings have, in recent years, come to play an increasingly important role especially for the EEC countries. In 1986, however, they seem to have stabilised. Of the EEC Member States, Denmark, the United Kingdom and Germany are particularly active, receiving large quantities in this way, while Iceland and Sweden are important extra-Community sources (see Table 10). Cod, herring and mackerel are the major species traded in this way.

Table 10

DIRECT LANDINGS

(tons)

	Landings abroad		Foreign landings received	
	1985	1986	1985	1986
United Kingdom	46 500	35 700	52 800	51 900
Denmark	11 300	8 900	186 300	184 400
Germany	14 500	12 400	42 200	45 100
Iceland[a]	95 500	112 800	-	-
Sweden	27 800	27 300	100	100

a) If all fresh fish exports are counted, the figures are 151 400 tons for 1986 and 163 600 tons for 1985. The figures shown cover direct landings plus exports in containers.

Over-the-side sales (or joint ventures as they are called on the American continent) constitute an important means of disposing of fish not readily in demand on the domestic markets in the United States, the United Kingdom and Ireland. Moreover, over-the-side sales may induce development of domestic harvesting capacity and in the long term eventually domestic processing. United States over-the-side sales to foreign fleets, principally Japanese and Russian, have shown spectacular increases over the past few years. From 0 in 1978, 255 000 tons in 1982 to 911 000 tons in 1985 and 1.25 million tons in 1986, the main species involved being Alaska pollack. The United Kingdom over-the-side sales to Eastern European processing vessels amounted to 171 200 tons in 1986, 11.4 per cent less than in the previous year. Irish sales, also to Eastern European vessels amounted to 26 000 tons, the same as in the 1985 season. The United Kingdom and Irish sales consist mainly of mackerel and herring.

b) <u>Frozen fish</u>

Attaining US$3 651 million in 1985, OECD imports of frozen fish set a new record and was, compared with 1984, up by almost 17 per cent (see Table 11). The main reason was a 20 per cent increase in Japanese imports of almost exclusively whole frozen fish. On the export side, OECD exports increased by only 6 1/2 per cent; hence, in the case of frozen fish, the trade balance has moved to the detriment of OECD Member countries. It should, however, be noted that these figures exclude United States trade in whole frozen fish.

Table 11

OECD IMPORTS & EXPORTS OF FROZEN FISH INCLUDING FILLETS, 1981–1985

(US$ million)

		1981	1982	1983	1984	1985
IMPORTS						
OECD	total a)	3 118	3 330	3 118	3 125	3 651
	fillets	1 265	1 287	1 346	1 236	1 376
U.S.	total	–	–	–	–	–
	fillets	677	665	747	662	717
EEC	total	1 091	1 091	969	919	1 080
	fillets	424	445	428	393	482
Japan	total	958	1 152	1 093	1 215	1 456
	fillets	22	32	33	36	37
EXPORTS						
OECD	total b)	1 962	2 038	1 929	1 931	2 057
	fillets	972	1 034	1 043	1 039	1 108
Canada	total	549	604	524	517	512
	fillets	252	278	238	249	218
Japan	total	146	142	124	163	137
	fillets	50	57	45	47	54
Iceland	total	246	233	254	224	268
	fillets	230	214	240	206	248
Norway	total	160	211	227	211	216
	fillets	148	143	150	137	140
EEC	total	623	674	620	600	669
	fillets	268	307	324	327	360
Denmark	total	192	211	223	224	228
	fillets	128	148	162	173	183

a) Includes, for United States, fillet imports only.

b) Does not include United States exports.

1986 exports of frozen fish were well up on 1985. The combined exports of the four major exporters, i.e. Canada, Norway, Iceland and Denmark reached around 678 000 tons, compared with 620 000 tons in 1985. All four countries experienced increases (as shown in Table 12). The main reason for the general improvement in exports is found in the increasing whole fish exports which were 20 per cent higher in 1986 for the four countries while exports of fillets were only marginally above 1985 levels.

Table 13 depicts trade in frozen groundfish by major importers and exporters, frozen groundfish being the most important product group for OECD products. It should be noted that small quantities of other species may be included and the figures shown include fillets only, unless otherwise stated. The data provides evidence of a continued strong demand in the two major outlets, i.e. United States and United Kingdom, these combined increased 5.5 per cent to 480 500 tons in 1986. On the export side it seems that only Norway experienced a decrease, although only marginal.

Table 12

EXPORTS OF FROZEN FISH

(tons)

		1986	1985
Denmark	whole	25 877	23 206
	fillets	100 007	93 332
	total	125 884	116 538
Iceland	whole	31 032	23 505
	fillets	115 385	115 388
	total	146 417	138 887
Norway	whole	100 200	85 100
	fillets	85 600	86 400
	total	185 800	171 500
Canada	whole	84 250	65 300
	fillets	75 400	72 600
	blocks	60 300	55 600
	total	219 950	193 500

Table 13

TRADE IN FROZEN GROUNDFISH

('000 tons)

	1984	1985	1986
Exporters			
Canada	140.9	128.2	135.7
fillets	86.3	72.6	75.4
blocks	54.6	55.6	60.3
Iceland	107.4	115.4	115.4
Denmark	83.0	93.3	100.0
Norway	94.2	86.4	85.6
Total	425.5	423.3	436.7
Importers			
United Kingdom	79.0	87.0	99.0
United States	339.0	368.5	381.5
fillets	195.5	216.9	214.1
blocks	143.5	151.6	167.4
Total	418.0	455.5	480.5

On the whole, trade in frozen tuna was up during 1986, boosted by increasing demand for canning material in the United States where imports reached 199 500 tons or 17 per cent more than the 170 300 tons imported in 1985. The United States import bill for frozen tuna amounted to

US$226.6 million, up 11 per cent. In Japan imports decreased marginally by 2 per cent to 133 000 tons but exports at 53 000 tons were 83 per cent above the 1985 level. The reason for this sharp increase was a doubling of exports of skipjack tuna. On the European scene, Italian imports were marginally down at 103 000 tons valued at L 180 billion. French imports increased to 19 400 tons (+13 per cent) while exports increased considerably, from 61 140 tons in 1985 to 82 400 tons in 1986, the main species being skipjack, now more important that yellowfin tuna.

Throughout 1986 prices for the major frozen groundfish species and presentations increased. On the United States market Alaska pollack blocks of fillets went from US$0.60/lb at the beginning of 1986 to just below US$1.00/lb at the end of the year. Similarly standard 16 1/2 lb cod blocks which traded at US$1.25 to US$1.30 per lb at the beginning of 1986 reached US$1.50 to US$1.60 per lb at year end. Other groundfish followed closely. Similar price increases were experienced on the European market. As regards frozen tuna, prices showed different tendencies depending on species and hence end use. Skipjack tuna prices (mainly used as canning material) fluctuated widely throughout the year with prices turning upwards towards year end. Prices of tuna for sashimi, of which supplies to the Japanese market amounted to 322 010 tons in 1986 (+11.0 per cent), decreased on the other hand.

c) *Fish, dried, salted or in brine and smoked fish*

Table 14 provides data on OECD trade in dried, salted or smoked fish, most of which is composed of trade in stockfish and klipfish. One should note that exports by OECD Member countries have been decreasing since 1981. A plausible reason for this would be that the major outlets, i.e. developing countries have experienced fluctuating export earnings and have had less funds to pay for imported goods. Iceland, Norway, Canada, Denmark and the Netherlands are the OECD Member countries most active in this trade on the export side, while the main outlets are Spain, Italy and Portugal. Outside the OECD area, Nigeria and Brazil constitute the major importing countries.

Table 14

OECD TRADE IN FISH DRIED, SALTED OR IN BRINE AND SMOKED FISH, 1981-1985

(US$ million)

	1981	1982	1983	1984	1985
Exports	1 286.0	987.2	884.6	786.7	779.6
Imports	841.6	804.8	800.9	751.1	890.5

The two major cured groundfish producers, i.e. Norway and Iceland both improved exports in 1986. However the product mix changed somewhat. Norwegian exports of salted fish, most of which would have been cod, at 21 300 tons, was down 11 per cent; exports of stockfish more than doubled to 12 900 tons, whole klipfish exports were down at 50 000 tons compared with 58 200 tons exported in the previous year. The improvement in the exports of stockfish was due to the reopening of Nigerian imports. Klipfish shipments to Brazil increased to 20 300 tons (+34 per cent) while the market in Zaire, previously the second largest outlet, took only 1 700 tons compared with 8 200 tons in 1985. Also due to the reopening of the Nigerian market, Icelandic stockfish improved considerably from 924 tons in 1985 to 7 200 tons in 1986 and thus regained "normal" levels. Exports of wet salted fish were up 3.9 per cent to 45 440 tons as all the Mediterranean EEC countries bought more. Icelandic exports of salted fillets at 7 150 tons more than doubled mainly due to a doubling of exports to Germany and Italy. Finally it should be added that Canada and Denmark, although relatively small exporters when compared with Norway and Iceland, had combined exports of 54 800 tons in 1986 [DK 11 260 tons, Canada 43 505 tons] compared with 51 700 tons in 1985 [DK 11 906 tons of cod products, Canada 39 769 tons of seafish.]

Trade in pickled herring an important export item for Norway, Iceland, Denmark and the Netherlands fared poorly in 1986. Combined exports of these countries reached 60 900 tons, compared with 73 300 tons in 1985. Major outlets for these exports continue to be the USSR and Germany, the latter having imported 12 000 tons of salted herring and 17 000 tons of herring preparations during 1986. Meanwhile, although herring is in abundance and hence is faced with decreasing prices it may be difficult to revive consumption when taking into account the low prices for mackerel and tuna products.

Italian imports of salted, dried or smoked products decreased considerably. They went from 44 600 tons in 1985 to 31 273 tons in 1986. The reason was that Italy imported almost 10 000 tons less of salted, dried cod and saithe, i.e. 20 900 tons, which was valued at L 169 billion. Spanish imports of dried, salted or products in brine totalled 28 710 tons, 35 per cent more than in 1985, valued at Ptas 10 780 million (+55 per cent on the 1985 figure).

d) <u>Prepared and preserved</u>

1985 saw a leap in trade of prepared fish and crustaceans; over 1984 figures, imports were up just less than 10 per cent to US$2 459 million, exports were up just above 10 per cent to US$1 757 million (see Table 15). Exports have thus regained the 1981 level. The United States and the EEC remain the largest importers reflecting, i.a. increased imports of low-priced products, such as tuna from LDCs.

Also in 1986 the major importers of canned tuna continued to import large quantities. United States imports, accounting for 107 100 tons valued at US$227.9 million set another record and was up by 10.6 per cent and 9.2 per cent respectively by quantity and value. By far the major source was Thailand supplying 65 per cent of the overall United States imports. In Europe, the three major markets, i.e. the United Kingdom, France and Germany had a combined import total of 94 400 tons, almost 20 000 tons more than in the previous year. The United Kingdom surpassed France, once the leading importer; United Kingdom imports reached 38 000 tons (up 58 per cent). French imports at 35 400 tons were 1 500 tons lower than in 1985 while German imports at 21 000 tons was up 50 per cent. These significant increases give evidence of a continued market dominance of these products from LDCs.

Table 15

OECD IMPORTS & EXPORTS OF PREPARED AND PRESERVED FISH AND CRUSTACEANS, 1981-1985

(US$ million)

	1981	1982	1983	1984	1985
OECD total imports	2 180	2 056	2 234	2 238	2 459
United States	379	413	468	493	627
Japan	184	214	216	258	308
EEC	1 167	1 030	1 124	1 035	1 090
France	265	263	270	234	249
United Kingdom	384	283	362	331	334
Germany	164	158	159	152	162
Belgium	117	108	104	85	87
OECD total exports	1 742	1 547	1 607	1 592	1 757
Canada	251	200	245	221	236
United States	60	34	29	24	100
Japan	519	455	406	438	426
Norway	154	148	204	185	197
EEC	426	448	477	466	536
Denmark	144	158	166	163	182
Netherlands	70	74	80	76	86

Exports of canned salmon for the two principal producer countries, i.e. the United States and Canada, were well up on the previous year. Their combined exports at 51 900 tons (United States 26 966 and Canada 24 900 for 1986) were up 47 per cent on 1985 exports of 35 200 tons (United States 21 887 and Canada 13 305). Average value of exports was marginally lower in 1986 implying that canned salmon continues to be in competition with other canned products such as tuna. Among the major importers, i.e. the United Kingdom, Australia and Belgium, only the United Kingdom increased its imports of canned salmon to any significant degree; imports reached 27 000 tons, up 42 per cent on 1985. United Kingdom imports seem to have regained the level which prevailed before the 1982 withdrawal of salmon from the market due to contamination of some canned products.

For the second consecutive year, Japanese exports of canned fish decreased. At 126 000 tons and a value of Y 36.3 billion, exports were down 23.8 per cent in quantity and 36 per cent in value. All types of exports were down however, the largest decreases were for canned mackerel 38 600 tons (-34.7 per cent) and canned sardines 55 900 tons (-19.2 per cent). Average export prices for these two products were also down considerably -14.3 per cent and -10.8 per cent for mackerel and sardine respectively.

All major EEC importers of canned fish (the United Kingdom, France and Germany) increased their imports; United Kingdom imports of 105 000 tons valued at £220 million were up 22 per cent by quantity and 20 per cent by value. French imports at 74 500 tons valued at FF 1 412 million were up 2.7 per cent by quantity and 1.6 per cent by value and German imports of 62 000 tons at DM 325.8 million were up by 17 per cent and 4.2 per cent respectively. However, while the larger imports into the United Kingdom and

Germany reflected considerable increases in canned tuna (United Kingdom +58 per cent, Germany 50 per cent) and canned salmon (United Kingdom +42 per cent), French canned fish imports were fairly stable with a minor increase in imports of canned sardine. For the main exporters, 1986 was a good year.

Denmark saw its exports of prepared and preserved fish increase by 5 300 tons to 73 500 tons with earnings up 20 per cent to DKr 2 283 million. Norwegian exports were marginally up in quantity to 33 500 tons while the value increased 8.2 per cent to NKr 696 million. Dutch exports at 24 200 tons and Dfl 146.7 million were up 9.4 per cent and 4.4 per cent respectively.

Portuguese exports of canned fish reached were probably somewhat up on 1985. The major product exported is canned sardine accounting for just over 80 per cent of the quantities exported. Main outlets are found within the EEC where Portugal enjoys a tariff quota of 5 000 tons for 1987, pending the full abolition of duties by 1996. Spanish exports of canned fish reached 19 100 tons (-36 per cent), at a value of Ptas 8 118 million (-22 per cent). In the meantime, Italy experienced difficulties; its exports of 9 705 tons were 23 per cent lower than in the previous year. This was due to a considerable fall of 32 per cent in quantities exported of canned sardines.

e) <u>Crustaceans and molluscs</u>

Also in 1985, imports of crustaceans and molluscs increased (US$5 568 million, re. Table 16). Crustaceans and molluscs now represent 37 per cent of total OECD imports of fish products, most of which, however, are imported from outside the OECD area. It is not likely that the OECD trade balance could be improved via culture of shrimps as the culture of these species almost exclusively takes place in warmer climates. Moreover, shellfish has its major outlet in the OECD area and more especially in the high income countries.

On the U.S import bill crustaceans and molluscs represent the most important item. In 1986, the United States imported 314 000 tons valued at US$2.4 billion, up 7.6 per cent and 17 per cent respectively. Of the total value of imports, shrimp (in all forms) accounted for about 60 per cent, totalling US$1 433 million and was up by 25 per cent on 1985. Other important shellfish products were lobster tails (US$232.4 million - -10.7 per cent); whole lobster (US$134.2 million - +16.1 per cent) and scallops (US$192.6 million - +30.9 per cent). For analogue products, i.e. products based on surimi, United States imports reached US$58.5 million, up 21.4 per cent on 1985.

Of the total United States shrimp imports of US$1 433.5 million, US$1 080.1 million (or three-quarters) were imported fresh/frozen with shell. In volume this represented 118 900 tons in 1986. Ecuador was the major supplier with 27 500 tons valued at US$273.6 million, followed by Mexico with 27 100 tons and US$270.5 million. This very sharp increase in supplies from Ecuador is attributed to a steady increase in aquaculture production. Imports of the second most important shrimp presentation, i.e. fresh/frozen peeled shrimp also increased. It attained 41 700 tons with a value of US$221.3 million, up 18.4 per cent and 27.6 per cent respectively. The major suppliers of peeled shrimps were India, Taiwan and Mexico.

Table 16

OCED IMPORTS & EXPORTS OF CRUSTACEANS AND MOLLUSCS, 1981-1985

(US$ million)

	1981	1982	1983	1984	1985
OECD total imports	4 335	4 720	5 203	5 288	5 568
Japan	1 964	2 053	2 028	2 099	2 285
United States	1 244	1 479	1 883	1 910	1 917
Spain	169	189	177	189	164
Canada	120	131	155	167	160
EEC	723	744	809	782	889
France	292	292	302	269	283
Italy	135	159	167	172	247
OECD total exports	1 458	1 519	1 605	1 539	1 682
Canada	255	279	313	279	342
United States	216	198	166	134	134
Japan	62	62	108	87	88
Australia	271	361	310	285	291
Norway	20	21	21	27	33
EEC	383	371	417	413	452
Denmark	89	82	102	105	124
Netherlands	74	62	71	60	61

Japanese imports of crustaceans and molluscs amounted to 653 000 tons and were thus well above the 565 000 tons imported in 1985. Total value of the 1986 imports reached Y 580 billion, marginally up on the previous year (Y 578 billion). Imports of all the major products increased; 225 800 tons of shrimp (1985: 193 600 tons) valued at Y 332.4 billion (1985: Y 340 billion); cuttlefish and squid 141 900 tons (1985: 124 700 tons) valued at Y 79.9 billion and octopus 107 000 tons (1985: 98 600 tons) at Y 59.0 billion. Major changes in supplies of shrimps were experienced; Taiwan and China which exported 37 825 tons and 18 724 tons respectively, up 74 per cent for Taiwan and 75 per cent for China on 1985 exports.

Also imports of shellfish into Europe were up. In the United Kingdom, imports reached £200 million, 38 per cent above the 1985 level, of which shrimps and prawns accounted for £90 million. The major suppliers were Denmark, Norway and Iceland, the United Kingdom market traditionally preferring cold water species. However, the United Kingdom also increased its imports of the warm water species, particularly from India. 1986 French imports of crustaceans and molluscs amounted to 131 800 tons valued at FF 3.4 billion. Compared with the previous year (118 800 tons and FF 2.56 billion) they were up by 10.9 per cent in quantity and 32.8 per cent in value. Also in France shrimps are of major importance; 1986 imports reached a value of FF 1.4 million, up by FF 249 million or almost 22 per cent. By volume imports of shrimps amounted to 35 670 tons, 9.6 per cent more than in 1985.

Shellfish imports into Germany were up considerably by quantity, they increased 31 per cent to 51 000 tons while the value at DM 324 million was up by 10.4 per cent. This development was due to a heavy increase in imports of mussels whereas imports of higher valued products were only marginally up. The opposite tendency was the case for Belgium where import value increased by 20 per cent to BF 7 006 million for a quantity of 50 533 tons, up 3.4 per cent on 1985. For Belgium, the increase in import value is thought to stem from higher levels of imports of shrimps.

Danish exports of crustaceans and molluscs fared well in 1986. At 75 000 tons valued at DKr 1.58 billion the value was up 19.4 per cent while the quantity was marginally up by 1.9 per cent. The most important items continued to be deep water prawns accounting for three-quarters of the export value. However it should be noted that these exports are almost entirely of Greenland or Faroe Island origin. The main outlets for Danish exports were France, Sweden and Italy, the latter being a major outlet for Norway lobster. Due to a considerable decrease in landings of prawns, Norway saw its exports decline 35 per cent by quantity. The value was down also but by only 5.5 per cent as average prices, especially for shrimps, increased considerably. The Netherlands exported 76 800 tons of shellfish valued at Gld 395 million, down in quantity but up in value, due to lower exports of mussels and increased exports of shrimps.

Australian total exports of crustaceans and molluscs amounted to A$462.5 million (+19.4 per cent) and represented 90.2 per cent of the overall exports. Of this prawn and shrimp exports increased to A$207.1 million, up 39.3 per cent from the previous year. The two major outlets continued to be Japan and the United States to which the former market received products valued at $A262.3 million, up 50 per cent while the United States market took A$119.4 and decreased by one-fifth.

f) Fishmeal

Table 17 provides an overview of major OECD trade in fishmeal. In addition to the exports by OECD countries, the world's largest exporters, i.e. Chile and Peru, reached a combined export figure in 1986 of 1.47 million tons compared with 1.32 million tons in 1985.

Table 17

OECD TRADE IN FISHMEAL

('000 tons)

	1985	1986
Exporters		
Iceland	159.9	191.5
Norway	173.7	95.8
Denmark	217.8	224.7
Japan	157.4	167.2
United States	31.4	34.9
Importers		
Germany	365.0	415.0
United Kingdom	236.0	235.0
Finland	230.3	197.0
Sweden	112.7	108.6
Italy	109.2	85.9
United States	231.6	168.1

During the first three quarters of 1986 fishmeal prices continued the upward trend which began in late 1985. At the beginning of the year fishmeal (64 per cent protein content) traded at around US$292 per ton at Hamburg, and reached US$339 per ton in September. However, at year-end, prices levelled off and were quoted at US$310 per ton. The main reasons for this development was a strong demand which was supported by lower price margins vis-a-vis soyabean meal - the main competitor. In this regard it should be mentioned that speciality markets for fish meal have emerged, e.g. food compounds directed towards the fish farming industry. This market is expected to advance considerably in the future.

g) <u>Fish oil</u>

Table 18 depicts major OECD countries' trade in fish oils. Although the table provides evidence of a fall in both exports and imports, world-wide production in 1986 was up by an estimated 14 per cent. This increase was mainly due to improvements in harvest and production in Chile and Peru. Combined exports of countries shown, in 1986, were at 506 000 tons compared with 678 200 tons a year earlier, while imports, which in 1985 reached 748 400 tons only attained 509 900 tons in 1986.

Table 18

OECD TRADE IN FISH OIL

('000 tons)

	1985	1986
Exporters		
United States	126.2	85.7
Iceland	126.6	97.9
Japan	250.8	225.7
Denmark	61.1	62.4
Norway	114.3	35.4
Importers		
Germany	249.0	170.0
United Kingdom	265.0	173.0
Netherlands	234.4	166.9

1986 was a catastrophic year for fish oil producers and exporters. The year began with prices of US$330 per ton, by August 1986 prices were down to US$138 per ton after which prices showed signs of recovery; at year end 1986 prices were US$ 213 per ton but still 35 per cent lower than in January. The reason lies in i.a. an increase in production in non-OECD countries, but as noted above imports by the major importers fell considerably. Fish oil was substituted in these countries by other oils, notably palm oil, in their production of fats and margarine. In addition, as noted in the section on processing, the fish oil industry is searching for alternative use of production.

VIII. OUTLOOK

In many OECD Member countries the fishing industry seems to be entering a phase of maturity. Most governments have now implemented and tested the broad lines of the management schemes which correspond best to the needs of their fishing industry. Refinement and tightening of certain rules are still to be expected, but by and large, no major changes in the management mechanisms should be expected in the near future. Thus, as in the past years, government interventions may be aimed at rationalising and modernising the fishing fleet as well as the processing sector.

New patterns are evolving in the composition of the fishing fleets. As some countries continue to disinvest, i.e. the Netherlands, Germany, others are acquiring larger vessels. In Europe, Norway and Ireland are the main examples of this kind of development, but the same tendency can also be found in the United States and Canada as these two countries continue to take better advantage of the resources inside their EEZs. This tendency to use large vessels, equipped to process fish at sea, will require more and more capital, hence large companies may gain more and more importance in parts of the fishing industry.

The role of the large companies involved in the fishing industry may be further enhanced by the development of new products requiring more and more sophisticated technologies and ever higher investments in the processing sector. Consequently, the fishing industry of OECD Member countries may become more and more internationalised through the involvment of large companies. Joint ventures and mergers may play an important role in the structuring of the industry. The same type of pattern may also evolve in aquaculture and some firms will probably be involved in these two areas of fish production.

More and more attention will continue to be devoted to markets. The governments are likely to be involved in trade negociations in GATT for the next few years, while companies, some quite large in that sector also, will be monitoring the markets and adjusting to this more and more closely. Some major questions are important for the next few years. They concern mainly the reaction of the consumers to the new products such as surimi, the demand for what were high-valued species, such as salmon, which are now produced through aquaculture. Finally, but for this question no answer is yet forthcoming, the industry will have to reflect upon the eventual interaction between the new processing technology and aquaculture.

All these changes will proceed gradually, leaps and bounds are unlikely. Hence, from one year to the next, one of the important factors for the development of the fishing industry will remain the general state of the economy which determines the demand for fish and fish products. Barring any recession, it should continue to grow as many new segments of consumers are discovering the virtues of fish. But also, because new quality products are offered in increasing numbers.

On the production side, it does not appear that overall changes are to be expected. Groundfish will be in tight supply on both sides of the Atlantic. This will be particularly the case for cod in the North-west Atlantic; therefore, prices are likely to remain firm. For its part, Iceland expects a shortage of capelin. Finally, the dependency of OECD Member countries on third world sources of molluscs and crustaceans is not expected to ease.

TABLES TO GENERAL SURVEY

NOTES

Exchange rates

Table 2 has been composed on the basis of the country chapters which follow and summarises production for 1985 and 1986. The conversion rates are those used in the OECD publication "Monthly Statistics of Foreign Trade" and give equivalent in US$ per national currency (ref. column 1).

Tables 4 to 6 have been extracted directly from the OECD data base on foreign trade, and refer to 1985 only. The conversion rates used in these tables are listed in columns 2 (imports) and 3 (exports) below, and also give equivalent in US$ per national currency.

Country	Monetary unit	1	2	3
Australia	Dollar	0.668	0.699	0.700
Belgium[a]	Franc	22.385	16.937	16.946
Canada	Dollar	0.720	0.732	0.732
Denmark	Krone	0.124	0.095	0.095
Finland	Markka	0.197	0.162	0.162
France	Franc	0.144	1.113	1.112
Germany	D.M.	0.461	0.342	0.344
Greece[a]	Drachma	7.144	7.194	7.210
Iceland[a]	Kronur	24.328	24.045	24.110
Ireland	Pound	1.346	1.066	1.067
Italy[a]	Lira	0.671	0.523	0.525
Japan[a]	Yen	5.934	4.165	4.187
Netherlands	Gulden	0.408	0.303	0.303
New Zealand	Dollar	0.497	0.497	0.494
Norway	Krone	0.135	0.117	0.117
Portugal[a]	Escudo	6.685	5.874	0.588
Spain[a]	Peseta	7.140	5.899	5.911
Sweden	Krone	0.140	0.117	0.117
Turkey[a]	Lira	1.483	1.905	1.905
United Kingdom	Pound	1.466	1.289	1.294
Yugoslavia[a]	Dinar	2.637	5.385	5.385

a) Per '000 national currency.

Table 1(a)/Tableau 1(a)

OECD FISHING FLEETS/FLOTTES DE PECHE DE L'OCDE

1976 & 1986

	0-49.9 GRT/TJB				50-99.9 GRT/TJB				100-149.9 GRT/TJB				150-499.9 GRT/TJB			
	1976		1986		1976		1986		1976		1986		1976		1986	
	No.	GRT/TJB	No.	GRT/TJB	No.	GRT/TJB	No.	GRT/TJB	No.	GRT/TJB	No.	GRT/TJB	No.	GRT/TJB	No.	GRT/TJB
Canada	35 310	na	34 135a)	na	41	na	666a)	na	114	na	145a)	na	273b)	na	294ab)	na
Finland/Finlande	10 000	13 000	110		57	3 950	70	na	5	642	15	na	2	674	-	-
Iceland/Islande	450	7 670	399a)	6 618a)	151	10 702	102a)	6 864a)	81	10 147	86a)	10 607a)	169	46 984	171a)	27 388a)
Norway/Norvège(d)	27 614	151 176	8 694	85 311	278	21 881	212	16 660	96	12 281	67	8 680	449	127 454	382	108 323
EEC/CEE:																
Belgium/Belgique	74	2 681	50	1 839	91	7 322	56	4 305	54	6 658	42	5 230	33	6 828	49	11 472
Denmark/Danemark	6 862	70 293	2 725	50 608e)	233	16 806	203	14 403	184	25 195	136	18 625	162	34 709	168	38 793
Germany/Allemagne	1 056	12 998	756	11 871	104	7 467	76	5 367	54	6 202	49	5 903	14	3 258	15	3 315
Italy/Italie	20 310	128 716	18 072f)	129 780f)	568	40 295	632f)	45 706f)	187	22 474	297fg)	42 710fg)	108	25 829	73fh)	44 069fh)
Portugal	4 144	39 474	6 410	44 684	185	12 215	248	16 754	56	4 146	2721)	133 313	102	29 643	na	na
Spain/Espagne(a)	13 784	109 240	14 906	116 204	1 138	81 485	1 114	80 419	605	74 294	563	69 967	1441	359 213	1 029	263 062
United Kingdom/Royaume-Uni	5 935	63 386	6 987	72 546	389	25 107	473	30 681	61	7 494	90	11 318	221	63 084	126	30 874

Table 1(a)/Tableau 1(a)

OECD FISHING FLEETS/FLOTTES DE PECHE DE L'OCDE

(Cont'd/Suite)

1976 & 1986

	500-999.9 GRT/TJB				+1 000 GRT/TJB				TOTAL (Vessels with engines) (navires à moteur)			
	1976		1986		1976		1986		1976		1986	
	No.	GRT/TJB	No.	GRT/TJB	No.	GRT/TJB	No.	GRT/TJB	No.	GRT/TJB	No.	GRT/TJB
Canada	1	582	-	-	-	-	-	-	36 112	na	na	na
Finland/Finlande	24	19 607	29a)	22 973a)	-	-	-	-	10 065	18 848	35 240a)	na
Iceland/Islande	67	47 895	85	61 674	5	6 846	31	44 306	875	95 110	450c)	110 588a)
Norway/Norvège(d)									28 509	367 533	9 469	324 954
EEC/CEE:											826a)	
Belgium/Belgique	1	555	-	7 072	-	-	-	-	253	24 044	197	22 846
Denmark/Danemark	3	2 146	10		-	-	2	2 767	7 444	149 149	3 244	132 268
Germany/Allemagne	34	30 245	9	7 805	32	81 079	5	15 903	1 294	141 250	910	50 164
Italy/Italie	33	22 471	na	na	21	29 714j)	na	na	5 168	194 969	19 074f)	262 265f)
Portugal	14	10 058			60	85 402					6 930	194 571
Spain/Espagne	86	56 063	59	39 509	98	137 229	78	109 727	17 152	817 524	17 749	678 888
United Kingdom/Royaume-Uni	61	43 212	5	3 263	36	45 038	1	1 478	7 239	117 360	7 703k)	150 160

NOTES TO TABLE 1(a)/NOTES CONCERNANT LE TABLEAU 1(a)

a) Figures refer to 1985/Les chiffres se réfèrent à 1985.
b) 150 grt and over/150 tjb et plus.
c) Excluding vessels of unknown tonnage/Non compris les navires de tonnage inconnu.
d) Decked boats/Navires pontés.
e) Between 5 grt and 49.9 grt/Entre 5 tjb et 49.9 tjb.
f) Figures refer to 1984/Les chiffres se réfèrent à 1984.
g) Between 100 grt and 199.9 grt/Entre 100 tjb et 199.9 tjb.
h) 200 grt and over/200 tjb et plus.
j) 900 grt and over/900 tjb et plus.
k) Including 21 vessels of unknown tonnage/Y compris 21 navires de tonnage inconnu.
l) 100 grt and over/100 tjb et plus.

Table 1(b)/Tableau 1(b)

OECD FISHING FLEETS/FLOTTES DE PECHE DE L'OCDE

1985 & 1986

	0-49.9 GRT/TJB				50-99.9 GRT/TJB				100-149.9 GRT/TJB				150-499.9 GRT/TJB			
	1985		1986		1985		1986		1985		1986		1985		1986	
	No.	GRT/TJB	No.	GRT/TJB	No.	GRT/TJB	No.	GRT/TJB	No.	GRT/TJB	No.	GRT/TJB	No.	GRT/TJB	No.	GRT/TJB
Canada	34135				666				145				294a)			
Finland/Finlande																
Iceland/Islande																
Japan/Japon																
Norway/Norvège(b)	8595	92574	8694	85311	215	12053	212	16660	66	8564	67	8680	391	110829	382	108323
Sweden/Suède(c)	297	5129	329	5300d)	176	13200	180	13500d)	33	4125	35	4400d)	41	7475e)	45	9000de)
EEC/CEE:																
Belgium/Belgique	65	2026	50	1839	59	3876	56	4305	42	5230	42	5230	50	11409	49	11472
Denmark/Danemark	2777	51524f)	2725	50608f)	204	14622	203	14403	133	18205	136	18625	163	37405	168	38793
France	19253	-	18489	-	381	-	392	-	102	-	113	-	170	-	151	-
Germany/Allemagne	801	12130	756	11871	75	5360	76	5367	46	5468	49	5903	17	3519	15	3315
Ireland/Irlande	1347	13965			144	9936			44	5072			35	7856		
Italy/Italie	18072g)	109780g)			632g)	45706g)			297gh)	42710gh)			73gj)	44069gj)		
Portugal	6 496	45 687	6 410	44 684	248	15 680	248	16 574	2751)	134 581	2721)	133 313				
Spain/Espagne	14906	116204			1114	80419			563	69967	-	-	1029	263062		
United Kingdom/Royaume-Uni	6979	77721	6987	72546	375	26106	473	30681	92	11290	90	11318	136	34470	126	30874

44

Table 1(b)/Tableau 1(b)

OECD FISHING FLEETS/FLOTTES DE PECHE DE L'OCDE

(Cont'd/Suite)

1985 & 1986

	500-999.9 GRT/TJB				+1 000 GRT/TJB				TOTAL (Vessels with engines) (navires à moteur)			
	1985		1986		1985		1986		1985		1986	
	No.	GRT/TJB	No.	GRT/TJB	No.	GRT/TJB	No.	GRT/TJB	No.	GRT/TJB	No.	GRT/TJB
Canada									35340			
Finland/Finlande												
Iceland/Islande												
Japan/Japon												
Norway/Norvège(b)	79	57776	85	61674	21	27300	31	44036	9367	314096	9469	324954
Sweden/Suède(c)									547	29929	589	32200d)
EEC/CEE:												
Belgium/Belgique	1	555	0		0		0		197	23096	197	22846
Denmark/Danemark	6	4519	10	7072	2	2767	2	2767	3285	129042	3244	132268
France	66	-	62	-	20	-	2	-	19 992	-	19227	-
Germany/Allemagne	7	6670	9	7805	6	18561	5	15903	952	51708	910	50164
Ireland/Irlande	7	3935							1577	40764		
Italy/Italie									19074g)	262265g)		
Portugal									7 019	195 948	6 930	194 571
Spain/Espagne	59	39509	0		78	109727	0		17749	678888		
United Kingdom/ Royaume-Uni	8	4724	5	3263	5	6381	1	1478	7595	160692	7703k)	150160

45

NOTES TO TABLE 1(b)/NOTES CONCERNANT LE TABLEAU 1(b)

a) 150 grt and over/150 tjb et plus.
b) Decked boats/Navires pontés.
c) Vessels used in fishing for at least 25 weeks per year if the value of fish sold from each vessel amounts to at least SKr 100 000/Navires untilisés pour pêcher pendant au moins 25 semaines par an si la valeur du poisson vendu par ce navire atteint au moins KrS 100 000.
d) Est.
e) 150 grt and over/150 tjb et plus.
f) Between 5 grt and 49.9 grt/Entre 5 tjb et 49.9 tjb.
g) Figures refer to 1984/Les chiffres se rapportent à 1984.
h) Between 100 grt and 199.9 grt/Entre 100 tjb et 199.9 tjb.
j) 200 grt and over/200 tjb et plus.
k) Includes 21 vessels of unknown tonnage/Y compris 21 navires de tonnage inconnu.
l) 100 grt and over/100 tjb et plus.

Table 1(c)/Tableau 1(c)

FISHERMEN/PECHEURS
1985 & 1986

	TOTAL		Full time/A plein temps		Part-time/Temps partiel		
	1985	1986	1985	1986	1985	1986	
Australia	na	na	na	na	na	na	Australie
Canada a)	83 791	85 049					Canada a)
Finland	6 994	6 950	2 101	2 100	4 893	4 850	Finlande
Iceland	6 420	na					Islande
Japan	na	na					Japon
New Zealand							Nouvelle-Zélande
Norway	29 559	29 981	22 460b)	7 099b)	22 619c)	7 362c)	Norvège
Sweden	4 678		3 349d)		1 329e)		Suède
United States	238 800						Etats-Unis
EEC:							CEE:
Belgium	1 266	1 258	1 266	1 258			Belgium
Denmark	18 815	9 000f)					Danemark
France	18 815	18 165					France
Germany	3 244	3 089	2 561	2 424	683	665	Allemagne
Greece	27 700	39 000		27 000		12 000	Grèce
Ireland	7 019		3 108		3 911		Irlande
Italy a)	54 879	na	-	na	-	na	Italie a)
Netherlands							Pays-Bas
Portugal	37 743	40 058					Portugal
Spain	101 158	na	101 158	na	-	na	Espagne
United Kingdom	21 948	22 181	16 151	15 918	5 797	6 263	Royaume-Uni

a) Figures refer to 1984 and 1985./Les chiffres se réfèrent à 1984 et 1985.

b) Sole or main occupation/Occupation unique ou principale.

c) Subsidiary occupation/Occupation secondaire.

d) Sole occupation/Occupation unique.

e) Main or subsidiary occupation/Occupation principale ou secondaire.

f) Est.

Table 2/Tableau 2

FISH PRODUCTION IN OECD MEMBER COUNTRIES/PRODUCTION DE POISSON DANS LES PAYS MEMBRES DE L'OCDE(a)(b)
1985 & 1986

('000 tons/tonnes)

	Weight type/ Type de poids	1985				1986				Index/Indice 1985=100				
		Food/ Alimentation	Industrial/ Industriel	Total quant.	Total val. Total $ mill.	$/ton	Food/ Alimentation	Industrial Industriel	Total quant.	Total val. Total $ mill.	$/ton	Food/ Alimentation	Industrial Industriel	Total quant.	Total val. $ mill.	$/ton
Australia(c)	***	150.6	4.6	155.2	348.4	2 246	147.1	4.8	151.9	395.9	2 606	98	104	98	114	116
Belgium(d)	***	44.6	-	44.6	70.3	1 576	39.0	-	39.0	74.6	1 913	87	-	87	106	121
Canada	***			1 347.2					1 368.7							
Denmark(d)	***	483.3	1 244.0	1 727.3	425.7	246	465.1	1 364.1	1 829.2	429.3	234	96	110	106	101	95
Finland(d)	***	59.8	58.2	118.0	48.4	410	53.6	63.8	117.4	46.8	398	90	110	99	97	97
France	**															
Germany	**	178.3	8.5	186.8	126.3	676	155.6	7.8	163.4	111.1	680	90	92	87	88	101
Greece	*	130.4	0.1	130.5	182.7	1 400	138.0	0.1	138.1	194.1	1 405	106	100	106	106	100
Iceland	*	686.4	993.9	1 672.8	-	-	810.9	839.7	1 650.6	-	-	118	84	99	-	-
Ireland(d)	**	167.3	25.5	192.8	38.5	200										
Italy(d)	**	395.6	-	395.6	1 123.0	2 839	413.6	-	413.6	1 300.0	3 142	105	-	105	116	111
Japan	***			12 197.0												
N'lands(d)	***															
N.Z'land(e)	***	166.0	-	166.0	-	-	161.0	-	161.0	-	-	97	-	97	-	-
Norway	***	868.7	1 202.6	2 071.3	604.8	292	907.5	934.7	1 842.2	673.6	366	104	78	89	111	125
Portugal	**	375.7	13.2	388.8	213.5	555	385.1	11.5	396.6	317.3	800	103	87	102	149	144
Spain	**	1 459.0	325.0	1 784.0			1 417.7	325.0	1 742.7			97	100	98		
Sweden(d)	**	156.7	41.4	198.1	91.7	465	136.6	36.7	173.2	90.3	523	87	89	87	98	112
U.K.(d)	**	673.4	88.4	762.1	474.8	623	652.4	58.1	710.5	519.9	732	97	66	93	109	117
U.S.	***	1 517.5	1 321.0	2 839.2	2 326.2	820	1 539.0	1 197.0	2 736.2	2 762.8	1 010	101	91	96	119	123

a) Fish production, i.e. national landings in domestic ports, unless otherwise stated/Production de poisson, c.a.d. débarquements nationaux dans les ports nationaux à moins qu'elle soit autrement indiquée.

b) Includes fish, crustaceans, molluscs, meal, etc./ Y compris poisson, crustacés, mollusques et farine, etc.

c) 1985 = financial year 1984-85/1985 = année fiscale 1984-85.
   1986 = financial year 1985-86/1986 = année fiscale 1985-86.

d) Est. 1986.

e) Figures for 1985 and 1986 refer to 1984 and 1985, respectively/Les chiffres for 1985 et 1986 se réfèrent à 1984 et 1985 respectivement.

(*) Live weight/Poids vif. (**) Landed weight/Poids débarqué. (***) Not specified/Non spécifié.

Table 3/Tableau 3

UTILISATION OF CATCH/UTILISATION DES PRISES
1985 & 1986

	Year/Année	Total	Non-human consump. non hum.	Human consump. hum.	Fresh, chilled/ Frais, sur glace	Frozen/ congelés	Cured/ Salés, fumés, séchés	Canned, prepared/ En conserve, préparés	Human consump. hum.	Fresh, chilled/ Frais, sur glace	Frozen/ Congelés	Cured/ Salés, fumés, séchés	Canned/ En conserve	
					'000 tons/tonnes						%			
Australia	1985	155.2	4.6	150.6	136.6	-	3.1	10.9	100	91	-	2	7	Australie
	1986	151.9	4.6	147.3	133.7	-	3.0	10.6	100	91	-	2	7	
Canada	1985	1 426.8	99.9	1 326.9	271.1	756.2	171.2	128.4	100					Canada
	1986	1 478.2	90.1	1 388.1	279.4	842.6	161.1	105.0	100					
Finland	1985	128.2	58.2	70.0	51.0	10.6	6.8	1.6	100	73	15	10	2	Finlande
	1986	127.5	63.8	63.7	44.2	11.6	6.3	1.6	100	69	18	10	3	
Iceland	1985	1 668.5	931.1	737.4	141.2	409.8	176.2	10.2	100	19	56	24	1	Islande
	1986	1 651.2	839.7	811.5	175.1	433.6	190.3	12.6	100	22	53	23	2	
Japan	1984	12 055.0	7 348.0	4 707.0	(2 249.0........)		1 630.0	828.0	100					Japon
	1985	11 453.0	6 919.0	4 534.0	(2 113.0........)		1 675.0	746.0	100					
	1986													
New Zealand	1985	145.1		145.1	5.8	134.1	1.5	3.7	100	4	92	1	3	N. Zélande
Norway	1985	2 071.4	1 216.8	854.6	134.2	471.9	229.6	18.9	100	16	55	27	2	Norvège
	1986	1 842.1	942.6	899.5	180.1	487.0	218.3	14.1	100	20	54	24	2	
Sweden	1985	225.9	51.1	174.8	-	-	-	-	100	-	-	-	-	Suède
	1986	200.5	49.7	150.8										
Turkey	1985	577.9	114.1	463.8	420.6	28.9	5.7	8.6	100	91	6	1	2	Turquie
	1986													
United States	1985	2 839.4	1 344.8	1 494.6	(944.2........)		31.7	518.6	100	(63........)		2	35	Etats-Unis
	1986	2 736.4	1 196.1	1 539.5	(1 056.3........)		27.2	456.0	100	(69........)		2	29	
EEC														CEE
Belgium	1985	44.7	1.9	42.8	38.8	3.4	0.4	0.2	100	91	8	1	0	Belgique
	1986	39.0	0.9	38.1	34.0	3.5	0.4	0.2	100	89	9	1	1	
Germany	1985	186.8	19.7	167.1	104.2	62.9	-	-	100	62	38	-	-	Allemagne
	1986	163.4	14.2	149.2	106.8	42.4	-	-	100	72	28	-	-	
Ireland	1985	222.2	38.2	184.0	80.7	89.4	13.9	-	100					Irlande
	1986	238.0	19.7	218.3	80.5	118.9	18.9	-	100					
Italy	1985	395.6	10.0	385.6	274.6	39.0	0	72.0	100	71	10	0	19	Italie
	1986	413.6	7.0	406.6	298.8	44.8	0	63.0	100	73	11	0	15	
Portugal	1985	388.4	13.2	375.2	212.0	60.0	50.6	53.0	100	57	16	13	14	Portugal
	1986	396.6	11.5	385.1	201.9	93.9	39.3	50.0	100	52	24	11	13	
Spain	1985	1 784.2	429.0	1 354.1	659.2	300.9	-	394.0	100	49	22	-	29	Espagne
	1986	1 742.7	441.0	1 301.7	635.0	272.7	-	394.0	100	49	21	-	30	
United Kingdom	1985	885.0	97.0	788.0	531.0	230.0	9.0	18.0	100	67	29	1	2	Royaume-Uni
	1986	717.0	70.0	647.0	437.0	186.0	10.0	14.0	100	68	29	1	2	

Table 4/Tableau 4

OECD IMPORTS OF FOOD FISH BY MAJOR PRODUCT GROUPS & MAJOR WORLD REGIONS/
IMPORTATIONS DE POISSON ET PRODUITS DE LA PECHE PAR PRINCIPAUX GROUPES DE PRODUITS ET PRINCIPALES REGIONS DU MONDE
1985

(US$ million c.i.f.)

	All fish/Tous poissons	%	Fish fresh, frozen, incl. fillets/Poiss. frais, sur glace, y compris filets	%	Fish dried, smoked/Poisson séché, fumé, salé	%	Crustaceans & molluscs/Crustacés & mollusques	%	Prepared & preserved/Préparés & conservés	%	Importateurs
Importers											
EEC	4 320	(29)	2 025	(34)	316	(35)	889	(16)	1 090	(44)	EEC
United States	4 202	(28)	1 583	(27)	74	(8)	1 917	(34)	627	(25)	Etats-Unis
Japan	4 610	(31)	1 767	(30)	251	(28)	2 285	(41)	308	(13)	Japan
OECD Total	14 879	(100)	5 960	(100)	891	(100)	5 568	(100)	2 459	(100)	Total OCDE
%	(100)		(40)		(6)		(37)		(17)		%
Origins											**Origines**
OECD	7 937	(53)	4 250	(71)	779	(87)	1 609	(29)	1 299	(53)	OCDE
Non-OECD (*)	6 937	(47)	1 708	(29)	111	(13)	3 958	(71)	1 160	(47)	Non-OCDE (*)
Comecon	460	(3)	115	(2)	3	(-)	271	(5)	71	(3)	Comecon
Africa	790	(5)	177	(3)	6	(1)	453	(8)	153	(6)	Afrique
America	1 953	(13)	431	(7)	11	(1)	1 383	(25)	127	(5)	Amérique
Asia	3 675	(25)	883	(15)	53	(6)	1 934	(35)	805	(33)	Asie
Oceania	47	(-)	20	(-)	a)	(-)	13	(-)	14	(-)	Océanie

a) Less than US$0.5 million/Moins de US$0.5 million.

(*) The total of the imports from the five non-OECD zones may not correspond to the global figure for non-OECD as a whole, since the latter also includes values from non-specified origin./Le total des importations en provenance des cinq zones non-OECD peut ne pas correspondre au chiffre global pour l'ensemble de la catégorie non-OCDE, celle-ci incluant aussi les valeurs d'origine non spécifiée.

Table 5/Tableau 5

IMPORTS OF FISH, CRUSTACEANS, MOLLUSCS & PRODUCTS THEREOF BY OECD COUNTRIES ACCORDING TO ORIGIN/
IMPORTATIONS DE POISSON, CRUSTACES, MOLLUSQUES ET AUTRES PRODUITS DE LA MER PAR DES PAYS MEMBRES DE L'OCDE SELON LEUR ORIGINE
1985

(c.i.f. prices in US$ million)

Import.country / Origin of imp.	Cdn.	U.S./E.U.	Jap./Japon	Aust.	N.Z.	EEC/CEE	Bel./Lux.	DK	Fra.	Ger./All.	Grc.	Irld.	Ita.	Nld./P.B.	U.K./R.U.	Port.	Spain/Esp.	Finl.	Icel./Isl.	Nor.	Swe./Suède	Tur.	OECD/OCDE
Canada	0	841	267	18	5	171	9	7	49	18	2	2	21	8	57	51	a)	2	a)	1	12	a)	1 378
U.S./E.U.	203	0	869	24	a)	174	12	10	46	8	1	2	13	12	69	6	2	a)	a)	1	10	a)	1 291
Japan/Japon	30	297	0	24	3	59	3	1	15	3	2	a)	7	4	24	a)	3	a)	a)	2	1	a)	427
Austr.	a)	122	181	0	2	12	a)	a)	6	1	a)	a)	4	a)	a)	a)	1	a)	a)	a)	a)	a)	317
N.Z.	1	118	41	40	0	7	a)	a)	2	a)	a)	a)	a)	a)	2	a)	a)	a)	a)	a)	a)	a)	209
EEC/CEE:	22	217	67	16	3	1 722	183	37	383	326	25	35	365	142	233	6	162	4	a)	13	58	1	2 414
Bel./Lux.	a)	5	1	a)	a)	69	0	2	18	9	a)	a)	1	28	10	a)	1	a)	a)	a)	a)	a)	76
DK	3	108	13	4	a)	549	36	0	81	170	5	2	108	37	110	5	19	2	9	9	46	a)	806
France	3	27	15	6	a)	207	30	1	0	36	1	a)	109	a)	21	1	48	1	a)	3	a)	a)	322
FRG/RFA	1	4	6	2	a)	146	16	12	33	0	1	a)	30	32	20	a)	a)	a)	a)	a)	1	a)	186
Greece/Grèce	a)	1	2	a)	a)	29	a)	a)	17	2	0	a)	8	a)	a)	a)	6	1	a)	1	2	a)	39
Ireland/Irlande	a)	1	3	a)	a)	70	2	1	28	10	1	0	a)	5	24	a)	2	a)	a)	a)	a)	a)	80
Italy/Italie	1	3	8	1	a)	64	4	1	25	14	11	a)	0	a)	4	a)	26	1	a)	1	a)	a)	112
N'lands/P.B.	7	34	16	2	a)	385	80	12	74	74	6	1	95	6	43	a)	21	a)	a)	2	7	a)	490
U.K./R.U.	5	33	2	6	3	204	14	8	106	13	1	24	14	23	0	a)	38	a)	a)	2	2	a)	302
Iceland/Islande	a)	224	22	a)	a)	187	2	9	22	38	9	a)	14	1	94	6	23	2	a)	3	4	a)	520
Norway/Norvege	7	157	71	6	a)	409	12	45	78	70	3	a)	79	5	117	23	16	23	1	0	68	a)	803
Portugal	5	11	6	1	a)	45	2	2	6	8	a)	a)	21	1	7	0	9	a)	a)	a)	1	a)	84
Spain/Espagne	3	43	116	1	a)	133	1	2	24	9	4	a)	89	1	3	40	0	a)	a)	a)	a)	a)	344
Sweden/Suède	a)	5	1	a)	a)	50	a)	35	3	8	a)	a)	1	1	1	a)	4	9	a)	9	0	a)	84
COMECON	16	25	173	8	a)	159	6	8	54	26	2	a)	45	4	15	4	50	1	4	13	2	a)	460
AFRICA/AFRIQUE	1	77	264	20	a)	363	11	2	197	14	10	1	98	3	27	3	56	1	a)	a)	a)	a)	790
AMERICA/AMERIQUE	35	1 237	241	7	a)	351	3	131	76	9	5	a)	114	2	12	6	60	a)	a)	2	11	a)	1 953
ASIA/ASIE	65	812	2 337	74	4	321	23	5	60	65	6	1	57	23	81	1	24	5	a)	2	13	a)	3 674
OCEANIA/OCEANIE	2	11	20	3	a)	12	a)	a)	a)	a)	a)	a)	a)	a)	11	4	a)	a)	a)	a)	a)	a)	47
World/Mondial	377	4 202	4 610	234	18	4 320	271	364	1 029	628	73	35	937	209	774	197	404	48	6	51	195	1	14 878

a) Value inferior to $500 000/Valeur inférieure à $500 000.

Table 6/Tableau 6

EXPORTS OF FISH, CRUSTACEANS, MOLLUSCS AND PREPARATIONS THEREOF BY OECD COUNTRIES, ACCORDING TO THEIR DESTINATION/
EXPORTATIONS DE POISSON, CRUSTACÉS, MOLLUSQUES & PRÉPARATIONS PAR DES PAYS DE L'OCDE SELON LEUR DESTINATION

1985

(f.o.b. prices in US$ million)

Export.country Destination of exp.	Cdn.	U.S./E.U.	Jap.	Aust.	N.Z.	EEC/CEE	Bel./Lux.	DK	Fra.	Ger./All.	Grc.	Irld.	Ita.	Nld./P.B.	U.K./R.U.	Port.	Spain/Esp.	Finl.	Icel./Isl.	Nor.	Swe./Suède	Tur.	OECD/OCDE
Canada	0	109	28	a)	1	22	a)	4	4	1	1	a)	1	7	5	4	5	a)	a)	5	1	a)	176
U.S./E.U.	839	0	269	101	96	171	5	74	20	2	1	2	2	30	37	11	33	1	203	130	4	1	1 859
Japan/Japon	234	685	0	176	85	73	a)	39	12	5	a)	3	4	8	3	5	117	1	19	53	2	2	1 453
Austr.	19	20	17	0	40	17	a)	4	a)	2	a)	a)	1	2	8	1	1	a)	a)	6	a)	a)	122
N.Z.	5	a)	9	1	0	3	a)	a)	a)	a)	a)	a)	1	a)	2	a)	a)	a)	a)	a)	a)	a)	18
EEC/CEE:	161	113	48	10	5	1 721	68	573	205	141	25	63	70	375	201	44	118	8	181	378	42	45	2 876
Bel./Lux.	9	6	2	a)	a)	184	0	38	31	17	a)	1	4	80	12	2	1	a)	1	11	a)	1	219
DK	6	7	a)	a)	2	24	a)	0	6	6	a)	1	1	10	4	1	1	7	11	42	29	a)	128
France	42	31	13	6	a)	435	26	110	0	41	17	25	28	78	110	6	23	a)	19	66	3	12	657
FRG/RFA	18	8	2	1	a)	338	9	183	37	0	2	9	14	74	11	8	8	a)	34	57	7	15	507
Greece/Grèce	1	1	2	a)	a)	30	a)	6	2	1	0	a)	13	6	1	a)	4	a)	7	3	a)	11	59
Ireland/Irlande	1	1	a)	a)	a)	30	a)	1	a)	a)	a)	0	a)	a)	28	a)	a)	a)	a)	a)	a)	a)	32
Italy/Italie	15	3	6	3	a)	335	1	98	105	27	5	a)	0	87	12	20	78	a)	11	69	1	5	546
N'lands/P.B.	9	9	4	a)	1	134	26	32	9	33	a)	5	6	0	23	1	1	a)	1	6	1	1	168
U.K./R.U.	60	47	18	a)	1	212	6	105	19	16	a)	22	4	40	0	7	2	4	98	113	1	a)	560
Iceland/Islande	a)	a)	a)	a)	a)	a)	a)	a)	a)	a)	a)	a)	a)	a)	a)	a)	1	a)	0	1	a)	a)	2
Norway/Norvège	2	a)	2	a)	a)	16	a)	11	a)	a)	a)	1	a)	a)	3	a)	a)	3	3	a)	8	a)	33
Portugal	17	2	a)	a)	a)	7	a)	2	4	a)	a)	a)	a)	a)	a)	0	39	a)	46	16	a)	a)	127
Spain/Espagne	a)	1	5	4	1	171	a)	23	64	2	3	5	24	20	30	11	0	a)	21	13	2	a)	229
Sweden/Suède	10	7	1	a)	a)	69	1	54	1	1	a)	2	a)	7	4	1	a)	4	4	68	0	8	171
COMECON	11	a)	a)	a)	3	43	a)	6	1	8	a)	3	4	4	21	3	5	a)	48	8	1	a)	122
AFRICA/AFRIQUE	11	1	52	2	1	100	1	2	12	3	a)	21	10	47	3	9	43	a)	2	35	a)	a)	254
AMERICA/AMERIQUE	34	31	30	a)	a)	16	a)	1	9	a)	a)	a)	a)	1	3	a)	4	a)	1	49	a)	a)	165
ASIA/ASIE	8	42	161	36	31	33	a)	5	8	4	a)	a)	a)	8	7	2	13	a)	1	7	a)	2	336
OCEANIA/OCEANIE	a)	1	55	1	5	5	a)	a)	5	a)	a)	a)	a)	a)	a)	a)	a)	a)	a)	a)	a)	a)	67
World/Mondial	1 358	1 014	685	332	268	2 604	78	849	363	194	33	100	123	375	337	95	389	17	533	817	73	60	8 249

a) Value inferior to $500 000/Valeur inférieure à $500 000.

AUSTRALIA

I. SUMMARY

The total quantity of fish, crustaceans and molluscs caught by the Australian fishing industry in 1985/86 fell by 2.2 per cent, while the value of recorded landings increased by 13 per cent. Major increases in the value of production of abalone (68.6 per cent), scallops (25.1 per cent) and prawns (18.9 per cent) were recorded. Rock lobster was the only commodity to fall in value of production (3.7 per cent).

Overseas markets for Australia's major fisheries exports were strong in 1985/86 and this situation is generally expected to continue in 1986/87. Landings of prawns, rock lobster and tuna are expected to decrease. In contrast production of oysters is expected to increase sharply in 1986/87, mostly as a result of expanded Pacific oyster production. Prices for prawns on the major market, Japan, are expected to be lower in 1986/87. Although the average return received by rock lobster fishermen is forecasted to increase in 1986/87, prices in the major rock lobster markets are not expected to rise. World prices for tuna are unlikely to improve in the short term because of the high supplies and large stocks of tuna in cold storage in Japan, South-East Asia, the United States and Italy, and continuing large catches of all tuna species world-wide. Although Japanese prices for sashimi tuna are expected to decline in the sort term, prices for high grade frozen and fresh sashimi are likely to be maintained.

Domestically the Government has moved further towards developing fisheries management regimes for the major Australian fisheries. At the same time, the Australian Government is proceeding with the Offshore Constitutional Settlement, which is designed to allow a single State, or Commonwealth or joint fisheries management arrangement and reduce problems of fisheries being divided into several jurisdictions. The Fishing Industry Policy Council of Australia remains the prime forum for consultation between industry.

II. GOVERNMENT ACTION

a) Resource management

Several domestic fisheries continued to be subject to pressure owing to increased fishing effort. In mid-1985, therefore, the Commonwealth Fisheries Act was amended to allow for the drafting of formal management plans for fisheries under Commonwealth control. A management plan for southern bluefin tuna was introduced on 1st December 1985 and for the northern prawn fishery on 1st March 1986. Several others are in the process of introduction during 1986. The details are as follows:

-- Most major elements of a new management plan for the limited entry northern prawn fishery have been progressively implemented since 1984. Major elements of the plan include a new joint industry/Government fishery management committee, a boat replacement policy, a voluntary buy-back scheme funded by a levy on industry, surrender provisions for replacement boats, permanent closures of prawn nursery grounds, seasonal closures of commercial banana prawn and tiger prawn grounds, substantially increased penalties for infringements of the management plan rules, extension of the fishery westward, and streamlined licencing and administrative arrangements. Following scientific advice that tiger prawn resources were being over-exploited action has been taken to reduce fishing effort by approximately 30 per cent from 1987 and to reduce fleet size by 40 per cent within three years.

-- Following the acceptance and implementation of the main recommendations from an industry/government task force, the Commonwealth is now in the process of transferring responsibility for the traditional area of the east coast prawn and scallop fishery to Queensland and NSW. Under the new arrangements the area of the traditional Trawl fishery adjacent to each State will be controlled as an extension of the respective State fishery. The Commonwealth will continue to be responsible for trawl resources seaward of the new state trawl fisheries, to the edge of the Australian fishing zone.

-- New management arrangements for the Bass Strait scallop fishery involving the establishment of two adjacent State zones (Tasmania and Victoria) and a central zone, managed jointly with the Commonwealth, came into force in July 1986. The two arrangements which were made under the Offshore Constitutional Settlement, provides for zones adjacent to Tasmania and Victoria to be managed under respective State laws, and the Central Zone to be managed under Commonwealth law.

-- The Chairman of the industry/government committee established to develop a management plan for its shark fishery off south eastern Australia circulated a discussion paper to relevant fishermen in December 1986 setting out longer term management options for the fishery. Industry comments will be considered by the committee in March 1987 to finalise recommendations to Commonwealth and State Ministers on long term management arrangements for the gillnet sectors of the fishery. Existing interim limited entry arrangements introduced in February 1986 will continue until the longer term arrangements are finalised. Management arrangements for the longline sector of the fishery are being examined.

-- The Torres Strait Treaty and arrangements between Australia and Papua New Guinea for joint management in the Protected Zone of the fisheries for prawns, Spanish mackerel, dugong, turtles and pearl shell, came into force on 15th February 1985. Discussions were held in November 1986 prior to renewing these initial arrangements which were due to expire on 15 February 1987. These renewed arrangements will be in force until 15 February 1990. An arrangement between the two countries for the management of the tropical rock lobster fishery in the Gulf of Papua, north of 9°S, came into force on 15 February 1985. On 15 November 1985, an arrangement for the management of the tropical rock lobster fishery in the Protected

Zone was put in place. Discussions between the two countries on renewal of these rock lobster arrangements, which expire on 15 February 1988 are scheduled for November 1987. Both countries are continuing research into the major commercial and traditional fisheries. A report of the proceedings of the Torres Strait Fisheries Seminar held in Port Moresby in February 1985 was published early in 1986.

-- A research programme on the northern pelagic shark resources started in 1984. The programme involves tagging, the collection of biological material and the analysis of fisheries data. Field work has been completed but the evaluation of results continues.

-- The North West Shelf deep-water crustacean resources (scampis, prawns and lobsters) continue to be managed under a Development Plan which started in 1985. The prime objectives of the Plan are to collect the biological information necessary for long-term management and to restrain fishing effort to appropriate levels.

-- The South-eastern trawl fishery, encompassing waters off New South Wales, Victoria, Tasmania and South Australia, is the principal source of Australian-caught fish for the major metropolitan markets of Sydney and Melbourne. A significant quantity of fish is processed and some exported. Demand for shore-based plant facilities has not increased with the expansion of the fishery after 1979, with a number of commercial processors suffering severe financial problems, largely in the face of competition from quality fish imported duty-free from countries such as New Zealand. The fishery is managed under a limited entry regime and there is at present a freeze on the issue of any new endorsements. Indications suggest that the fleet capacity in the fishery is sufficient to take the available resources. Management arrangements are intended to contain fishing effort and provide a basis for reduction over time. The boat replacement policy was finalised on 2nd June 1986 and provides for the unitization of fleet capacity. A mechanism has been introduced that requires a compulsory forfeiture of capacity units upon alteration of fishing capacity.

-- The management plan based on the allocation of individual transferable catch quotas to Australian fishermen active in the southern bluefin tuna fishery and implemented on 1st October 1984 has continued in 1986/87. The total catch quota has not changed from the 14 500 tons permitted in 1984/85 but in view of continued scientific concern at the depletion of stocks, Japanese and Australian industry developed an arrangement whereby the former has provided financial assistance to enable Australia to refrain from taking 3 000 tons of that quota. In its first season, the plan achieved its objectives of restricting total catch to a scientifically acceptable level and of encouraging Australian fishermen to target on larger fish.

-- Following significantly increased interest in longlining for yellowfin and bigeye tuna off the East coast of Australia to supply sashimi markets in Japan and the United States, the introduction of management measures for this fishery is now being considered with a view to promoting the orderly development of the commercial fishery. Detailed management arrangements will not be finalised until extensive consultations have been held with the relevant State Governments as well as the commercial and recreational fishing interests involved.

The agreement governing the joint gillnetting venture in northern waters of the Australian Fishing Zone (AFZ) between KKFC Pty Ltd and commercial fishing interests of Taiwan expired in October 1986. In accordance with the decision of the fishing interests of Taiwan to cease gillnetting in Australian waters, the agreement was not renewed.

In 1986, the Seanorth Pty Ltd joint venture involving Australian and Thai fishing interests completed its first year of operation. Only trawlers were deployed, as the introduction of the gillnet length restriction led to a decision by the company not to gillnet in Australian waters. Steps are being taken to amend the agreement covering the venture to allow for the operation of up to 12 stern trawlers with at least one foreign vessel being replaced by an Australian vessel during each of the remaining two years of the agreement. During these two years the annual catch quota will be 10 000 tons, and, subject to availability, the company will annually purchase a minimum of 1 000 tons from Australian fishermen. The company is also required to make maximum use of Australian shore facilities and spend on Australian goods and services an amount not less than 10 per cent of the value of the catch of foreign vessels licenced under the joint venture agreement and Australian vessels introduced to replace those vessels.

Surveillance of the AFZ is primarily carried out by Royal Australian Air Force P3 Orion long range maritime aircraft and Royal Australian Navy "Fremantle" and "Attack" Class patrol boats. The surveillance effort by these vehicles is programmed to monitor the activities of licenced foreign fishing vessels and to provide a high probability of detection of illegal foreign fishing activity.

In addition to being subjected to routine surveillance by Defence force vehicles, licenced foreign vessels are required to participate in a comprehensive logbook programme and to comply with a radio reporting system whereby regular position and catch reports are passed via coast radio to the Federal Sea Safety and Surveillance Centre (FSS&SC). In addition, those licenced foreign vessels which are on quotas are required to undergo pre- and post-fishing inspections. Under the radio reporting system, vessels are also required to indicate their intention to enter or depart from AFZ, details of movements within the zone and times of commencing or ceasing fishing.

The FSS&SC which is administered by the coastal Protection Unit of the Australian Federal Police is the coordinating authority for all matters of national coastal surveillance interest including fisheries, quarantine, immigration, search and rescue, pollution, marine parks and customs requirements. Patrolling and enforcement action against Australian fishermen in Commonwealth waters is undertaken on behalf of the Commonwealth by State fisheries authorities under a reimbursement programme.

b)   Financial support

Financial support to the Australian fishing industry is provided by several means, some of which are direct while others are indirect. Non-financial support is also provided, primarily in the form of management of the industry and the collection, analysis, research and dissemination of information to fishermen and interested individuals.

The long standing financial means of support include the Primary Industry Bank of Australia, the Commonwealth Development Bank, the Australian Industry Development Corporation, taxation concessions, the Export Market Development Grants Scheme, the Fishing Industry Research Trust Account (FIRTA) and the Fishing Industry Development Trust Account (FDTA) and the research through the Commonwealth Scientific and Industrial Research Organisation and the Bureau of Agricultural Economics.

During 1986, FIRTA was altered and expanded with more emphasis on development funding. The new Fishing Industry Research and Development Trust Account will be funded up to a ceiling of one per cent of the gross value of fisheries production, which is a significant increase in funding.

The fishing industry has achieved and retained exemption from the full Federal excise duty on distillate. In May 1986, the excise was 18.3 cents per litre.

During 1986, the Innovative Agricultural Marketing Programme was commenced. This scheme will involve expenditure of A$25 million over five years beginning 1986-87. Although fish products are included under the Scheme there has been no specific allocation for fisheries. The Government-based scheme is aimed to stimulate innovation in the marketing of agricultural products, with the emphasis on export marketing and associated production, processing and development activities for agricultural products.

A$3 million was spent on buyback programmes during 1985-86. A further A$3 million has been allocated to all Australian fisheries for buyback purposes provided certain conditions are met. These conditions require that an effective management plan be in place to control fishing capacity and effort and that fishermen must be willing to contribute, via a levy, to the cost of the scheme.

c)   Economic efficiency

The main thrust of the Commonwealth Government's fisheries policy during 1986 was to improve the economic performance of the industry through the development and implementation of management arrangements for various fisheries. The fisheries most affected were northern prawn, east coast prawn, south-eastern trawl, east coast tuna, Bass Strait scallop and Great Australian Bight fisheries. A buyback programme operated in both the northern prawn and the east coast trawl fisheries.

The south eastern trawl fishery saw the introduction of a transferable unitisation system based on boat volume and engine power in an effort to control fishing capacity. The management responsibility for the Bass Strait scallop fishery was transferred from the Commonwealth to the Tasmanian and Victorian Governments in early 1986. The southern shark fishery became a limited entry fishery based in gillnet limits.

All management arrangements are subject to modification to improve the economic efficiency of the industry and ensure the conservation of the resource.

In each case the need to promote economic efficiency has been an important consideration behind the mechanisms chosen. Output controls through individual transferable quotas have been implemented in the southern bluefin tuna fishery and the Tasmanian and South Australian abalone fisheries. Such controls place an effective restraint on the catch while minimising the extent to which the actual fishing operation is regulated. Individual fishermen are thus free to choose the most economically efficient methods of taking their quota. Furthermore, since quotas are transferable, fishermen are able to adjust the size of their operation if they wish.

The management regime for the northern prawn fishery operates through input controls since output controls were not regarded as feasible. The introduction of "units of capacity" which are calculated by aggregating several factors of production, for example, vessel tonnage plus power, was chosen as the most appropriate control for the northern prawn fishery. The resulting units can be traded and used quite flexibly to allow fishermen to vary the size, power, etc. of their vessels as long as the total number of units of capacity in the fishery does not increase. The industry funded buy-back scheme aims to reduce capacity by buying out units mainly during periods of economic downturn. The surrender provisions which require fishermen replacing boats to surrender a proportion of the units required for that boat aim to reduce capacity in time of economic buoyancy when buy-back activity is slow and costly.

d) Bilateral arrangements

The Australian Government continued its policy of granting restricted access to the AFZ to foreign fishing vessels. A new agreement with the Government of Japan was signed in October 1986 and permits access to specified areas of the AFZ by up to 290 Japanese tuna longliners for the twelve month period from 1st November 1986.

In view of continued scientific concern at the depleted status of southern bluefin tuna stocks, conservation of the species was inevitably a major factor in determining Australia's attitude to continued Japanese access to the AFZ; in particular to those AFZ waters south of latitude 34°S in which the Japanese have taken the bulk of their AFZ catch of southern bluefin tuna.

Australia, Japan and New Zealand were unable to reach agreement on quota levels but an arrangement developed by the Australian and Japanese tuna industries will result in reduced catches by the fleets over the next three years. Japanese industry has accepted an annual quota of 19 500 tons, compared with its previous 23 150 tons.

Australian industry has agreed to refrain from taking 3 000 tons of its 14 500 ton quota. Australian fishermen have already made a significant contribution to the stock conservation objective by establishing a quota which was approximately two-thirds of pre-management catch levels.

An agreement with an Australian company, KKFC Pty Ltd, which acts as agent for commercial fishing interests in Taiwan, was renewed in 1986. The agreement sets down the terms and conditions under which fishing vessels from Taiwan may operate off the north and north-western coasts of Australia during the twelve months from 1st August 1986. Under the terms of the agreement a total of 50 pairs of trawlers may take 15 000 tons of demersal stocks. A gillnet length restriction introduced in 1986 to reduce the incidental catch of dolphins led to a decision by the fishing interests of Taiwan to discontinue gillnetting in Australian waters.

Under the head agreement between the Government of Australia and the Republic of Korea, and the second annual subsidiary agreement signed in December 1986, a total of 10 Korean vessels will be permitted to engage in squid jigging in a designated area off Tasmania, Victoria and South Australia. A proportion of the catch may be landed in Australia. The Koreans will pay an access fee based on the estimated market value in Korea of the quota allocated to their vessels.

e) Other Government Action

In July 1985, the Commonwealth Government agreed that the Commonwealth/State agreement on control of marine fisheries (the "Offshore Constitutional Settlement" or OCS) should proceed in the manner originally envisaged for it, i.e. that there should be formal legal arrangements between the Commonwealth and one or more States whereby a fishery is to be managed under a single law (Commonwealth or State) from the coast to the outer limits of the Australian fishing zone. Discussions were held in late 1985 between Commonwealth and State fisheries authorities and with representatives of the fishing industry about the future pattern of jurisdiction, and recommendations arising from those discussions are to go before the respective Governments. It is now expected that the first of the legal arrangements will take effect by mid-1987.

Negotiations on maritime boundaries with Papua New Guinea and France were concluded sometime ago but are continuing with the governments of the Solomon Islands and Indonesia.

f) Structural adjustment

In the southern bluefin tuna fishery, the transferability of quota resulted in a high level of adjustment within a relatively short period. It assisted over half of the tuna operators previously engaged in the fishery to shift to alternative fisheries or industries by the sale of their quota entitlement. In the northern prawn fishery, it is yet too early to determine how effective the buy-back scheme will be, but the scheme aims to remove about 30-40 per cent of total capacity over a number of years. In 1985-86 a reduction of 10 per cent in capacity was achieved.

g) Sanitary regulations

Health regulations relating to fish operate at both Australian federal and State level, and cover fish intended for consumption on both the domestic market (local production and imports) and the export market. The Department of Health, through the co-ordinating work of the National Health and Medical Research Council, publishes a Model Food Act and Food Standards. These are only legally binding as and when they are adopted under relevant pure food acts of the various Australian State and Territory Health authorities. They have prime application at the point of retail sale. Compositional standards for fish and fish products, as well as requirements for design, construction and operation of processing plants, retail establishments, labelling, additives, methods of analysis, etc., are covered by State and Territory Acts and Regulations. Health regulations relating to export of fish are administered by the Department of Primary Industry under the Export Control Act and Fish Orders and cover product standards, the design, construction and operation of export processing plants.

III. PRODUCTION

a) Fleet

The Australian Bureau of Statistics does not collect statistics on the number of commercial fishing vessels. There were however 4 612 vessels with Commonwealth licences in 1985/86 compared with 4 973 in 1984/85.

b) Operations

Development in deeper waters adjacent to Victoria, South Australia and Tasmania began to prove fruitful with the identification of what could be large resources of deepwater species, in particular orange roughy.

c) Results

Total landings by Australian vessels in 1985/86 were estimated at 152 000 tons live weight, valued at A$592 million. Production of shellfish increased by 13 per cent while finfish increased by 17 per cent.

IV. PROCESSING AND MARKETING

a) Handling and Distribution

Domestic handling and distribution of fish and fish products is a matter for private enterprise supplemented by controls operated by the individual State Governments.

b) Processing

During 1984/85 approximately 88 per cent of edible marine produce caught by commercial fishermen was used in fresh and frozen form; 7 per cent was canned, 2 per cent cured and 3 per cent reduced to meal.

c)   Domestic Market

Seafood consumption in Australia was estimated at 17.3 kg (live weight) per person in 1983/84, the latest year for which figures are available. The average annual wholesale price received by fishermen at Melbourne and Sydney markets increased in Melbourne in 1985/86, to reach A$2.14/kg (+15.6 per cent) and decreased in Sydney to A$2.39/kg (36.6 per cent).

d)   External Trade

The value of Australia's imports of edible marine produce in the year ending June 1986 was A$352.8 million, an increase of 11 per cent over the 1984/85 figure. Edible marine exports were valued at A$512.5 million, an increase of 22 per cent. Whole fresh and frozen fish imports totalled 10 572 tons and were valued at A$18.4 million in 1985/86, a decrease of 13.8 per cent in quantity over the preceding year. New Zealand supplied 44 per cent of these imports followed by the United States (19 per cent) and Malaysia (15 per cent). Imports of chilled and frozen fillets decreased slightly, the major suppliers being South Africa (33 per cent), New Zealand (27 per cent) and Japan (10 per cent). Fish finger imports decreased by 25 per cent to 2 561 tons with 45 per cent coming from South Africa and 36 per cent from New Zealand. Imports of smoked fish decreased by 11 per cent to 3 387 tons, South Africa having supplied 63 per cent. Canned fish imports increased to 15 571 tons, the main products being salmon (8 268 tons), sardines (2 550 tons) and tuna (2 510 tons). Salmon was imported mainly from the United States (53 per cent) and tuna mainly from Thailand (53 per cent) and Japan (32 per cent). Norway (31 per cent), Thailand (24 per cent) and the United Kingdom (13 per cent) were the principal sources of sardines.

Australia's main exports in 1985/86 were frozen prawns (13 109 tons valued at A$207 million), frozen rock lobsters (5 486 tons valued at A$146 million) and abalone (4 262 tons valued at A$87 million). Of these exports, Japan took 88 per cent of the prawns, 27 per cent of the rock lobsters and 54 per cent of abalone. The United States took 66 per cent of rock lobster exports. France took 41 per cent of the scallops exported.

V. OUTLOOK

Prospects for the fishing industry in 1986-87 are mixed. Relatively strong export returns resulting from the depreciation of the Australian dollar are likely to partly offset the effect on total returns of lower landings of a number of important products.

While returns from fisheries exports have increased as a consequence of the depreciation of the Australian dollar, many of the overseas markets may be weaker than in 1985-86. In Japan (one of Australia's main fisheries export markets -- the other being the United States), the strength of the Yen against other currencies is expected to reduce economic growth as a result of a slowdown in the export sector. This in turn is likely to reduce growth in demand for high value seafood products. However, the appreciation of the Yen against the US dollar will continue to attract high levels of supply to Japan from the main exporting countries.

Growth in US demand for seafood is also expected to level off as a consequence of a slow down in economic growth.

In response to the significant increase in the number of boats and level of fishing effort in recent years in prawn fisheries and other fisheries in waters off southern and eastern Australia, controls on catching capacity and/or catch have been introduced, or are under consideration. Development of new fisheries management arrangements is aimed at ensuring protection of the resources and maintenance of viable and efficient fisheries. Close government/industry liaison is a feature of the formulation of new fisheries management plans.

BELGIUM

I. SUMMARY

Total landings of fish in 1986 by Belgian fishermen in national ports fell by 4 650 tons to 34 942 tons, a reduction of 12 per cent.

Despite this considerable decrease in the quantity of catches, the value of catches in Belgian and foreign ports increased, due to the larger catches of the more expensive demersal species, by BF 185 million to BF 3.33 billion, an increase of 6 per cent.

About 2.6 per cent of total fishery products, i.e. 812 tons, were withdrawn from the market in order to maintain withdrawal prices. This constituted 1 000 tons less than in 1985, when 1 815 tons, or 5.1 per cent of total catches, were put into intervention.

The official price in Belgium for diesel oil for sea fishing fell sharply, by 45 per cent. This considerable reduction in the largest cost items combined with a large increase in income from double-outrigging trawling has led to this being an excellent year for this sector.

Imports of fish-based products for human consumption increased in volume by some 7 per cent. Despite the proportional increase of 6 per cent in exports, Belgium remained a net importer of about 100 000 tons of fishery products.

Annual fish consumption per capita in 1986 was around 16 kilogrammes.

II. GOVERNMENT ACTION

The multi-annual guidance programme for the fishing fleet provides for continuing adjustment of the fleet's production capacity to catch possibilities. The main features of the programme are:

-- Maintaining the fishing fleet at its 1982 level for the duration of the programme, while at the same time restructuring and modernising the fleet to ensure a good rate of return;

-- Maintaining the fishing industry's small owner-operated units;

-- Maintaining the fleet at around 200 vessels, and, within these limits:

i) Strictly limiting the total engine power of large beam trawlers (over 300 hp) to 65 000 hp, with average engine power not exceeding 1 200 hp;

ii) Limiting the total engine power of "Eurokotters" (small beam trawlers of 70 grt and 300 hp) to around 5 000 hp;

iii) Limitation of the fishing fleet's total engine power to 96 000 hp.

-- In renewing and modernising the fleet, priority, in line with the programme's objectives, shall be given to:

i) A coastal fishing fleet (small trawlers and shrimpers);

ii) Multi-purpose fishing vessels, not equipped with a beam trawl.

In order to respect quotas, the quantities authorised by the EEC, a number of Royal Decrees introducing conservation measures came into force:

i) Under Royal Decree of 28th May 1986, providing for conservation and management of sole reserves, it was forbidden:

-- In 1986:

- To spend more than 265 days at sea with a fishing vessel in the European fishing area.

-- During the period 1st June to 15th September 1986:

- To fish with a vessel equipped with a beam trawl in ICES area VIId;

- To fish in the ICES area IVc with a vessel of over 66 grt equipped with a beam trawl;

- In the ICES area IVa,b, for the total weight of the catch of a fishing vessel to include more than 1 per cent of sole.

ii) Under Royal Decree of 27th November 1986 introducing measures for conservation and management of cod reserves in the North Sea, it was forbidden:

-- During December 1986:

- In the ICES area IV, for the total weight of catch of a fishing vessel equipped with a beam trawl to include more than 15 per cent of cod.

Notwithstanding the active negotiation of quotas under EEC Regulation 170/83 Article 3, it was necessary in about ten ICES sectors to stop the fishing of certain species.

Regarding assistance for the development of fisheries ventures in Third World countries, training and further training schemes for managerial and supervisory staff (non-specialised) have been widened to include similar courses in French. Thus, in 1986 two groups of fisheries officials, one French, one English-speaking, pursued intensive, multidisciplinary studies, mostly in Belgium, but also in neighbouring countries.

In accordance with the Royal Decree of 5/5/73 financial assistance was granted for experimental fisheries (semi-pelagic cod-fishing).

III. PRODUCTION

a) <u>Fishing fleet</u>

In 1986, 15 new trawlers joined the fleet while 15 vessels were permanently withdrawn, laid up, scrapped or lost at sea, so that in 1986 the total number remained at 197. Six new units are due to join the fleet in 1987, replacing the vessels permanently withdrawn. In 1986 total engine power fell from 97 006 hp to 96 810 hp (-0.2 per cent) and total tonnage decreased from 23 096 grt to 22 846 grt (-1.1 per cent).

The total number of fishermen available at 31st December fell from 1 266 in 1985 to 1 258 in 1986. Thus, since 1980 the number would seem to have stabilised.

b) <u>Operations</u>

In 1986, the price of fuel fluctuated between BF 12.35 per litre at 1st January to BF 6.65 per litre at 31st December (-46.2 per cent). The lowest price (BF 5.40 per litre) was recorded at the beginning of August. The official weighted average price per litre fell from BF 13.97 in 1985 to BF 7.72 in 1986, a fall of 44.7 per cent in nominal terms.

Double outrigging trawling continued to gain ground. However, despite the addition of 17 units, bringing the fleet to 179 vessels, the number of landings fell from 4 136 in 1985 to 3 746 in 1986 (-9 per cent). Moreover, the average number of days at sea rose by 9 per cent to 6.1, so that the total number of days at sea was stabilized at 23 000. But catches are not likely to equal the 1985 level, 22 000 tons. Despite the 8 per cent fall in catches, to 20 200 tons, there was an increase in value of 11 per cent. The "outrigger trawl" sector thus accounted for 65 per cent of landings at national ports (in 1985: 62 per cent), and even 77 per cent of total value (75 per cent in 1985). Although only 21 per cent of catches consisted of sole (1985: 19 per cent), this single species accounted for 58 per cent of this sector's gross revenue (1985: 51 per cent). Thus the average value of catches per day at sea rose from BF 92 800 in 1985 to BF 104 600 in 1986 (+13 per cent).

Thirteen more vessels were added to the fleet equipped with otter trawls, bringing the number up to 85 units. Revenue per day at sea rose from BF 60 800 in 1985 to BF 65 500 in 1986 (+7.7 per cent). However, the fishing effort in terms of days at sea increased from 8 360 days in 1985 to 8 437 days in 1986 (+1 per cent).

Despite the good herring catches, these vessels went over prematurely -- as in 1985 -- to other types of fishing, owing to rather low prices and market uncertainty.

Shrimp catches and prices were very poor, with a gross revenue per day of BF 20 000 (+3 per cent).

c) Results

Despite the increased fishing effort in terms of weighted average engine power (+4.2 per cent), landings at Belgian ports, down by 5 per cent in 1985, fell by a further 11.7 per cent to 34 942 tons in 1986. The catch of pelagic fish (herring) fell by 84 per cent, while the supply of crustaceans was down by 13 per cent. Catches of demersal species accounted for 92 per cent of the total.

After an average annual increase of 12.8 per cent the total value of the first hand sale on Belgian markets increased from BF 2 863 million in 1985 to BF 3 097 million in 1986 (+8.2 per cent). In terms of value the leading species were sole (BF 1 448 million), cod (BF 360 million), plaice (BF 306 million) and anglerfish (BF 133 million).

Landings in foreign ports in 1986 totalled 4 025 tons (including 3 923 tons of demersal fish) compared with 5 036 tons in 1985 (-20 per cent). Plaice and cod were the main species, with 2 558 tons and 360 tons respectively. The total value of these catches was BF 235 million, or BF 41 million less than in 1985 (-14.8 per cent).

The total value of catches thus rose quite substantially to BF 3 332.5 million (+6.1 per cent) while total landings were down 12.7 per cent from 44 620 tons in 1985 to 38 967 tons in 1986.

In 1986, about 2 per cent of catches, or 812 tons, were withdrawn from the market, 1 000 tons less than in 1985. These withdrawals mainly consisted of plaice (from 881 tons to 301 tons) and whiting (233 tons).

IV. MARKETING

On the domestic market, the average price rose by 23 per cent (1985: +23 per cent) to BF 99 per kilogramme. This increase was mainly due to a shift in the breakdown of the species available. The market share of the relatively cheap pelagic species ($\pm$ BF 10 per kilogramme) continued to decline in terms of quantity compared with the relatively expensive demersal species and crustaceans (BF 101 and BF 92 per kilogramme respectively). But the average price of white fish, apart from sole, remained around BF 60 per kilogramme. The price of cod fell (-14 per cent) in response to an increase of 35 per cent in supply. While the small size classes showed a fall in price of about 10 per cent, the large classes increased by 8 per cent. Although the average price of plaice remained steady at BF 43 per kilogramme, prices converged for the different size classes: good increases were recorded for small sizes (+18 per cent) while the other sizes followed a diametrically opposite trend. Sole prices soared (to BF 425 per kilogramme in December), with the average price (BF 321 per kilogramme) up 20.7 per cent on 1985.

V. INTERNATIONAL TRADE

Total imports of fish and fishery products with the exclusion of fish meal and oil increased from BF 16 009 million in 1985 to BF 19 438.8 million in 1986 (+21 per cent) while quantities imported increased from 127 489 tons in 1985 to 137 154 tons in 1986 (+8 per cent).

In quantity, total exports excluding fish meal and oil totalled 35 348 tons, an increase of 15.5 per cent. However, in value terms, exports rose from BF 4 521 million in 1985 to BF 6 448 million in 1986, an increase of 43 per cent.

CANADA

I. SUMMARY

Ten years have passed since the extension of Canadian fisheries jurisdiction to 200 miles. Since Canada took control over harvesting efforts within the 200-mile limit, stocks have been rebuilt. In addition to phasing out much of the foreign fishing activity inside the Canadian zone, restrictions have been applied to Canadian fishing to assist stock recovery.

The improved health of the stocks, combined with larger allocations to Canadian vessels as opposed to foreign vessels, have provided marked benefits to the Canadian fishing industry. Foreign activities are restricted primarily to "straddling stocks" which overlap the Canadian fishing zone and to species not currently utilised substantially by the Canadian industry.

In 1986, commercial fish landings in Canada were valued at C$1.30 billion, compared to C$1.11 billion in 1985 and C$390 million in 1976, prior to the extension of Canadian fisheries jurisdiction to 200 miles. Of the 1986 landings, 66 per cent was caught in the Atlantic, 29 per cent in the Pacific and 5 per cent in the non-coastal provinces. These landings were processed into C$2.88 billion worth of fish products. Exports increased from C$1.86 billion in 1985 to C$2.42 billion in 1986. Imports also increased from the 1985 level of C$496 million to C$616 million in 1986.

II. GOVERNMENT ACTION

a) <u>Resource management</u>

A fishing management plan, developed each year by the Canadian Government, sets out the Total Allowable Catches (TACs) and allocations among the different fleet sectors and prescribes the permissible gears, seasonal constraints and other key regulatory controls. On the Atlantic coast, Canada's Groundfish Management Plan sets out the TACs for all major groundfish species and stocks. There are 12 major groundfish species and 44 stocks involved, including 7 key transboundary or straddling stocks. The Northwest Atlantic Fisheries Organisation (NAFO) sets TACs for these 7 transboundary stocks as well as for 3 stocks which occur outside the Canadian 200-mile fishing zone and allocates quotas to contracting parties, including Canada.

Groundfish available to the Canadian fishing industry in 1986 was about 975 000 tons, a slight decrease from 1985. Canadian allocations of cod decreased by 45 000 tons due to decreases in TACs on the Scotian shelf and the Gulf of St. Lawrence. The haddock stocks in NAFO Division 5 also decreased in 1985 but this was compensated partly by increases in 4VW haddock. Redfish allocations overall increased by 5 000 tons and Greenland halibut by 9 000 tons.

The herring TAC in NAFO Division 4T in the Southern Gulf of St. Lawrence was increased to 43 375 tons in 1986 from 32 500 tons in 1985. The inshore fishery was managed by seasonal quotas in the Spring and by a global quota with weekly closures in the fall. A three-day extension to the inshore fishing season in September effectively raised the 1986 TAC to 50 464 tons, of which the inshore sector was allocated 40 362 tons and the purse seine fleet 10 106 tons. Total catches in 4T in 1986 were 56 494 tons, representing a 59 per cent increase over 1985 herring landings. In Sydney Bight (NAFO Division 4Vn), the herring TAC for purse seine vessels was increased to 4 200 tons from 3 500 tons in 1985. For the Northern Gulf of St. Lawrence (Divisions 4RS), the 4R TAC was increased to 17 000 tons from 10 000 tons in 1985. The 4S fixed-gear quota remained unchanged. Purse seine vessels were allocated 1 000 tons of herring to conduct an experimental fishery in 4S, but vessel owners declined to participate in the fishery. Since 1983, only a limited herring gillnet fishery was permitted along Newfoundland's east and southeast coasts. However, in 1986, with improvements in stock abundance and recruitment, the fishery was opened to purse seine and bar seine gears which were assigned quotas of 4 000 tons and 3 700 tons respectively. In the Bay of Fundy and south and eastern shores of Nova Scotia (NAFO Divisions 4WX), the herring quota for purse seine vessels was lowered to 92 600 tons from 100 000 tons in 1985. Fixed gear vessels were allocated an allowance to 13 000 tons, up from 10 000 tons in 1985.

Based on strong market projections for frozen female capelin in Japan, the Atlantic capelin quota was increased to 85 000 tons in 1986, from 40 200 tons in 1985. Catches in 1986 totalled 69 118 tons compared to 35 085 tons in 1985. The landed value of the fishery more than tripled in 1986, increasing to C$20.9 million from C$6.3 million in 1985.

The TAC for Gulf of St. Lawrence shrimp stocks was set at 12 100 tons in 1986, a reduction from the 14 500 tons in effect in 1985. The reduction of the TAC was due to reduced level of exploitation proposed for the Esquiman Channel. For the first time in 1986, the 40 mm minimum mesh size regulations in the Gulf shrimp fishery was enforced in the Gulf.

The quota for Canada's Atlantic bluefin tuna fishery, which is conducted in inshore waters, remained unchanged in 1986 at 700 tons or 1 521 fish. The quota is assigned in tonnage by the International Commission for the Conservation of Atlantic Tunas and then converted by Canada for management purposes into numbers of fish. Catches in 1986 were low for the fourth year in a row. Only 436 fish were caught, of which 243 were landed offshore by two Japanese longliners under charter to a Nova Scotia company.

The mid-shore snow crab fishery season in the Gulf of St. Lawrence was again extended by several days. As in 1985, the total quota of 26 000 tons was caught, but the extension ensured that fishermen had adequate time to remove all their gear from the water, and prevented dumping of catches at sea and wastage of the resource. In addition to the existing crab fishery in Division 3KL, off the coast of eastern Newfoundland, supplementary licences were again available in 1986 to certain groundfish fishermen in Area 2J (southern Labrador), 3PS, (southern Newfoundland) and 3K.

A four month test fishery for offshore clams on the Scotian shelf was carried out between January and April, 1986. Positive findings resulted in DFO granting two licences and associated enterprise allocation of 15 000 tons of Stimpsons Surf clams to each of two fishing enterprises for fishing on the Banquereau Bank on the Scotian Shelf for a three year period. In addition, permits are being issued to continue exploratory fishing for offshore clams outside Banquereau Bank.

In 1986, Greenland agreed to adopt an Atlantic salmon quota of 850 tons for 1986 and 1987 through the North Atlantic Salmon Conservation Organisation (NASCO). Since almost one-half of salmon taken off West Greenland is of Canadian origin, the reduced quota level will continue to ensure that Canada's conservation efforts since 1984 are not wasted. Canada continued to work through NASCO to attempt to control foreign interceptions of Canadian-origin Atlantic Salmon. Through NASCO, Canada also agreed to close its commercial salmon fisheries in Newfoundland on 15 October instead of the previous closure date of 31 December, in order to reduce its interception of Atlantic salmon originating from United States rivers. The commercial Atlantic salmon fishery in New Brunswick, Nova Scotia and Prince Edward Island had been closed in 1985 and 1986. In 1986, a voluntary licence buy-back programme was offered to all Maritime commercial salmon fishermen. In New Brunswick, fishermen were eligible to receive between C$15 000 and C$45 000 depending on previous years' salmon landings, while in Nova Scotia and Prince Edward Island, minimum and maximum payments were C$8 000 and C$22 000 respectively. The tough measures put in place in 1984 for the recreational fishery for Atlantic salmon were continued in 1985 and 1986 with some modifications. The grilse-only restriction was continued (salmon less than 63 cm weighing 1.35 kg and 2.70 kg). However, the angling season for Atlantic salmon in the Maritime provinces and the province of Newfoundland and Labrador was extended by two weeks in 1985. Starting in 1986, all salmon hooked and then released did not have to be counted toward the two-fish-a-day bag limit. However, anglers had to stop fishing for salmon once they had released twice their daily bag limit. Mandatory tagging of all salmon harvested was extended to the Newfoundland commercial salmon fishery in 1986.

In northern British Columbia, sockeye salmon returns were approximately one week later than anticipated and about 50 per cent of the forecast magnitude. Returns on the west coast of Vancouver Island were also well below predictions for the second year in a row. However, Fraser River returns were above expectations for both early and late run fish. A strong late run to the Adams River was predicted but actual returns exceeded expectation, resulting in a seine fishery at the mouth of the Fraser for the first time in 15 years. For coho, 1986 was a good year for trollers off the west coast of Vancouver Island and the northern coast of British Columbia, but only average in the Strait of Georgia, the reverse of 1985. The Georgia Strait sport catch of coho was above average but below 1985. Expectations for chum salmon in the north were average with the exception of record returns in the Bella Coola region. South Coast chum returns were larger than expected. Returns of pinks and chinooks were average or better in the north but chinook returns in the Barkley South area were well below early 1980 levels.

The Salmonid Enhancement Programme is yielding significant economic benefits as it moves closer to the goal of rebuilding British Columbia's salmon stocks to historic levels. In 1985, 14 per cent of the salmon catch (21 per cent by value) in commercial, sport and food fisheries was as a result of all enhancement facilities on the coast. All indications are that the SEP contribution has increased significantly and that by the 1990s, SEP could contribute up to a third of the Pacific salmon catch.

The Pacific Salmon Commission, created to implement the Pacific Salmon Treaty signed by the US and Canada in 1985, continues to present recommendations to both governments to prevent overfishing and provide for optimum production and to provide benefits equivalent to the production of salmon originating in each country's waters. Annual meetings of the Commission lead to recommendations with respect to adjustment of existing fishery plans, establishment of new ones, and an enhancement and research programme. In making recommendations, the Commission reviews information from the governments, the panels, and technical committees assigned to the Commission. Negotiations are still continuing to resolve the fishing arrangements for the Yukon River as set out in the Pacific Salmon Treaty.

The roe herring fishery of British Columbia is one of the most intense and lucrative in Canada. Since 1981, a system of area licencing of roe herring fishermen has distributed the fishing power of the fleet and has limited the number of vessels converging on a particular opening. Due to the continuing herring stock decline, two of the five usual fishing areas were not opened for fishing in 1986, namely the west coast of Vancouver Island and the Gulf of Georgia. The total roe herring fixed quota for 1986 was 12 400 tons. Stock strength is expected to increase for the 1987 season.

b) Economic efficiency

For the most part, 1986 was a good year for the fishing industry. Harvests were good, prices were high and inventories moved quickly. In Atlantic Canada, 65 per cent of total landings and 80 per cent of landed value was accounted for by the inshore component of the industry. The offshore fleet, comprising mainly of large trawlers over 100 feet, accounted for the remaining 35 per cent of total landings and 20 per cent of landed value.

The Resource-Short Plant Programme (RSPP) continues to provide a specific amount of offshore caught fish to designated seasonal fish processing plants in the five eastern provinces during the off-peak fishing season. As in 1985, the 1986 allocations for the RSPP were 11 000 tons of cod, 6 000 tons of redfish and 8 000 tons of turbot. In 1986, agreements were signed by the Minister of Fisheries and Oceans and the RSPP industry consortia in Newfoundland and in the Maritimes/Quebec delegating the administration of RSPP to those consortia for the three-year period 1986-88.

The enterprise allocation system continues to operate in the Atlantic offshore groundfish industry. Each company is allotted a certain tonnage of fish which it may catch in specific areas over a one year period. The three-year trial project, implemented in 1984 to extend enterprise allocations to mobile gear vessels under 19.8 metres (65 feet) in length along the west coast of Newfoundland, was continued in 1986. In addition, the enterprise allocation system was extended to the offshore lobster fishery in 1985 and the offshore clam fisheries in 1986, all on an experimental basis.

Canada's first factory freezer trawler, owned by National Sea Products (NSP), was put into operation in 1986. The company has described its first year as a success. The licence is one of three potential licences; the other two are reserved for Fisheries Products International and for a company or consortium from the remaining offshore groundfish companies. These licences are available for a five-year introductory period. NSPs licence was granted provided that the company adhere to certain conditions concerning the species harvested and where it is harvested. A socio-economic impact study was implemented in 1986 and will continue in 1987 to assess the effect that this trawler, the Cape North, is having on the community of Lunenberg, Nova Scotia, on the plant workers both at the NSP plant and on the vessel itself, and on the fishing industry in general.

British Columbia's Pacific salmon stocks currently support commercial fisheries valued at C$500 million, employing 16 000 in the harvesting and processing sectors. Commercially produced salmon is sold in a wide variety of product types around the world. The recreational fishery, involving 328 000 tidal water anglers, generates total expenditures of C$250 million and over 3 800 person-years of employment annually. The recreational salmon fishery plays a key role in the province's important and growing tourism industry, attracting 50 000 non-residents to British Columbia annually. There is potential for this sector to grow significantly in the future. The province's Native population of 63 000 relies on salmon fishing as an important source of food. Although salmon dominates the West Coast industry, roe herring and halibut are other important species.

c) Bilateral arrangements

Canada continues to strive for cooperation in fisheries conservation, particularly with respect to stocks which straddle the 200-mile limit. In 1986, Canada had bilateral fisheries treaties in force with Cuba, the EEC, the Faroe Islands, France, the German Democratic Republic, Japan, Norway, Poland, Portugal, Spain (expired 10th June 1986) and the USSR. Most of the foreign allocations of Canadian managed stocks under these treaties are surplus to the needs of the Canadian fishing industry. Canada continues to place emphasis on cooperation to conserve fishery resources outside the 200-mile limit as a condition of access under these treaties to Canadian resources.

In 1986, in Atlantic Canada, total allocations of Canadian managed stocks and stocks determined by the Northwest Atlantic Fisheries Organisation and under Canadian management totalled 1.68 million tons. Of this amount 264 165 tons (16 per cent) were allocated to foreign countries. These allocations were comprised mainly of redfish, Greenland halibut, capelin, silver hake squid and some cod. Atlantic over-the-side sales in 1986 amounted to 4 404 tons comprised mainly of mackerel and herring but also some cod and gaspereau.

The 1986 Pacific hake fishery, the only West Coast foreign allocation, included Polish and Soviet vessels fishing off the west coast of Vancouver Island. The catch by foreign vessels was 23 764 tons. The catch by Canadian vessels sold to the foreign vessels was 30 147 tons with a landed value of C$5.4 million for Canadian fishermen.

d) Other Government action

The Coastal Fisheries Protection Act establishes the legal parameters may be granted port privileges and may enter Canadian fisheries waters to engage in fishing.

Canada strictly enforces its 200-mile fishing zone. In 1986, on the Atlantic coast, there were numerous cases of unauthorised entry and fishing within the Canadian zone. Several of these resulted in the payment of fines. In 1986, a total of C$296 232 in fines and C$50 485 in forfeitures was assessed in relation to foreign fishing infractions. Maximum penalties on conviction for several types of infractions under the Coastal Fisheries Protection Act were substantially increased in December of 1986 as follows: increase in maximum fines from C$25 000 on summary conviction and C$100 000 on indictment for foreign vessels fishing without a licence to C$150 000 and C$750 000 respectively. Maximum fines for other offences have increased from a maximum of C$5 000 on summary conviction and C$25 000 on indictment to C$100 000 and C$500 000 respectively. Amendments to the Coastal Fisheries Protection Act constitute one element in a series of measures intended to provide greater protection of Canadian fishery resources. Included in the overall strategy are the acquisition of a helicopter for offshore surveillance work, increased aircraft patrols, re-arming of Atlantic offshore fishery patrol vessels and boarding parties, development of an electronic licencing system for fishing vessels and 100 per cent observer coverage of foreign fishing vessels. In addition, allocation of fish to foreign fleets will now be based primarily on the cooperation provided to Canada in the management and conservation of fisheries resources.

In 1986, the Minister of Fisheries and Oceans appointed the Atlantic Fisheries Licence Appeal Board to provide for third party review of fisheries licencing decisions. The Department published a review paper on fisheries licencing policies and subsequently undertook consultations with industry on licencing issues.

The Fishing Vessel Insurance Plan (FVIP) of the Federal Government continues to provide insurance coverage at cost for about 9 000 vessels 90 per cent of which are under 45 feet in length. Many are located in remote communities which are not accessible to private insurers.

Under the federal-provincial Economic Regional Development Agreements (ERDAs), the three Fisheries Subsidiary Agreements signed in 1984 continued in 1986. The Nova Scotia agreement was for C$50 million (C$35 million federal and C$15 million provincial), the New Brunswick agreement was for C$45 million (C$25 million federal and C$20 million provincial), and the Prince Edward Island agreement was for C$10 million (C$7.5 million federal and C$2.5 million provincial). All agreements expire in 1989. Programmes within these fisheries subsidiary agreements are: Resource Development, which includes aquaculture; Market Development; Infrastructure; Harvesting and Quality Enhancement.

In 1986, an approved federal policy for the management of fish habitat was tabled in the House of Commons by the Minister of Fisheries and Oceans. This policy calls for a net gain in the productive capacity of fish habitats for Canada's fisheries resources through the conservation, restoration and development of those habitats. This policy will now be implemented by the federal government through its direct administration of the habitat provisions of the Fisheries Act, through negotiated federal-provincial agreements, through integrated resource planning with other economic resource sectors, and through the rehabilitation of degraded habitats and the enhancement of natural productive capacity.

In 1986, Canada continued its programme announced in 1985 to clean up its share of North American acid rain emissions. The goal is to cut sulphur dioxide emissions from Canadian sources by 50 per cent in provinces east of the Saskatchewan-Manitoba border by 1994. Maximum safe levels for deposition are estimated to be less than 20 kg wet sulphate per hectar per year. The Department of Fisheries and Oceans estimates that 14 000 lakes in Canada are acidified and 150 000 more are endangered. The Canada-US Joint Report of the Special Envoys on Acid Rain, released in January 1986, acknowledges that acid rain is a serious transboundary problem for both countries and the report contains the central recommendation that the United States sponsor a US$5 billion cost shared emissions control technology development programme with industry. The budget presented by the President of the United States in January 1987 does not reflect this commitment. Canada is continuing to pressure the United States into taking definitive actions leading to controls.

In November 1985, Canada proposed discussions with the US on the possibility of cooperation in the management of two Gulf of Maine transboundary stocks, haddock and herring, in order to rebuild them from their seriously depleted states. In February 1986, the US counter-proposed that consultations at the technical level be initiated through existing scientific meetings. This will provide an opportunity to start the rebuilding process with technical and scientific talks. Improving resource levels for transboundary stocks on a cooperative basis is moving slowly but this subject is still high on the Canada-US fisheries agenda.

Canada has significant aquaculture potential. However, other nations with similar cold water aquaculture resources have far outstripped Canada's production. The aquaculture industry has been developing in all regions of Canada, notably with salmon, trout and oysters in British Columbia; trout in the Prairies and Ontario; trout and lobsters in Quebec; and salmon, mussels and oysters in the Atlantic Region. Estimated aquaculture production in Canada in 1986 was about 7 000 tons, valued at about C$17.5 million. By 1995, there could be sales potential for 36 000 tons of Canadian aquaculture products. Canada has the resources to meet this potential provided that markets are developed, special technology is adopted and federal and provincial policies on aquaculture are developed and clarified. Fisheries Ministers have agreed that the lead development role rests with the private sector, but that governments can provide the appropriate investment climate and regulatory framework to foster industry growth. They have agreed to increase the federal-provincial coordination to ensure orderly development of the aquaculture industry. Ministers also agreed to negotiate bilateral memoranda of understanding on aquaculture development, in order to establish the framework for federal-provincial involvement in each province. Key provisions of these agreements include: "one-stop: licencing and leasing of commercial aquaculture ventures by the provinces; and federal-provincial cooperation to promote the orderly development of the industry. Significant progress was made in 1986 and agreements are at various stages of negotiations in all provinces. An agreement with the Province of Nova Scotia was signed in March 1986.

The Fisheries Prices Support Board bought C$3.8 million worth of Canadian canned mackerel in 1986 to help meet the food fish requirements of the Canadian International Development Agency (CIDA) for food aid and development programmes. Processors were paid C$21.10 (compared to C$20.25 in 1985) for each case of 24 tins of 397 g delivered to Fisheries Prices Support

Board warehouses. The product is sold at cost to CIDA by the Fisheries Prices Support Board.

The Atlantic Fisheries Technology Programme has approved approximately 60 projects valued at C$3.2 million. These projects aim at improving product quality and value through programmes designed to upgrade handling, storage and off-loading methods and equipment; reducing the cost of harvesting and processing through the promotion of methods and procedures to improve productivity and efficiency; expanding the commercially harvestable resource base through innovative new approaches such as surimi and aquaculture development; and transferring known technology and techniques not currently in use in the east coast fishery, the intent being to encourage the industry to advance technologically and to improve its international competitiveness. Plans are underway to implement three Atlantic Technology Transfer Government/Industry Consultation Workshops in 1987. These are the Atlantic Surimi Workshop, the Atlantic Aquaculture workshop and the Atlantic Fisheries Technology.

Under a new privatization initiative, fish promotional activities were taken over from the Department of Fisheries and Oceans by the Canadian Seafood Advisory Council, representing the East Coast fishing industry and the Freshwater Fish Marketing Corporation, and the Fisheries Council of British Columbia.

During 1986, the Department of Fisheries and Oceans undertook an intensive policy review to fulfill its mandate to coordinate Canada's policies and programmes concerning the oceans. The major outputs of this review were:

i) An inventory of federal ocean-related programmes and an overview of the oceans economy;

ii) Consultations with the private sector;

iii) Development of proposals to strengthen Canada's ocean management regime.

The Department of Fisheries and Oceans, in September 1986 sponsored Oceans forum, where representatives from the oceanic industries, universities and government exchanged views. An interim private sector group is preparing recommendations for the Minister of Fisheries and Oceans concerning the establishment of a permanent National Oceans Advisory Council. A national Oceans Strategy is being developed with the overall objective of securing maximum economic, scientific and strategic benefits from Canada's three oceans. The national Oceans Strategy will set goals, outline approaches and put forth proposals for Cabinet consideration in the Spring of 1987.

e) Sanitary regulations

Under the authority of the Fish Inspection Act, the Federal Department of Fisheries and Oceans continued to place emphasis on the need for good sanitary conditions and practices to ensure a good quality final product. The Department continued its efforts in 1986 to consolidate and clarify regulations and increase information to the industry on construction, equipment and operation of processing establishments, vessels, cold storages, freezing facilities and handling methods.

Section 15 of Part I of the regulations requires that establishments which process fish shall be registered, and under section 17, Part 1, they shall maintain sanitary processing and operating conditions consistent with good manufacturing and food handling practices. Facilities and operations are inspected regularly and standards enforced.

With regard to vessels, unloading, handling and storage, these operations and equipment are covered by schedules III, IV and V of the regulations. Considerable effort was directed to assist the industry in upgrading these operations in 1986.

The national quality improvement programme has been revised to reflect a national consensus between the Federal and Provincial governments. The changes to the programme reflect the move from mandatory grading of raw material and final products to a focus on quality control and its possible linkage to the plant registration programme. Efforts are being made to include the dockside grading system in regulations to permit its voluntary use by industry. There will be no initiatives with respect to final product grading until the quality control linkage to plant registration can be evaluated.

III. PRODUCTION

a) Fleet

Statistics on fishing vessels and fishermen are not yet available for 1986. The total number of commercial fishing vessels in Canada excluding freshwater fisheries, decreased from 37 326 in 1984 to 35 240 in 1985. The decrease was for the most part in vessels less than 10 grt. The total number of commercial fishermen in Canada increased from 83 791 in 1984 to 85 049 in 1985, largely due to an increase in the number of Pacific fishermen to 18 580. The number of sea fishermen on the Atlantic coast declined slightly to 58 402 in 1985. For inland provinces, there were 8 067 freshwater commercial fishermen active in 1985.

b) Landings

Canadian landings in 1986 totalled 1.48 million tons with a landed value of C$1.30 billion compared with 1.43 million tons valued at C$1.11 billion in 1985.

Groundfish landings on the Atlantic coast increased in volume by only 1 per cent since last year but increased in value by 23 per cent. The 1986 landings of 772 800 tons were valued at C$362 million.

Atlantic coast landings of pelagic and other finfish rose from 258 680 tons in 1985 to 276 100 tons in 1986. Total landed value increased by 55 per cent from C$50 million to C$77 million. The significant increase in landings was due to a large increase in the capelin catch. The significant increase in value was largely due to improved prices for herring and capelin.

Landings of shellfish on the Atlantic coast increased in volume and value from 145 240 tons to 162 460 tons and C$319 million to C$407 million respectively. Scallop landings climbed by 20 per cent to 55 800 tons and value increased by 19 per cent to C$72 million. Lobster landings increased 9 per cent in volume and 22 per cent in value, while shrimp landings increased by 33 per cent in volume and 57 per cent in value. Crab landings decreased by 6 per cent but had a significant 42 per cent increase in value.

Total landings on the Pacific coast remained relatively stable between 1985 and 1986. Total Pacific fisheries landings were 221 800 tons valued at C$380 million. Salmon catches dropped slightly to 107 565 tons but increased slightly in landed value to C$255 000. Herring landings declined by 36 per cent and value decreased by 30 per cent. Pacific groundfish landings improved between 1985 and 1986, particularly for redfish, halibut and hake. Shellfish landings and values decreased slightly in 1986.

In 1986, total freshwater landings increased to 45 000 tons from 44 000 tons in 1985. The total landed value increased by 9 per cent to C$67 million.

IV. PROCESSING AND MARKETING

a) <u>Utilisation</u>

Out of a total landed catch of 1.48 million tons in 1986, 1.40 million tons were sold for human consumption. As in 1985, approximately 19 per cent of the catch was marketed as fresh fish. The share of frozen products increased from 53 per cent to 57 per cent of the total catch and increased in quantity by 11 per cent. A significant 18 per cent decrease in production of canned products lowered its share of total production from 9 per cent in 1985 to 7 per cent in 1986.

Total Canadian production increased by 4 per cent to 813 560 tons and the value of production increased by 16 per cent to C$2.9 billion in 1986. The value of Atlantic coast production was C$2.0 billion, a significant increase of 24 per cent from the C$1.6 billion reported in 1985, largely due to higher prices for fresh and frozen groundfish products. The value of Pacific coast production decreased slightly by less than 1 per cent to C$719 million. The marketed value of freshwater fish products was C$128 million, a 5 per cent decrease from 1985.

b) <u>Marketing</u>

Domestic sales of fishery products is estimated at 169 000 tons (edible weight) for 1986. This corresponds to a per capita consumption of 7.9 kg.

In general, fish export prices were good in 1986, and total Canadian exports increased significantly from C$1.86 billion in 1985 to C$2.42 billion in 1986 with an increase in volume of 6 per cent. Both the volume and the value of exports increased to the United States, Japan, the EEC, other European countries and Central and South America. Canada's largest export market is the United States followed by Japan and the EEC. The gains in the value of exports were primarily in whole or dressed frozen salmon and unspecified sea fish, fresh and frozen cod fillets, frozen cod blocks, salted or dried cod, canned salmon and fresh or frozen crab, lobster, shrimps and prawns.

Total Canadian imports increased by 12 per cent in volume and 24 per cent in value between 1985 and 1986. In 1986, 152 371 tons of fish products were imported valued at C$616 million. The United States is the most important supplier of fish imports to Canada. In 1986, the EEC was the second most important supplier followed by Japan and Central and South America.

V. OUTLOOK

Canada continues to look forward to improved resource availability in many species, improved world markets and prices for fish products and considerable potential for aquaculture development. Future improvements will depend largely upon continued efforts to conserve and enhance the resource base within 200-miles as well as international cooperation in the management of those stocks which migrate beyond Canadian fisheries jurisdiction.

DENMARK

I. SUMMARY

Total 1986 landings by Danish fishermen in Danish ports increased by 5.9 per cent to 1 829 164 tons; the catch of fish for human consumption decreased by 3.8 per cent to 465 029 tons from 483 372 tons in 1985, whereas the catch of fish for other purposes increased by 9.7 per cent to 1 364 135 tons.

The landed value of the catch by Danish fishermen in Danish ports increased by 0.8 per cent mainly due to higher prices of fish for human consumption. The value increased despite lower prices on landings of fish for purposes other than for human consumption.

Danish exports of fish and fishery products in 1986 increased by 10.5 per cent in value from 1985 despite a drop of 0.1 per cent in quantities exported.

II. GOVERNMENT ACTION

a) Resource management

The general framework of the Danish resource management was the Common Fisheries Policy (CFP) of the European Economic Community.

As a complement to the CFP, a series of national management schemes were operated to try to achieve continued fishing opportunities whilst, at the same time, ensuring that Danish quotas allocated under the CFP were not exceeded.

Following consultations with the industry, a seasonal division of the Danish quotas was established.

In 1986, a regulatory system was supplemented with catch quotas per vessel per trip for cod, haddock and saithe. In the pelagic fishery this principle was used for herring and mackerel.

Access to the fisheries for herring and sprat in the Skagerrak and Kattegat was limited by a system of notification, allowing vessels to fish for only one of the above-mentioned species in either the Skagerrak or Kattegat during one week and prohibiting these vessels from fishing in other waters during that same week.

The fisheries for herring and mackerel by purse seiners and the shrimp fishery in Greenland waters were managed by using a licence system with individual vessel quotas.

Finally, a number of regulatory measures were operational in the coastal fisheries.

On a national level the minimum fish sizes for cod, haddock, saithe and plaice were increased.

The surveillance of the national fishing zones was carried out by four fisheries inspection vessels assisted by four rescue vessels, and controlled by the Ministry of Fisheries.

b) Aid

In 1986, several financial aid programmes were in effect, the purpose being to improve efficiency and utilisation of resources in the fishing sector.

Pursuant to a law originally enacted in 1978, grants were extended for structural measures, in accordance with EEC Regulation 355/77, aimed at developing or rationalising plants dealing with the processing or storage of fish and fish products for direct human consumption. The grants are limited to a maximum of 25 per cent of project costs. In 1986 DKr 4.5 million of national funding was expended and national grants were given as 6 per cent of project costs.

According to a law enacted in 1984, grants were provided on account of the specified costs of equipment installed onboard fishing vessels with the purpose of promoting productivity in handling catches for human consumption. The grants were limited to a maximum of 25 per cent of project costs. Furthermore, grants were available to cover a maximum of 40 per cent of the expenditure for installations resulting in improved fuel utilisation. According to the same Act, grants are provided for the development of aquaculture and construction of new fishing vessels. Grants are given as 5 per cent of project costs. In 1986, DKr 20.3 million was expended for these purposes.

Under the provisions of the annual Appropriation Acts, grants of DKr 3.8 million were allocated in 1986 for measures promoting experimental fisheries. Grants were also provided for the purpose of engaging consultative services to fishing, for which an amount of DKr 1.8 million was appropriated.

To encourage modernisation of the fleet, the state-guaranteed Royal Danish Fisheries Bank (Kongeriget Danmarks Fiskeribank) provided loans of up to 70 per cent of the construction costs of new fishing vessels and up to 60 per cent of the cost of purchasing second-hand vessels. The Bank also granted loans to an amount of up to 60 per cent of the cost for the purchase of processing plants and machinery. The interest rate for the loans corresponded to the market rate of interest, and the repayment is made over 10 to 30 years. By the end of the year, loans granted by the Royal Danish Fisheries Bank amounted to DKr 318 million, paid out in bonds and DKr 113 million paid out in cash.

c) Economic efficiency

In 1985, a set of guidelines on the adjustment of the fishing capacity to the fishing possibilities was adopted, which resulted in a substantial reduction in the number of licences granted to insert new fishing capacity in the fleet.

The main elements of the 1986 guidelines were:

i) Only to allow insertion of the same capacity which was withdrawn from the fleet;

ii) With the following exceptions:

-- Vessels under 5 grt;

-- Modernisation which increases the capacity by less than 15 per cent;

-- Vessels which are used exclusively to fish non-critical stocks, e.g. stocks for which there were no regulatory measures (quotas).

It is expected that the guidelines, with small modifications, will also be applied in 1987.

In implementing an EEC directive on fleet capacity, grants have been given for the permanent withdrawal of vessels from fisheries within EEC waters. DKr 40 million have been appropriated for the period 1984-1986.

d) <u>Sanitary regulations</u>

General regulations concerning hygiene, i.e. regulations concerning the purchase and sale of, and the catching, storing, carrying, freezing, preserving, or otherwise processing or manufacturing, for commercial purposes, of fish and fish products are laid down in the Fisheries Act of Quality Control with Fish and Fish Products No. 167 of 12th May 1965, as amended by the Acts No. 139 of 24th April 1975 and No. 560 of 30th November 1983.

More specific regulations are laid down in a number of departmental orders.

III. PRODUCTION

a) <u>Fishermen and fleet</u>

From 1985 to 1986 there has been no change in the number of vessels with engines but there was an increase in tonnage of 2.5 per cent, indicating a reorientation of the fleet towards larger vessels.

b) <u>Results</u>

Total landings by Danish fishermen in Danish ports increased by 5.9 per cent to 1 829 164 tons. Landings for direct human consumption decreased by 3.8 per cent to 465 029 tons whereas landings for other purposes increased by 9.7 per cent to 1 364 135 tons.

In value terms, Danish landings in Danish ports increased by 0.8 per cent to DKr 3.5 billion, of which landings worth DKr 2.8 billion were for direct human consumption and landings worth DKr 0.7 billion for other purposes. The average value of fish for human consumption increased 7.8 per cent whereas the average value of fish for other purposes decreased by 16.9 per cent. Codfish amounted to 51.8 per cent of the total value of landings for human consumption, flatfish 14.3 per cent and herring and mackerel 8.5 per cent. There was an increase of 17.5 per cent on the price of codfish, a decrease of 6.7 per cent on the price of herring and mackerel and an increase of 12.3 per cent on the price of flatfish.

c)   Aquaculture production

Total production of freshwater trout amounted to approximately 24 000 tons in 1986. Saltwater production of trout reached approximately 3 600 tons and the production of eel was approximately 200 tons. It is expected that the production of large trout and eel will increase considerably over the next few years (see Table 1).

Table 1

AQUACULTURE PRODUCTION, 1986

Freshwater trout	Approx. 24 000 tons
Saltwater trout	Approx. 3 600 tons
Eel	Approx. 200 tons

IV. MARKETING

At the end of 1986, 2 907 fishermen, or around 90 per cent of all owners of fishing vessels, were members of the Danish Fishermen Producers' Organisation, which was set up in 1973 according to the EEC regulation for the Common Market Organisation for fishery products. The Organisation is nationwide and members are guaranteed certain minimum prices for their landings of fish. Two new organisations were applying minimum prices in 1986. The Purse Seiners' Producers' Organisation, which had ten members at the end of 1986 and the Skagen Fishermen's Producers' Organisation, which had 54 members, were set up in 1985.

V. EXTERNAL TRADE

a)   Imports

The quantity of imports increased from 454 151 tons in 1985 to 493 582 tons in 1986. The value rose from DKr 4 009 million in 1985 to DKr 4 901 million in 1986, corresponding to an increase of 22.2 per cent.

The most important import item in 1986 was whole fish (fresh/chilled or frozen), which amounted to 26 per cent of total imports, measured in value.

Shellfish amounted to 24 per cent of total imports, canned or prepared fish amounted to 20 per cent and fillets (fresh/chilled or frozen) to 10 per cent.

b) Exports

Total exports of fish and fish products decreased from 804 589 tons to 803 893 tons in 1986, corresponding to a decrease of 0.1 per cent. The value rose from DKr 10 142 million to DKr 11 210 million in 1986, an increase of 10.5 per cent.

The most important fish export item, measured in quantity, was fishmeal, of which 224 661 tons was exported at a value of DKr 815 million, a decrease of 8.9 per cent in value terms from 1985 caused by a drop in prices.

Exports of frozen fillets amounted to 100 448 tons at a value of DKr 2 327 million, giving an increase in value terms compared with the previous year of 18.9 per cent. This increase was mainly due to a rise in exports to other EEC countries.

Canned or prepared products rose from 68 208 tons in 1985 to 73 477 tons in 1986 and the corresponding value from DKr 1 912 million to DKr 2 283 million, an increase of 19 per cent, higher than the 16 per cent increase recorded in the previous year.

The most important markets are to be found in the EEC, in which around 70 per cent of the total Danish exports of fish and fishery products take place. The most important markets are Germany (20 per cent); the United Kingdom (13 per cent); France (12 per cent), Italy (12 per cent) and Benelux (8 per cent), including fishmeal.

Outside the EEC, the most important markets are the United States (6 per cent); Sweden (7 per cent); Switzerland (5 per cent) and Japan (5 per cent).

VI. OUTLOOK

On the basis of the new quotas allocated to Denmark for 1987, a further decrease in landings of codfish is expected, which will lead to difficulties in obtaining sufficient supplies for the industry. Low prices for herring and fish for industrial purposes are expected.

FINLAND

I. SUMMARY

The overall number of fishing licences issued was about 803 900, yielding Mk 18.2 million.

The total commercial catch was 117 500 tons. Of this amount, 53 600 tons was used for human consumption and 63 800 tons for other purposes.

The Government appropriation for different subsidy measures was Mk 22 million.

The total amount of fishing capital covered by fishery insurance increased by 9.1 per cent from 1985. The Government's share of indemnifications increased by 24.6 per cent.

Imports of fish products decreased by some 50 000 tons and by Mk 24 million. Imports for human consumption increased by 3 714 tons and by value Mk 73 million. Exports decreased considerably by 4 470 tons.

II. GOVERNMENT

A total of 625 801 ordinary fishing licences (Mk 20 each) were issued yielding Mk 12.5 million. This revenue was used to finance management of fish stocks, scientific research and extention work in the field of fisheries. In addition, 178 104 special licences (Mk 32 each) authorising ice-fishing without permit from the owner of the fishing area in question, were issued yielding Mk 5.7 million. This amount will be reimbursed to the owners. Compared with the previous year, there was a slight decrease in the number of ordinary licences whereas the number of special licences was about the same.

Finland negotiated the following catch quotas with the Soviet Union and Sweden; negotiations are based on bilateral reciprocal agreements with the two countries:

-- Finland is entitled to catch 50 tons of salmon with an unlimited by-catch of sea trout and cod in the Soviet economic zone. The Soviet Union is permitted to catch 4 600 tons of herring with the bycatch of sprat limited to 2 300 tons in the Finnish fishing zone. The agreement covers only the Baltic main basin excluding the Gulf of Bothnia and the Gulf of Finland.

-- Finland and Sweden still agreed not to allocate any specific quotas for salmon due to the weak condition of spawning stocks, but reserved, as in 1985, 5 tons of salmon for both countries to cover cases where fishing gear of either country had drifted by winds or currents into the foreign zone. Herring and cod fishing is free for both countries in traditional extent.

In 1986, the Soviet Union exhausted its quota while Finland's salmon catch in the Soviet zone was 47.5 tons.

In traditional fishery a few hundreds tons of herring and cod were caught by Finnish and Swedish fishermen in each others fishing zones. In this fishery no quota or licence system is applied.

Seventeen loans intended for use in connection with fish handling, freezing and storage, plant and equipment, as well as transport facilities, were granted by private banks under the scheme of interest rebates paid by the Government. The loans amounted to Mk 6.3 million which was Mk 3.3 million more than in 1985. The Government appropriation for interest rebates was Mk 334 845. The amount granted was allocated as follows:

-- Three loans for building or purchasing
   of fish handling and storage establishments   Mk   396 600

-- Four loans for expanding and repair of
   establishments   Mk 3 981 250

-- Three loans for fish handling machinery
   and equipment   Mk 1 080 000

-- Seven loans for acquisition of transport
   facilities   Mk   868 000

The rate of interest was 6.75 per cent.

Fishermen received 255 loans from private banks for fishing vessels and equipment, amounting to Mk 12.9 million. This was 38 loans and Mk 1.9 million more than in 1985. The rate of interest was 5 per cent, the same as in the previous year. The loans provided are indicated in Table 1. The interest rebates paid by the Government totalled Mk 1.2 million which was Mk 0.4 million less than the year before.

Table 1

	No. of loans	Mk
Vessels:		
Purchasing	27	3 061 000
New construction	8	499 500
Engines and equipment	25	1 891 000
Small boats:		
Purchasing	47	2 828 555
New construction	8	1 023 400
Engines and equipment	33	672 950
Fishing gear	64	1 655 878
Other	43	1 279 700
Total	255	12 911 983

Fishing insurance activities were maintained by six fishery insurance associations and one private insurance company operating in Aland County. The main part of the indemnifications comes from the Government. Claims for loss and damage of up to Mk 3 000 receive 50 per cent from the Government and 25 per cent from the association. Claims exceeding Mk 3 000 receive 90 per cent and 5 per cent respectively. Only commercial fishermen are entitled to insure their vessels, gear and equipment under this scheme, which applies to fisheries in the Baltic Sea region.

The overall coverage of current insurance increased by 9.1 per cent from 1985 to Mk 227.4 million. The number of accidents increased from 871 to 927. The total claims increased by Mk 0.3 million compared with the previous year. At the end of 1986, the situation was as follows:

Number of units insured	2 988
-- Trawlers	301
-- Small boats	798
-- Gear and equipment	1 591
Total claims from accidents	Mk 7.6 million
Total indemnifications	Mk 6.3 million
-- Government's share	Mk 4.9 million

III. PRODUCTION

The registered fishing fleet in 1986 consisted of 550 units, which was 18 more than in 1985. Compared with the previous year, the number of salmon boats increased by 10 and that of trawlers by 8 units.

The situation at the end of 1986 is given in Table 2.

Table 2

Type	No.	Registered in 1986	Taken off register in 1986
Salmon boats	288	20	10
Trawlers	262	15	7
Total	550	35	17

The total commercial catch in 1986, according to preliminary estimates, was 117 415 tons, about 10 000 tons less than in 1985. The decrease was mainly for herring and cod.

Some 53 600 tons were utilised for human consumption, which was around 6 000 tons less than in 1985. About 63 800 tons were used as fodder in fur farms, an increase of 5 500 tons compared with 1985.

The number of full and part-time fishermen decreased somewhat in both groups. The situation was as follows, the figures of 1986 being preliminary.

	1985	1986
Full-time	2 101	2 100
Part-time	4 893	4 850

Fish farm production, consisting mainly of rainbow trout decreased slightly from 10 000 tons to 9 500 tons. However the value was the same as in the previous year, about Mk 210 million, due to an increase in prices.

IV. MARKETING

The support scheme followed the lines agreed upon in 1985 in which the price system of all categories of herring and sprat for the food industry was unified. In this context the minimum price was increased and the food industry was granted subsidies in order to maintain a steady price level for raw materials. This subsidy was extended to freezing, canning and semi-preserved products.

The budget appropriation for direct price support was Mk 10 million and the subsidy for the food industry, Mk 7 million, the same as in the previous year. Altogether 8 900 tons of fish for human consumption were supported by Mk 9.3 million and 25 000 tons for fodder by Mk 7 million. Price support to raw materials for the food industry varied from Mk 1.40 to Mk 1.45 per kg and that for fodder from Mk 0.28 to Mk 0.31 per kg.

Table 3

	Minimum price Mk per kg	Target price Mk per kg	Subsidy Mk per kg		
			Fishermen	Industry	Total
Herring and sprat for food industry					
1.1–31.07	2.15	2.55	0.40	1.00	1.40
1.8–31.12	2.15	2.60	0.45	1.00	1.45
Herring for fodder					
In the Gulf of Bothnia (only north of 62° lat)					
1.1–15.05	0.69	0.97	0.28	–	0.28
1.8–31.12	0.64	0.95	0.31	–	0.31
Elsewhere					
01.1–15.5	0.64	0.92	0.28	–	0.28
16.5–31.7	0.59	0.90	0.31	–	0.31

Altogether, the budget appropriation for support to the fishing industry in 1986 was Mk 22 million (in 1985 Mk 22.7 million), of which direct price support consisted of about 80 per cent.

The appropriation for transporting herring from production areas was Mk 2.6 million. Altogether 29 787 tons were subsidised by Mk 1.8 million the average support being Mk 0.06 per kg.

In 1986 there was no appropriation for exports of fish products.

In order to promote domestic fish production an appropriation of Mk 1.6 million was fixed in the 1986 budget. Sales promotion by publishing booklets and advertisements was financed from this appropriation by Mk 0.5 million.

Total imports of fish and fish products decreased by 49 442 tons and Mk 23.7 million. However, imports for human consumption increased by 3 714 tons and Mk 73 million. The increase occurred in all product groups except in fresh and chilled products. The import of canned products increased most, by 1 694 tons. Imports of fish meal decreased by 21 357 tons and that of fish waste by 33 360 tons.

Exports decreased considerably from 6 717 tons in the previous year to 2 247 tons. The decrease occurred mainly in the fresh and frozen product groups. This might be due to a decline in the cod fishery in the Finnish fishing zone and the recently introduced strict application of customs procedure in the Community to ensure conformity with Community tariff legislation concerning trout.

V. OUTLOOK

Fish production in Finland has reached the extent where all commercial species are fully exploited. Catches are expected to remain on the same level as in recent years. It seems that a greater part of the herring catch will be used as fodder in fur farms. However, human consumption is expected also to remain on a high level (30 kg/capita). New management measures to maintain wild salmon stocks are expected to emerge both in national and international contexts. Last year Finland introduced restrictions in coastal and river fisheries for salmon.

FRANCE

I. SUMMARY

In the area of legislation and regulations, a first decree has been published for implementing the Act of 22nd May 1985, amending the Decree on sea fisheries of 9th January 1952. This implementing Decree, dated 27th August 1986, provides for penalties for sea fishing offences. As regards the rules governing authorisation for aquaculture, a revision of the Decree of 22nd March 1983 is being prepared.

Concerning investment, the government, in accordance with its planning commitments and the multiannual guidance programme, has sustained its financial contribution in favour of fishing fleet modernisation.

Sea fishing activity in 1986 has been stable with regard to tonnage landed and its performance in terms of value has again improved compared with the previous year. Distant water fishing has, however, run into difficulties, as has the tropical tuna industry. Even though landings have, for the first time, risen over the 100 000 ton mark, the tropical tuna industry has suffered from lower world market prices.

II. GOVERNMENT ACTION

a) Resource management

The measures taken are under the jurisdiction of the European Economic Community (see chapter on the EEC).

b) Financial aid to sea fishing and aquaculture

In 1986 the Government (Secretariat of State for the Sea) allocated FF 96.82 million in capital aid for fishing vessel construction and modernisation, equipment to improve the marketing of fishery products and pilot investment in aquaculture.

In the sea fishing sector, public financial aid is granted subject to the interministerial circular of 14th January 1983 as amended by the circulars of 17th September 1985 and 1st July 1986.

i) Owner-operated fleets

Aid for renovating the fleet

In 1986 State subsidies amounted to FF 46.23 million for 56 investment operations, of which 50 involved vessel construction and 6 involved modernisation of vessels. The total cost of these operations was FF 262.302 million, giving an average State subsidy rate of 17.62 per cent.

The investment operations, broken down by coastal region, were as follows:

Nord-Pas de Calais	3 trawlers
Haute Normandie	1 trawler
Basse Normandie	4 trawlers
Brittany	22 trawlers
	1 crabber
	1 longliner
	1 net setter
Pays de Loire	7 trawlers
	2 vessel modernisations
Poitou-Charente	4 trawlers
Aquitaine	1 trawler
	2 vessel modernisations
Languedoc-Roussillon	1 seine netter
	2 crabbers
	2 tuna vessel modernisations
Guyana	2 shrimp trawlers

All of this investment was covered by loans granted by the Crédit Maritime Mutual at a State-subsidised interest rate of 5 per cent.

Aid for purchasing second-hand vessels

The scheme was amended in order to make it more attractive, especially to young persons wanting to own their first fishing vessel. It is better that young owners should prove their ability before going heavily into debt in purchasing a new vessel.

They can take advantage of loans at split State-subsidised interest rates of 5 per cent for one-third of the loan and 8.75 per cent for the other two-thirds.

In 1986 the State paid FF 42 million to finance lower interest rates on these loans.

ii) <u>Industrial fishing fleet</u>

In 1986 the Government contributed capital aid to the "semi-"industrial fishing fleet. It shared in the cost of building twelve 24- to 38-metre trawlers, half of them for Brittany, at a cost in subsidies of FF 33.42 million and two shrimpers for Guyana at a cost in subsidies of FF 1.52 million.

The Government also awarded FF 4.8 million in subsidies for alterations and sundry equipment, bringing its total aid to FF 39.74 million.

iii) <u>Onshore investment</u>

In the area of onshore investment in sea product marketing and storage, Government (the Secretariat of State for the Sea) action reflected its commitments partly under Regional/State Plan contracts and partly under the multiannual guidance programme. In carrying out the latter, the FEOGA acts alongside the Government to achieve the following aims:

- streamlining operations and obtaining efficient market distribution of catches and products (e.g., fish auctions);

- ensuring regular market supplies in terms of both quantity and quality (e.g., refrigeration units).

In 1986 the State participated in the financing of 22 investment programmes in France and its overseas territories' coastal regions. The aims pursued were as follows:

- fullest and most rational use of landing areas where auction premises are being built or renovated (as in Fécamp);

- speeded-up sales organisation through direct transfer of catches from vessel to auction-place, by means of entrance and exit construction and automation of handling operations;

- improved product quality (for example, through temperature control in buildings).

The State contributed to these investments (amounting to roughly FF 36.4 million before tax) in the form of capital subsidies amounting to FF 7 299 915.

iv) <u>Aquaculture</u>

The main feature of government aid arrangements for aquaculture investment is the extensive decentralisation accomplished in 1983 in favour of the regions (aid to businesses) and Departments (infrastructure). The State is sustaining its incentives to development by means of:

- planning contracts between the Government and the regions spanning the period 1984-1988;

- direct intervention in fields outside the scope of decentralisation (pilot projects, hydro works, Overseas Territory investment);

- action by special funding agencies (Fonds Interministériel de Développement et d'Aménagement Rural).

Capital aid in 1986 -- excluding research -- provided by the Government and regional bodies amounted to approximately FF 16 million.

Loans at State-subsidised interest rates to aquaculture amounted to about FF 70 million in 1986.

A limited number of farming programmes received government aid. The main activities to benefit were:

Shellfish
- revival of common oyster and scallop production
- modernisation of Pacific oyster and mussel production methods
- development of clam farming

Crustaceans   stepping up prawn production in France
                    and tropical Overseas Territories

Fish   boosting production of trout and bass.

v) <u>Market organisation, marketing and sea product promotion</u>

Aid for fish market organisation, supplements that granted by the EEC under Council Regulation 3796/81 of 29th December 1981.

State aid in this area is handled by the "Fonds d'intervention et d'organisation des marchés des produits de la pêche maritime et des cultures marines" (FIOM). It takes the following forms:

-- support for experimental fishing programmes: aid to persuade certain fleets to move to other fishing areas or to fish other species with a view to improving long-term market supplies. State aid is applicable to vessels excluded from EEC Regulation 2909/83 of 4th October 1983, namely vessels under 24 metres in length. The FIOM partly underwrites the guarantee for fitting out survey vessels. In 1986 aid under this heading, initiated in 1976, came to an estimated FF 17 million;

-- training for producers and senior employees of producer organisations: aid is aimed at providing technical and economics training for the heads and members of producer organisations. Essentially FIOM provides 80 per cent of study travel costs. In 1986 aid under this heading, initiated in 1976, came to an estimated FF 1.9 million;

-- local guarantee funds: aid intended to encourage the setting up and operation of local guarantee funds designed to offer partial security to agencies granting loans for the purchase of fishery products fished by members of producer organisations. The FIOM aids the organisations in setting up or supplementing the funds up to a limit of 50 per cent of the sums engaged. In 1986 aid under this heading, initiated in 1980, came to an estimated FF 3 million;

-- underwriting subordinated loans: investment incentive aid to marine wholesale and processing co-operatives. Co-operatives obtaining subordinated loans may ask a financing co-operative to provide guarantees. The FIOM, along with producer organisations, contributes to the financing co-operative's underwriting reserves up to a ceiling of 11.25 per cent of the underwritten loans. In 1986 aid under this heading, initiated in 1980, came to an estimated FF 1 million;

-- sea product promotion: aid for increasing consumption of sea products at home and abroad and for selling off products subject to periodic surges in supply. In association with the "Comité national de propagande pour la consommation des produits de la mer" (PROMER), the FIOM mounts radio and television advertising campaigns, takes part in national and international fairs and exhibitions and publishes informational literature, including a periodical. The FIOM, drawing on a quasi-tax, entirely finances these operations. In 1986 aid under this heading, taken over by the FIOM in 1976, came to an estimated FF 29.7 million;

-- commercial and technical innovation: aid to foster the technological advances needed for upgrading methods of onboard and onshore processing of fish products with the goal of improving or diversifying the marketing of these products. The FIOM assumes part of the costs of preliminary studies for technology research. In 1986 aid under this heading, initiated in 1980, came to an estimated FF 3 million;

-- study of sea products: aid meant to finance trade studies aimed at gaining better insight into the sea product market and distribution network. The FIOM finances all of these studies. In 1986 aid under this heading, initiated in 1976, came to an estimated FF 4.5 million;

-- compensation for weather-enforced lay-up: aid in favour of the setting up and running of unemployment funds which pay compensation to their members whenever bad weather stops vessels putting to sea. The FIOM subsidises the funds on a Franc-for-Franc basis according to what income is received from employer and worker contributions. In 1986 aid under this heading, initiated in 1976, came to an estimated FF 6.5 million.

c) Economic viability

i) Fishing fleets

Implementation of the multiannual guidance programme with respect to the renewal and modernisation of small fishing fleets, as provided for by EEC Regulation 2908/83, continued in 1986.

France has been especially watchful where the stability of the overall strength of the fleet is concerned, and has sustained its attempts to match catch capacity to available resources.

ii) Onshore investment

In 1986 the port master plan experiment started the year before was both continued on a national scale and endorsed for the future by EEC Regulation 4028/86 of 18th December 1986 on measures to improve and adapt structures in the fisheries and aquaculture sector. Title VIII of the Regulation indeed stipulates that investment projects must form part of overall "product flow" programmes.

d) Bilateral arrangements

Bilateral arrangements within the scope of the Treaty of Rome are negotiated, concluded and supervised by the Commission of the European Communities on the Communities' behalf. Other agreements are a matter purely for France and concern access to the EEZ surrounding the following Overseas Territories: French Polynesia, New Caledonia and the Kerguelen archipelago. These agreements are not published in the French Official Gazette.

e)   Other government measures

With regard to fishing vessel management, eligibility for a government capital subsidy requires membership of a "managerial group".

Managerial groups are State-approved companies, including co-operatives, which supply technical and financial management service for member fishing vessels.

A circular setting out the terms for obtaining State approval was issed in 1986. It completes the legal framework providing for stricter fishery vessel bookkeeping, accounting transparency and, indirectly, compulsory auctioning of the catches and products of member vessels.

Regarding the organisation of fishery product markets, the French Government, pursuing the policy initiated in 1985 and in virtue of a decision of the Commission of the European Communities dated 9th October 1985, has abolished all market support measures (withdrawal-price support, aid for storage and processing).

f)   Sanitary regulations

The salubrity of shellfish farming areas, the fitness for human consumption of oysters, mussels and other shellfish and the standards of hygiene for the transport, sale and packaging of shellfish are regulated by:

-- the Decree of 20th August 1939, as amended by Decrees No. 48-1324 of 25th August 1948 and No. 69-578 of 12th June 1969;

-- Order of 12th October 1976;

-- Order of 6th January 1977.

Sanitary standards for the installation, operation and upkeep of premises and plant, handling, processing and storage of fish products and hygiene requirements for personnel are regulated by:

-- Order of 1st October 1973: fishing vessels;

-- Order of 2nd October 1973, as amended by Order of 30th July 1982: preparation and processing establishments;

-- Order of 3rd October 1973, as amended by Order of 25th October 1975: wholesaling premises;

-- Order of 4th October 1973: retailing premises.

The conditions for importing marine or freshwater fishery products for human consumption into France and the formal requirements for sanitary certificates are regulated by:

-- Order of 3rd November 1982: list of customs checkpoints for animal foodstuffs;

-- Order of 25th July 1986 (text only published in 1986): new certificate forms and list of marine fish permitted to enter as fillets, slices or mince.

The quick or deep freezing, defreezing, transport and storage of fishery products are subject to the general regulations covering animal foodstuffs:

-- Order of 1st February 1974, as amended by Orders of 9th July 1975, 11th May 1981 and 20th June 1984: transport;

-- Order of 26th June 1974: quick and deep freezing, defreezing;

-- Order of 18th June 1980, as amended by Order of 2nd August 1984: cold stores.

Standards of conformity for fishery products are regulated by:

-- Order of 21st December 1979, as amended by Order of 17th December 1984: microbiological standards;

-- Order of 9th March 1981: banning the consumption of marine and freshwater fishery products preserved or prepared with unauthorised substances.

III. PRODUCTION

a) <u>Fishing fleet</u>

Its size in units continues to decline, but its power and tonnage remain stable.

b) <u>Canning, preserving and processing</u>

The French fishery product processing industry is divided into three sectors:

i) <u>Preserves and semi-preserves</u>

Output for 1985 totalled 101 370 tons and centred on three main species:

```
Sardines      27 240 tons
Mackerel      25 900 tons
Tuna          41 760 tons
```

Domestic consumption was estimated at 184 567 tons, meaning that France had to import 78 377 tons. Exports were calculated at 8 076 tons.

The canning sector (turnover: FF 2.8 billion) is marked by the size difference among the 33 canning companies (4 800 people employed). Seven of these companies account for 74 per cent of turnover.

ii) Quick and deep frozen products

The quick and deep frozen fishery product market is a complex one as statistics deal with items that may be used as such (for example, quick frozen individual fillets), processed by the canning industry or even by the deep freeze industry itself (preparation of precooked dishes or breaded portions from slabs, fillets or slices).

Tables 1 and 2 nevertheless break down the output in 1985 of quick and deep frozen products into the amounts used for direct consumption and for industrial processing. The distinction is made in Table 1 between deep and quick freezing. Quick frozen products are those which are frozen ultra-fast and marketed in containers or packages that enclose them completely.

It should be further noted that:

-- the market in industrial fillets and quick frozen fish slices continued to grow in 1985;

-- the market in breaded fish portions or fingers has remained steady. The increase in household consumption of these items in 1985 matched that of other forms of presentation;

-- the market in whole quick frozen fish for immediate consumption declined in 1985 after having risen in 1983 and 1984;

-- supermarket sales of quick frozen fish-based prepared dishes, for the most part cooked in France, soared by 35 per cent in 1985. They are the largest-selling item in the prepared dish field.

It should be noted that, because of the increasing problems encountered by French deep-sea fishing vessels in catching species for quick freezing on their usual fishing grounds, processors -- except in the last-mentioned case -- are relying more and more on imported fish.

iii) Salting, curing, smoking

On the whole, this sector is expanding.

Salting and curing: Some forty firms are active in this field. They have replaced the old practice, now almost totally abandoned, of salting on board ship. In 1985 the industry processed almost 20 000 tons of fish, mainly herring (18 000 tons) for a yield of about 8 000 tons of products (mainly salt herring and salted or cured herring fillets).

Smoking: Also expanding, this industry processes, in the main, imported fish (salmon, herring) and employs from 1 500 to 2 000 people depending on the season. The output of smoked salmon (nearly 9 000 tons) practically covers domestic requirements.

c) Catches

Total landings increased in 1986. The quantity (up by 2 per cent) and value (up by 14 per cent) of fresh products rose moderately.

Table 1

FROZEN PRODUCTS FOR THE PROCESSING INDUSTRY

Species	Whole	Filleted(a)
A. FISH		
Anchovy	-	-
Sea bass	16	-
Cod	26	12 125
Alaska hake	-	-
Conger eel	156	-
Sea bream	24	-
Haddock	-	1 237
Smelt	1	-
Halibut	-	95
Gurnard	467	-
Herring	967	-
Pollack	82	-
Saithe	440	1 981
Dab	30	-
Ling	12	-
Anglerfish	198	-
Mackerel	9 220	-
Blue whiting	-	869
Whiting	105	157
Hake	9	-
Plaice	9	-
Kerguelen fish	-	-
Skate	252	-
Goatfish	-	-
Brown cat shark	202	-
Sardine	4 381	-
Sole	2	-
Sprat	1 220	-
Squale	-	-
Pout	45	-
Tuna	82 724	-
Turbot	1	-
Miscellaneous fish	3 318	-
TOTAL	103 907	16 464

Species	Tonnage
B. CRUSTACEANS	
Crab	33
Whole shrimps	1 096
Peeled shrimps	-
Lobster	-
Rock lobster (whole)	-
Rock lobster (tails)	-
Norway lobster	21
Other	-
TOTAL	1 150
C. MOLLUSCS	
Cuttlefish	8 978
Squid, octopus	2 016
Scallop	34
Mussel	-
Carpet shell	-
Other	67
TOTAL	11 095
A+B+C GRAND TOTAL	132 616

a) Most of which is frozen, including imported fillets, used in the production of portions and breaded sticks, fish "croquettes" and ready-to-cook dishes.

Table 2
CHILLED AND FROZEN PRODUCTS FOR DIRECT HUMAN CONSUMPTION

Species	Chilled Whole	Chilled Sliced fillets or blocks	Frozen Whole
A. FISH			
Anchovy	-	-	-
Sea bass	109	-	65
Cod	167	444	-
Alaska hake	6	1	-
Conger eel	132	30	6
Sea bream	8	61	5
Haddock	15	526	-
Smelt	-	-	2
Halibut	-	385	-
Gurnard	72	-	-
Herring	5	15	-
Pollack	5	105	-
Saithe	901	920	32
Dab	160	14	-
Ling	215	253	-
Anglerfish	62	46	79
Mackerel	157	1	-
Blue whiting	-	-	-
Whiting	528	1 043	25
Hake	73	82	63
Plaice	32	-	-
Kerguelen fish	-	-	288
Skate	5	58	20
Goatfish	-	-	-
Brown cat shark	7	104	25
Sardine	1 244	95	16
Sole	45	51	15
Sprat	267	30	102
Squale	-	-	12
Pout	128	-	15
Tuna	602	-	21
Turbot	-	-	25
Miscellaneous fish	306	174	882
TOTAL	5 251	4 438	1 698
Whole + fillets		9 689	
Breaded portions and sticks	-	23 979	-

Species	Frozen	Deep-frozen
B. CRUSTACEANS		
Crab	132	90
Whole shrimps	91	2 327
Peeled shrimps	436	200
Lobster	119	2
Rock lobster (whole)	2	1
Rock lobster (tails)	-	137
Norway lobster	448	102
Other	-	-
TOTAL	1 228	2 859
C. MOLLUSCS		
Cuttlefish	390	1 735
Squid, octopus	1 855	60
Scallop	47	12
Mussel	1	-
Carpet shell	50	-
Other	56	-
TOTAL	2 399	1 807
A+B+C GRAND TOTAL Sea products	37 295	6 364

The tropical tuna catch, on the other hand, declined in value by 9 per cent, despite a rise in volume of nearly 20 per cent.

d)   Aquaculture

Traditional oyster (110 000 tons) and mussel (50 000 tons) farming account for by far the largest share of output.

Some more recent ventures have grown in importance, such as rope-farming of mussels at sea (1 000 tons), salmon farming at sea (700 tons) and cockle farming (500 tons).

Table 3

FRENCH AQUACULTURE PRODUCTION
(CONTINENT AND OVERSEAS TERRITORIES)
1986

Molluscs	
Oysters	110 000 tons
Mussels	50 000 tons
Clams	500 tons
Crustaceans	
Prawns	150 tons
Fish	
Sea trout	650 tons
Coho salmon	80 tons
Bass	60 tons
Sea bream	10 tons
Turbot	10 tons
Reptiles	
Green turtles	50 tons

Farming areas and beds are located as follows:

-- oysters and mussels: Atlantic coast, Channel, Golfe de lion;

-- clams: Atlantic;

-- sea trout, coho salmon, turbot: Atlantic, Channel;

-- bass, sea bream: Mediterranean, warm water locations

-- prawns: Atlantic, French Pacific territories;

-- turtles: Reunion.

IV. PROCESSING AND MARKETING

a) <u>Handling</u>

In 1986 the Secretariat of State for the Sea pursued its objective of modernising unloading and handling facilities.

It should be emphasised that the operations financed during the year, demonstrate the extent to which investment in unloading and handling equipment is being integrated with building projects (for example, auction and storage facilities). Efficiency in terms of design and seaport siting is greatly enhanced (as measured by the economic return on onshore investment) compared with unplanned action.

b) <u>Processing</u>

In 1986 investment (FF 112 150 000 before tax) with State participation (FF 11 666 225) highlighted the following priorities:

-- modernisation of existing plant;

-- expansion of facilities for perfecting and developing new products.

Most enterprises showed their determination either to foster innovation or find new outlets using conventional production techniques. Investment programmes in this area comply with French guidelines as approved by the European Economic Commission.

c) <u>Domestic market</u>

i) <u>Promotion campaigns</u>

The FIOM budget for promoting consumption of fishery products totalled FF 28.9 million (FF 24.4 million of which was allocated to the domestic market and FF 4.5 million to the foreign market) in 1985, and FF 29.7 million (FF 24.5 million and FF 5.2 million allocated to the domestic and foreign markets respectively) in 1986. The FIOM took part in trade fairs or exhibitions and radio and TV campaigns more particularly advertising oysters, scallops, white tuna meat, quick frozen breaded fish and quick frozen items generally.

A consumer impact survey was conducted in connection with these campaigns. It revealed that the sales revenues generated by the advertising were double or quadruple the cost of each campaign.

ii) <u>Market organisation</u>

Market organisation for sea fishery and aquaculture products is governed by Council Regulation (EEC) No. 3796/81, as recently amended by Regulation No. 2315/86 of 21st July 1986.

In France the agency for market organisation is the FIOM (Fonds d'intervention et d'organisation des marchés des produits de la pêche maritime et des cultures marines), founded in 1975 and reorganised under Decree No. 83-1031 of 1st December 1983.

The FIOM, which as from 1986 desisted from all national market support action, is the correspondent of the European Agricultural Guarantee and Guidance Fund (FEOGA).

d) Foreign trade

The quantity of fishery product imports in 1986 remained steady at 551 234 tons, close to the 1985 figure of 551 513 tons, although their value rose by 5 per cent. The quantity of exports rose by 9 per cent and their value by 1.7 per cent compared with 1985.

The trade balance cover rates for fishery products in 1986 were 37 per cent in quantity and 33 per cent in value, a rise in quantity of 3 per cent over the 1985 rate.

The unfavourable trade balance for fishery products remained sizable, however, reaching FF 6.3 billion. It was mainly due to the high cost of a limited number of imported items.

i) Fresh fish

This market is supplied mostly by domestic output. It has a cover rate in value of about 74 per cent.

ii) Frozen and processed fish

For lack of adequate domestic output, this sector has traditionally depended on imports, especially of raw material for the processing industry.

In 1986 imports of frozen fish continued to rise in both quantity -- 151 171 tons as against 135 688 tons in 1985 -- and value (up 34 per cent).

The major imported item was frozen cod, 16 180 tons of which arrived in 1986 (including 14 183 tons of fillets).

Exports rose in quantity by 33 per cent in 1986 but did not increase in value compared with 1985.

The frozen fish market is the only one where the export-import ratio has worsened -- from 41.11 per cent in 1985 to 29.47 per cent in 1986.

In the field of salted, dried and smoked fish, 1986 bore out the downward import trend noted in previous years -- imports fell from 17 404 to 7 140 tons. The foreign trade cover rate by value simultaneously rose from 30.6 per cent in 1985 to 33 per cent in 1986.

iii) Canning

The export-import ratio in the sector was a very low 10.7 per cent. Because of special agreements concluded with countries belonging to the old

French Union, France is a heavy importer of canned tuna and sardines from countries like Senegal, Ivory Coast and Morocco.

    iv) <u>Crustaceans and molluscs</u>

The export-import ratio (10.7 per cent in 1986) remained low because of prawn and shrimp imports accounting for 14.8 per cent of the overall deficit.

*
* *

This deficit, which could be termed as structural, was due for the most part to a few species.

Among fish, imports of salmon, cod, sole and other kinds of fish accounted in value for 14.5, 8.9, 9.7 and 2.3 per cent respectively of the deficit (or, taken together, 35.4 per cent of the deficit).

As for crustaceans and molluscs, imports of prawns, mussels, crab and scallops made up nearly 45 per cent of the overall deficit.

The deficit thus resulted from imports of a dozen or so products with a high market value. The demand for these items is strong but, mainly for reasons of climate, French production is weak or non-existent and cannot satisfy it.

V. OUTLOOK

The 1985 Law on sea fishing sweepingly altered the existing legislation and defined a course of action in line with Community fisheries policy.

The provisions of the new Community regulations on structural investment aid in the marine fishing and aquaculture sector, particularly those applying to vessels over 33 metres in length, should strengthen the position of the industry and stimulate a revival of offshore fishing.

Given the sustained domestic demand for fishery products, the stableness of prices on the fresh fish market and lower energy costs, business should almost certainly prosper.

GERMANY

I. SUMMARY

In 1986 the German fishing industry finally seems to have overcome its depression caused by its very difficult and at times turbulent process of adaptation to the changed general conditions of maritime law which had lasted ten years. The reduction of capacities in deep-sea fisheries came to a halt at 14 units (5 freezer trawlers and 9 wetfish trawlers with a total of 23 200 grt). Cutter fisheries showed a satisfactory propensity to invest with a view to improving the age structure of the fleet.

In 1986, catches of the German sea fishery again declined by approximately 13 per cent to 176 000 tons (catch weight).

In terms of quantity, imports rose by 2 per cent, while exports decreased by 5 per cent.

II. GOVERNMENT ACTION

a) <u>Resource management</u>: See the chapter on the EEC.

b) <u>Financial support</u>

The Federal Government spent little less than DM 9.0 million -- i.e. about DM 1.0 million more than in 1985 -- on investment promotion. DM 5.9 million of this sum were spent on grants for structural measures (mainly grants for construction, reconstruction and purchase in cutter fisheries), DM 0.3 million on subsidies for reduced interest rates and DM 2.7 million on loans to cutter fisheries.

Loans granted by the Federal Government for the adaptation of capacities in sea fisheries, i.e. for the temporary laying-up of fishing vessels, amounted to DM 9.4 million, which is about DM 2.8 million less than in 1985. About DM 2.6 million of this sum will be refunded by the EC on German application. Grants for the reorientation of the fishing activity towards new fishing grounds and new fish species were no longer available.

In 1985, cutter fisheries were promoted by the four coastal Laender, with about DM 5.7 million in total, DM 2.8 million accounting for grants and DM 2.9 million for loans. Financial support in 1986 for which the corresponding data are not yet available, is expected to be similar.

III. PRODUCTION

a) <u>Fleet</u>

At the moment, the German deep-sea fishing fleet still consists of 14 units with about 23 200 grt, 5 of which are freezers, 4 are older and 3 relatively new wetfish trawlers of traditional style as well as 2 small trawlers. The reduction in capacity has not continued since early 1986.

The cutter fleet, too, maintained its number with approximately 650 units and about 24 000 grt. The construction of more cutters along with the breaking-up of older craft has improved the unsatisfactory age structure of the cutter fleet to a considerable -- but not yet sufficient -- extent.

b) <u>Operations</u>

In 1986, the fishing opportunities allocated to Germany in both EC waters and waters off third countries amounted to 214 000 tons and 127 000 tons respectively. The following quantities of the overall quota of 341 000 tons -- 2 000 tons less than in 1984 -- were allocated to the two branches of fisheries: about 221 000 tons to deep-sea fisheries (of which ca. 182 000 tons were accounted for by traditional species) and some 120 000 tons to cutter fisheries (of which 115 000 tons were accounted for by traditional species). Although these quotas seemed to be satisfactory in quantity, a considerable part of the catching opportunities they promised could, however, not be realised economically. Depending on the fishing ground or fish species, this was due to special biological/hydrological reasons, to an inadequate situation in proceeds as well as to a lack of abundance of fish resulting from overfishing.

c) <u>Results</u>

In 1986 the landings of German deep-sea fisheries declined, in comparison with the previous year, to 67 000 tons by just under 36 000 tons (live weight) or by about 35 per cent, while earnings derived from these catches decreased to DM 122.0 million by DM 36.6 million or by about 23 per cent. The continuously poor fishing conditions off Greenland caused the production of frozen fish to decline by about 37 per cent and the landings of wetfish by about 30 per cent.

With some 94 000 tons (live weight) the landings of German cutter fisheries in German harbours in 1986 were well up by about 6 000 tons or 7 per cent on the figures for 1985, with proceeds amounting to DM 115.5 million, which is about DM 5.0 million or ca. 4.5 per cent more. Landings in foreign ports rose from 15 000 tons to about 16 200 tons. Shrimp catches were also declining.

As to freshwater fisheries, the preliminary data for carp and trout suggest an average harvest and are expected to be satisfactory on the whole.

IV. MARKETING

a) Handling

In 1986, German fishing vessels landed a total of about 176 000 tons (live weight) of fishery products, including crustaceans and molluscs in national and foreign harbours. This meant another decline of 15 per cent as against the previous year. The share of quantities landed in German ports dropped from 93 per cent to 91 per cent, whereas landings abroad -- mainly in Denmark -- rose to 16 200 tons (+12 per cent). Owing to a smaller fleet size and lasting unfavourable fishing opportunities, in particular off Greenland, landings of wetfish continued to decline. Landings by the German freezer fleet were also decreasing to a considerable extent after another reduction in capacity (1985: 74 100 tons; 1986: 45 100 tons). If it had not been for the addition of two vessels suited for pelagic fisheries (herring and mackerel), the situation would have been even more unfavourable. A substantial increase in landings was recorded for mussel fisheries in the year under report (+37 per cent).

Domestic landings by the German fleet of sea fish for human consumption decreased by more 21 per cent (from 140 000 tons in 1985 to 106 000 tons). Hence, the declining tendency of German sea fish landings continued in 1986. The main causes were lower catches of redfish off Greenland as well as cod in the North Sea in connection with the earlier-mentioned reduction of capacities in large deep-sea fisheries.

Wetfish landings by German trawlers in Germany further decreased in 1986 (--13 per cent). Thus they accounted for just under 18 per cent of the average landings in the years 1977-1979. In 1986 landings of sea fish by the cutter fleet were also again declining (--6 per cent); the decline in this area was not so serious as in the preceding year, in spite of the fact that the external conditions had become even less favourable.

The main places for the marketing and landing of fresh fishery products are the sea fish markets of Cuxhaven and Bremerhaven. The sea fish market of Hamburg had to cope with a 15 per cent decrease in landings in 1986. The auction turnovers at the 3 sea fish markets approximately reached last year's level. The quantities landed at the three sea fish markets and offered for sale was (product weight):

Bremerhaven	42 983 tons
Cuxhaven	31 443 tons
Hamburg	2 559 tons

In contrast to domestic landings, direct supplies by foreign trawlers and cutters from the Community and from third countries experienced an increase in 1986 on the previous year (+7 per cent). These were predominantly landings of redfish and saithe from Iceland, the Faroe Islands, the United Kingdom and France. In 1986, the number of foreign importers landing their catches in the ports of Cuxhaven and Bremerhaven came to 199 (1985: 221). Container shipments are gaining more and more importance, in particular since a direct link was established from Iceland to Bremerhaven. The share of wetfish sales of foreign origin in the auction turnovers of the sea fish markets rose from 48 per cent in 1985 to 51 per cent in 1986. Direct landings by foreign fishing vessels were not able to compensate for the declining supplies of fresh sea fish. Direct landings of redfish by foreign vessels rose considerably as against the previous year and continued to be a significant factor of supply.

The German market as well as the processing industry are dependent on fresh raw materials. In this connection it is of special importance that the market is supplied continuously. The shortage of supply brought about a further reduction in the market withdrawals for both German and foreign landings (--55 per cent and --81 per cent, respectively).

Domestic landings of frozen fish by the German deep-sea fishery with about 45 000 tons (live weight) were down 36 per cent on the previous year; its fillet blocks, which are shock-frozen at sea, once formed the basis for the manufacture of ready-to-serve fish products of the industry and of deep-frozen products for big consumers. The decline in domestic production has led to an ever increasing dependence on imports. The overall result also includes the catches of two vessels which are specialised in pelagic fishing (herring and mackerel) and landing their catches whole-frozen.

b) <u>Utilisation</u>

The form in which landed wetfish and frozen fish was offered for sale remained almost unchanged in comparison with previous years. At the auctions, the majority of landed wetfish were offered gutted with head. Wetfish was, however, rarely sold in this form to the consumer; they were mostly filleted onshore or processed into ready-to-cook products. In 1986, this form of presentation did not lose significance. A large part of the frozen-at-sea products was already suitable for final consumption, particularly at restaurants (single fillet, interleaved). The produced fish blocks, on the other hand, constituted an important raw material for the processing industry onshore. They were further processed -- in frozen condition -- into final products, such as fish-fingers, fillets, etc.

c) <u>Demand</u>

It is estimated that the consumption of fishery products in 1986 will show a rising tendency.

d) <u>External trade</u>

In 1986, the importation of fishery products increased by approximately 2 per cent in terms of quantity; its value was DM 24 million higher than that of the previous year. Fish meal imports again rose distinctly in quantity, but receded markedly in value. Another increase was recorded for fresh products, frozen fillets and canned products. Fish oil imports were down on the previous year's level by nearly one-third.

The import volume of wetfish slightly increased in comparison with the previous year (+2 per cent, product weight). In terms of quantity, herring is still the most-bought fish species, followed by redfish and saithe. Imports of whole redfish, which were predominantly sold at the sea fish markets as direct landings by foreign fishing vessels and as container supplies, again showed an increase. A positive tendency was also reported for saithe.

As in previous years, Denmark continued to be the most important supplier, followed by France. Among the third countries, Iceland must be named as the main exporting country. As to frozen products of sea fish, including herring, fillet imports rose distinctly in the year under report, whereas imports of whole products increased to a minor extent only. The Netherlands were able to maintain their leading position for frozen herring. The share of frozen sea fish fillets in imports increased as against the previous year; saithe from Norway remained the main product. In the year under report, frozen blocks of saithe were increasingly bought by the fish industry for processing purposes, and the market would certainly have absorbed even larger quantities if they had been available.

The export of fishery products in terms of product weight reached about 307 000 tons, which is approximately DM 721 million in value. These figures represented a decline in quantity and value of 5 and 6 per cent respectively. Wetfish accounted for 14 000 tons and frozen products for 37 000 tons. Cod as wetfish or deep-frozen fillet, which was predominantly sold to Community member states, was still the most important export product in terms of quantity. Exports could only be increased for wetfish and for prepared and preserved products.

V. OUTLOOK

The German market dependence on imports of fishery products has grown stronger in 1986. The decline of landings by the German fleet resulted in the German market and the processing industry being no longer continuously supplied with raw materials from domestic landings. This decline was only balanced to a certain extent by an increase in direct supplies from abroad. Without these supplies the German fish processing industry would be faced by very negative consequences.

In 1986, the German fishery was able to make approximately the same profits as in the previous year. This seems remarkable in the face of decreased landings. It became possible as a result of another rise in prices. As fuel costs were lower than in the previous year, the situation of fisheries should have eased off a little.

Further development of the German fish industry will decisively depend on its access to economically important fishing grounds as well as on the supply of the market, at reasonable prices, with raw materials from other EC member states and from third countries.

It remains desirable to strengthen self-sufficiency. This could be achieved in the coming years by renewing harvesting capacities.

GREECE

I. SUMMARY

In 1986 fish production amounted to about 138 000 tons; the major part of this production (about 112 000 tons) included fresh products from sea-water fisheries.

In general, production increased and 1986 production data were taken from estimates made by the Central Fisheries Service (this Service has greatly improved statistical information in the fisheries sector) (1).

The basic improvements within the fisheries sector for 1986 were the following:

-- Replacement and modernisation of the fishing fleet at a rapid rate.

-- Carrying out programmes on the development of aquaculture through construction of intensive aquaculture units and hatchery stations.

-- Elaboration of draft decrees pertaining to the protection of fish production.

-- Implementation of research and training programmes.

II. GOVERNMENT ACTION

a) <u>Resource management</u>

   i) <u>Fish stocks</u>

The following measures were taken in 1986 for the rational exploitation of fish resources and the conservation of fish stocks.

-- The Presidential Decree No. 143/86 governing the issuance of new fishing licences for trawlers was published. By the same Decree the activities of "inshore trawl" fishing gear are restricted in certain sea areas.

-- The Presidential Decree No. 144/86 pertaining to the restrictions on fishing and marketing octopus and the prohibition of fishing activities in sensitive sea areas was issued.

-- A Ministerial Decision on experimental fisheries by mean of "Gagava" fishing gear was issued, aimed at collecting data and amending regulations in force for this fishing gear.

-- As regards environmental protection, a joint Ministerial Decision pertaining to the "quality of surface waters for certain uses" was issued. According to this Decision, a systematic control of the waters where fish of certain categories are found, such as crustaceans and echimoderms, will be carried out.

ii) Licencing arrangements

On the basis of EEC arrangements with third countries, 13 fishing licences to Greek vessels were granted for fishing activities in Senegal, Guinea-Conakry and Guinea-Bissau. Other vessels employed in distant-water fisheries operated in Nigeria, Gambia, Sierra Leone and the Persian Gulf under private agreements.

b) Financial support

In 1986, pursuant to the programme on measures and incentives within the framework of fisheries development (Productivity Programme), the following credits for financial support were allocated, Dr 206.6 million by the Public Investment Programme for the following activities.

-- National contribution to EEC programmes.

-- Replacement and modernisation of fishing vessels (not included in EEC programmes).

-- Abolition of inshore trawl fishing gear and purchase of gillnets.

-- Aquaculture in fresh and sea water.

-- Establishment of refrigerators and cold stores.

From the above-mentioned credits an amount of Dr 175 million was contributed by Greece to the EEC programmes for financial support and particularly for the activities financed within the framework of the following Regulations.

-- 2908/83/EEC - Fishing fleet reorientation and development of aquaculture.

-- 2909/83/EEC - Encouragement of experimental fisheries and joint ventures.

-- 83/515/EEC - Directive concerning the adjustment of productivity.

A sum of Dr 69.5 million was allocated from the Regular Budget for dismantling fishing vessels, payment of insurance premium on sea voyages, damages and research on fisheries.

The following activities were supported by the Public Investment Programme:

-- Dr 64.5 million for studies on fisheries development (research, studies on new fishing ports, artificial reefs, etc.).

-- Dr 240.2 million for construction, such as the establishment of fishing ports, modernisation of existing one, hatchery stations, etc.

Out of the activities already financed, the modernisation of 7 fishing ports was included within the framework of EEC Regulation No. 355/77 pertaining to the development of the conditions of processing and marketing fish products.

Within the framework "Reclamation Projects for Fish Breeding Stations" an amount of Dr 17.3 million was paid for carrying out 35 reclamation projects on lagoons. In 1986 within the framework of the Mediterranean Integrated Programmes, the Programme concerning Crete was approved by the EEC amounting to Dr 1 091.4 million.

c) Bilateral arrangements

Negotiations for the conclusion of bilateral agreements, under Community policy, with Tunisia, Algeria, Syria, Albania and Egypt were carried out to ensure fishing grounds and sponge fishing ground outside Greek waters. The conclusion of bilateral agreements with Egypt, Syria and Tunisia may soon be made.

d) Other actions

To achieve a rational distribution of fishing activities, the registration of the total fishing fleet will come into force in order to select which vessels will operate on the Greek fishing grounds.

The absence of scientific data prevents the adoption of protection measures within the fisheries sector. A solution to this problem is pursued by means of experimental fisheries, training of fishermen and fisheries research.

In 1986:

-- 6 fishing licences for trial fishing were granted;

-- The research programme on fishing of swordfish began;

-- The research programme on fishing large mackerel was carried out in collaboration with the Italian Institute "Communale Biologia di Marina" and the University of Crete;

-- A post-training seminar in cooperation with the Faculty of Biology of the University of Crete came to an end;

-- A "special training programme" of Ichthyologists for agricultural counsellors employed under Regulation No. 2966/83 (EEC) was established.

III. PRODUCTION

a) <u>Fishing fleet</u>

There is a total of 20 276 vessels engaged professionally in fisheries and 56 engaged in sponge fishing. The number of vessels according to type of fishery is distributed as follows:

	1985	1986
Distant-water fisheries	38	48
Offshore fisheries	865	826
Coastal fisheries	14 832	19 346
Sponge fisheries	68	56

During the past year a considerable increase in the number of professional vessels taking part in coastal fisheries has been observed. This is due to both the natural increase of professional vessels (about 5 per cent) each year and mainly to the implementation of strict restrictions to non-professional and game fishing under the Presidential Decree 373/85. Many fishermen who had a non-professional fishing licence for vessels but who used professional fishing gears, changed their vessel licence to a professional one when the new legislation came into force. This concerns small fishing vessels of 9 metres in length.

It should be noted that the restriction on game fisheries has contributed to the withdrawal of many fishing gears.

Most coastal fishing vessels are less than 15 years in age. Sponge fishing vessels are from 8 to 10 metres in length and are usually over 30 years in age.

The off-shore fishing vessels although only constituting 4 per cent of the total fishing fleet, represent about 30 per cent of the total capacity of the fleet. This category includes trawlers, purse seines and small vessels, mainly trawlers having licences for purse seines. About 46 per cent of the off-shore fishing vessels are 20 years in age.

The distant-water vessels are transporter-freezer vessels which are satisfactorily equipped and divided into two categories, i.e. fishing activities and shrimp activities. Half the distant-water vessels are over 20 years of age.

i) <u>Replacement and modernisation of the fishing fleet</u>

In 1986, 45 new vessels were built (replacing old ones) of which 34 were financed from National and Community sources and the remaining 11 from National sources. The replacement of 66 fishing vessels was also approved as well as their inclusion in the National and Community Programmes on financial aid. The purchase of 14 second-hand off-shore fishing vessels was also approved as well as replacement of engines in 19 off-shore vessels. Four in-shore trawls were decommissioned by National aid granting.

ii) <u>Fishermen</u>

The fisheries sector employs approximately 39 000 fishermen, most of whom are engaged in coastal fisheries. A high percentage of fishermen owning vessels smaller than 9 metres fish for one season, mainly from Spring to Autumn.

In the off-shore fisheries there is a shortage of workers and, as a result, vacancies are filled by foreign workers.

b) <u>Results</u>

In 1986 there was an increase in fish production. Production from overseas fisheries can be attributed to the number of vessels in operation as well as the potentialities provided for in fisheries agreements between the EEC and third countries.

In Greek territorial waters the increased quantities are mainly due to the reappraisal of production data (1) but it can be said that there is an upward trend in production coming from this sector.

Finally, marine aquaculture is a sector now being developed and there has been growth in production over the last two years while fresh water aquaculture and oyster culture has remained the same.

c) <u>Aquaculture</u>

One of the main targets of the Greek fisheries policy is the development of aquaculture, with special emphasis on marine culture. In 1986, production from aquaculture was 2 320 tons. Of this total 2 000 tons came from fresh waters, the main species bred are trout, eel and carp.

Government hatcheries produce trout spawn which is supplied to trout farms. The total trout production has stabilised to cover all domestic needs. A carp breeding farm at Ioannina came into operation to enrich the Ioannina Lake. There is a great interest, particularly from the private sector, in establishing units of intensive eel culture and fresh water crabs.

Marine culture is under development and there are 5 units engaged in the culture of golden bream and bass. However, the relevant infrastructure is lacking at present (euryhaline species hatcheries).

The establishment of a network of hatcheries is scheduled together with an Aquaculture Research Centre and training of personnel in this sector. Individual investors show a special interest in units of intensive euryhaline species culture, mainly those vertically integrated but also in fattening units of golden bream, bass and sea shrimp.

In 1986 the following projects progressed in the aquaculture sector:

-- The construction of the carp-breeding hatchery in Ioannina Lake by the Municipal Enterprise of Ioannina Lake. This station began to operate by producing 4 carp species;

-- The construction of a carp breeding hatchery in Mikri Prespa Lake; this is expected to begin operations in 1987;

-- The construction of the model training and pilot carp breeding unit in Psathotopi of Arta was completed; this is expected to begin operations in 1987.

d) <u>Lagoons</u>

Lagoons belong to the Greek State and they are leased out either to fishing cooperatives according to priority at a rent representing 25 per cent of fish production, or to individual by auction.

Approximately 90 per cent of the lagoons have been conceded to fishing cooperatives but recently local government or development companies also obtained leases. Exploitation is made by extensive culture, the main species caught are common grey mullet, eel, golden bream, bass, etc. In 1986 production reached approximately 2 500 tons.

IV. PROCESSING AND MARKETING

The improvement in conditions of processing and marketing is one of the main aims of the fisheries development policy. The problems met are the limited development of the processing industry and the lack of competitiveness of the final product due to high cost and lack of systematic organisation in marketing fish products, resulting in increased transport/conservation costs.

a) <u>Processing</u>

The construction of three new processing units has been scheduled within the framework of efforts for the promotion of the processing sector. There is great interest in the modernisation of existing units and favourable prospects for tuna processing, in combination with the implementation of the tuna fishing programme for tuna destined for processing.

b) <u>Marketing</u>

Modernisation of the seven existing fishing ports is to be carried out in two stages with the purpose of rationalising the marketing-distribution cycle of catches. Two new fishing ports were constructed in Messologi and Calymnos and construction at Preveza, Argolida Valley and Volos are under way.

Until 1986 there were special Community premiums for the processing of Mediterranean sardines and anchovies. Following the accession of Spain and Portugal, a special premium for the support of sardine producers was accorded within the framework of the special system to bring into line the prices practised in the Community of ten and in the two new Member States of the Community.

V. OUTLOOK

The main targets of the Fisheries Development Programme are to increase fish production and improve fishermen's income. Special emphasis is laid on the development of fisheries, aquaculture and processing/marketing of fish products with:

-- The 5-Year Programme (1983-1987) on Economic and Community Development;

-- The Specific Sectoral Programme (1983-1987) on Fisheries for the processing and marketing of fish products in the framework of Regulation 355/77/EEC as 1987 is the last year for the implementation of these programmes;

-- The Fisheries Programmes in the framework of the IMP;

-- The Programme-Contracts signed by the Ministry of Agriculture;

-- The specific regional/local programmes.

The following activities have been scheduled for 1987:

i) <u>Sea fisheries sector</u>

-- Continuation of efforts for searching new fishing grounds off Greek territorial waters;

-- Implementation of programmes on experimental fisheries (continuation of research in tuna and swordfish fishing);

-- Redrafting of regulatory decrees within the framework of updating fishing legislation;

-- Continuation of efforts for registering the fleet total, capacity, etc.;

-- Continuation of temporary or final decommissioning of fishing vessels.

ii) <u>Aquaculture sector</u>

-- Extension of the fish hatcheries network;

-- Continuation of aid granting for the construction of intensive pisciculture units, giving priority to sea culture;

-- Aid-granting to semi-intensive breeding units in State hatcheries;

-- Elaboration of hydrobiological studies on the lakes of Mornos, Polygitos, Tavropos and Vegoritis;

-- Elaboration of a final study on the initiation of projects for the construction of the experimental eel-culture unit in Psathotopi - Arta;

- Elaboration of a study for the determination of sites within the country suitable for aquaculture.

iii) <u>Transport, processing and marketing of fish products</u>

- Completion of the construction of fishing ports in Preveza, Calymnos and the Argolida Valley;

- Completion of the modernisation of fishing ports in Piraeus, Salonica, Alexandroupoli, Kavala, Patra, Chalkis and Chios;

- Study on the fishing ports of Paros and Nea Michaniona;

- Study on shellfish cleaning and sanitation in Salonica;

- Study and construction of two ell gathering and selection centres.

iv) <u>Cooperative organisations</u>

- Continuation of aid granting to cooperative fishing organisations through increased aids, technical and administrative support, establishment and operation of producer organisations, etc.

v) <u>Sector of research and training</u>

- Evaluation of the research programmes results;

- Experimental fisheries and fishermen's training in tuna fishing;

- Programme on training of fishing cooperative organisation staff;

- Promotion of two programmes concerning training of staff in research institutions;

- Implementation of a "specific programme" for the newly employed personnel in the fisheries sector;

- Participation and facing of "environmental" items, etc.

NOTES AND REFERENCES

1. The data on fish production and fishing fleet for 1985 and 1986 are considerably different compared with the corresponding data of the previous years. It is likely that differences will occur in data of the previous years after the new control. Thus it is not recommended to proceed with inter-annual comparison for the time being until the results of the new estimates are extended to cover the past years as well.

ICELAND

I. SUMMARY

The quantity of fish, shellfish and crustaceans landed by the Icelandic fishing fleet decreased by 1 per cent compared with 1985. The total 1986 catch amounted to 1 656 000 tons. In constant prices, the value of the catch increased by around 13 per cent.

In 1986 there was a slight decrease in the number of decked fishing vessels but an increase in total tonnage.

The volume of exports of fish products increased by some 3 per cent -- in value terms (current prices) by 37 per cent, reflecting increased exports of fresh fish, in particular, and higher prices.

The management measures introduced at the beginning of 1984 were continued in 1986, with some changes and modifications in light of past experience. However, a difference in approach resulted as some regulations in 1985 became law in 1986.

The export levy on fish products was abandoned and the fund and transfer system within the sector changed radically and simplified.

II. GOVERNMENT

The conservation and management measures for the demersal fishery, adopted in 1984, were continued in 1986, but with some modifications based on experience gained. There was a considerable change in the approach of making the fishery policy. In December 1985, the Althing (Parliament) passed a new law valid for a two-year period (1986-1987). The feeling had been that measures should be in force for periods longer than one year at a time. A major change from earlier policy is that some of the most important and controversial points are now written into the law.

The main reason for resorting to this extensive quota system was to secure sufficient reduction in catch rates in view of the deterioration in the condition of major stocks such as cod and thus build up the stocks. The year-classes of cod and some other stocks from 1977 through 1982 were either poor or, at best, below average. For example, the 1982 year-class of cod is the poorest on record. Secondly, adverse hydroclimatic conditions for some years have resulted in a reduction in weight/age ratio of most fish stocks by 10-15 per cent. These adverse conditions reversed in mid-1984 and have since given favourable results in improvement in the weight/age ratio of fish.

As the effects of these management measures gradually unveiled themselves, demands for greater flexibility emerged. Consequently, in Devising the management rules for 1986, without however abandoning TACs and licencing systems, effort was made to avoid rigidity and to make a greater leeway for vessel-owners.

Briefly, the main rules adopted for the demersal fishery in 1986, to be continued in 1987, were as follows:

-- Vessels over 10 grt can choose between a catch quota and an effort quota. If a catch quota is chosen, the vessel is allocated individual catch quotas for each of the five most important demersal species. These quotas are allocated on the basis of their average share of the catch during 1981-1983 and later adjustments. The catch quotas are freely transferable to another vessel within the same fishing firm or to another vessel based in the same fishing port. Transfers of catch quotas to vessels in other ports are subject to the approval by the Ministry of Fisheries. A change from 1985 is that vessels were able to transfer 10 per cent of unused quotas in 1986 to the following year, or use 5 per cent of 1987 quotas in 1986. This made way for greater flexibility.

-- If an effort quota is chosen, the vessel can fish for a certain number of days during each of the specified periods of the year. The number of days and periods vary from one group of vessels to another, depending on size and activity. Effort is measured in days at sea. Vessels fishing under effort quotas have a maximum cod catch limit, but can fish an unlimited amount of other species. Effort quotas are not transferable. A change from 1985 is that vessels could build up a higher catch quota during 1986 for the following year, if they did well under the effort quota system. This new device was intended to give room for fishermen to improve their fishing rights. A share of the TACs is allocated to vessels below 10 grt. On the other hand, longer periods of stops have been introduced, since the number of these vessels has increased considerably.

Other conservation and management measures, such as minimum mesh sizes for trawls, gill nets and seines remained unchanged. Quota regulations with regard to herring, nephrops, scallop, inshore shrimp and capelin were unchanged apart from an annual revision of the TACs based on stock and abundance assessments. Deep-water shrimp was not under quota, but closure of areas for limited time periods was introduced for the first time.

The following is a brief description of existing rules for some stocks of fish and crustaceans, other than demersal.

i) <u>The capelin fishery</u>

In 1981, an individual vessel quota system was introduced. Each vessel is allocated a share of the total quota. Two-thirds of the total is divided equally between the vessels and one-third on the basis of the carrying capacity of the vessels. From 1980 to date, no new vessels have been allowed to take part in the capelin fishery. The individual quotas are issued for each capelin season.

ii) <u>The herring fishery</u>

The total quota was, until 1986, divided between purse-seiners and drift-netters, which have formerly participated in the fishery. In 1986 each vessel was given a share in the total quota, independent of the way of fishing. The quotas are transferable, subject to approval by the Ministry of Fisheries.

iii) The lobster fishery

Access to the lobster fishery is restricted to vessels which have formerly participated in that fishery. Every successful applicant obtaining a lobster fishing licence is allocated a certain share of the total catch quota, based on the performance of the vessel during the past three seasons. The catch quotas have not been transferable.

iv) The inshore shrimp fishery

The shrimp fishery is, to some extent, based on the exploitation of regional stocks. TACs and quotas are established for each stock. Fishing licences are issued to local applicants provided their vessels are of a certain size. The licences are issued for one season at a time and are not transferable.

v) The scallop fishery

Like the inshore shrimp fishery, the scallop fishery is based on regional stocks. The management system is practically identical to that of the shrimp fishery discussed above.

From February 1986, loans have been made available to vessel-owners holding a fishing licence to replace older vessels with new ones of similar size and fishing potential under certain conditions. Thus it is expected that the rate of renewal will be faster in the next few years than the previous years. No new entries are permitted.

No Government subsidies were granted to the fishing industry.

III. PRODUCTION

The number of decked registered fishing vessels as of 1st January 1987 totalled 819 units -- 112 000 grt -- compared with 826 units totalling 110 600 grt in January 1986. In addition, roughly 600 open motorized vessels landed fish for sale, the majority of which, however, should be counted as recreational fishing boats, as weekend and evening fishermen are numerous.

Total landings decreased slightly from 1 680 000 tons in 1985 to 1 665 000 tons in 1986 or by 1 per cent. The main contributor to this decrease was capelin, as the catch fell by 93 000 tons, from 993 000 tons in 1985. The catch of demersal species increased by 52 000 tons mainly due to improved landings of cod (+38 000 tons). The improved cod catch is attributed to an increased weight/age ratio as well as better than anticipated returns of the 1979 and 1980 year-classes of cod.

The general cost/earning ratio of the fishing fleet improved considerably in 1986. This development can be traced to higher catch rates, favourable prices for fish on markets abroad, lower prices of oil and oil-related products, as well as lower rates of interest.

Processing aboard increased consideraly -- both of demersal species and shrimp. Seven trawlers have now the capacity to fillet and freeze their catch aboard, and six trawlers to freeze part of their catch -- mainly whole redfish, Greenland halibut and shrimp. Several other vessels, a few capelin purse seiners included, have been equipped with freezing facilities mainly for shrimp.

The ownership structure in the fishing industry remained virtually unchanged. All Icelandic fishing vessels are now privately owned or operated by private companies.

One of the major structural changes in the management of fisheries in 1986 was that the export levy of 5.5 per cent on most products was abandoned and the fund and transfer system within the sector was simplified.

IV. PROCESSING AND MARKETING

There was a marked increase in the exports of fresh fish on ice, especially in containers. In fact, markets for fresh fish were very favourable in 1986 on the whole. In all, 112 800 tons of demersal fish (nominal weight) was landed directly abroad or exported in containers in 1986, an increase of 18 per cent over the previous year. Freezing of fish declined slightly in spite of a larger catch of demersal species. On the other hand, there was a considerable increase in the production of salted fish.

Market prices for processed fish products (other than meal and oil) improved during the latter part of the year, although the rate of exchange of the United States dollar is a cause for concern as the United States is the most important market for frozen fish products. Inventories at the end of the year (except for stockfish) were at their lowest for many years.

The U.S. market was the most important one in 1986 for fish products, but its share decreased. On the other hand there was a large increase in exports to Great Britain, France, Germany and, in particular, Japan.

Table 1

	% of total visible trade	
	1985	1986
Imports:		
EFTA	22.0	21.6
EEC	49.5	50.8
Eastern Europe	8.8	6.5
North America	7.2	7.3
Other	12.5	13.8
Exports:		
EFTA	14.2	16.6
EEC	39.3	44.1
Eastern Europe	7.8	5.5
North America	27.3	22.0
Other	11.4	11.8

The main markets for imports and exports (services excluded) are shown in Table 1. The share of international trade in GDP was around 45 per cent in 1986. The share of fish and other marine products, in value terms was around 78.9 per cent of total visible exports in 1986 one of the highest ratios in the last two decades.

V. OUTLOOK

Revised estimates indicate that the 1983 year-class of cod is fairly strong and the 1984 year-class reasonable, whereas the 1985 and 1986 year-classes are below average, according to 0-group surveys.

The prospects for the catch of capelin in the 1987/88 season are not as good as in the season 1986/87. Results so far suggest that the capelin catch in the season 1987/88 will be considerably lower than in the previous two seasons. It is expected that the 1987 catch of species other than demersal, will be similar to that of 1986.

IRELAND

I. SUMMARY

While there was an improvement in prices obtained by fishermen for demersal species in 1986, the pelagic market continued to be slightly uncertain with weakness in the herring market still causing concern. Operational costs continued to be high.

II. GOVERNMENT

a) <u>Policy</u>

As in previous years, the Government's development plan for the fishing industry was continued with the aim of expanding the industry in the national interest. Programmes included grants and subsidised loans for the purchase and modernisation of fishing vessels and for aquaculture projects, education and training programmes for fishermen and a harbour development programme.

b) <u>Management</u>

Various licensing arrangements were continued and extended during the year to ensure compliance with EEC quotas and to secure an equitable distribution of those quotas among the various components of the fleet.

c) <u>Aid</u>

The aid scheme introduced in 1985 was continued in addition to the grant and loan schemes outlined at a) above.

III. PRODUCTION

a) <u>Exploratory fishing and gear technology</u>

Programmes were mainly geared to encouraging more of the medium-sized fleet (80' to 90' l.o.a.) to exploit 200-400 metre offshore waters for high value species such as hake, monk, megrim, prawn, squid and associate catches. These extensive grounds are off the South-west, West and North-west coasts including the South Irish Sea. Monofilament longlining for hake was expanded up to 5 boats at one stage during the year working grounds from the South-west as far as North-west Scotland.

Further work was undertaken to improve the landed condition of scad as a continuation of the B.I.M.s programme (commenced in 1985) aimed at introducing better trawling and handling practices.

Catching methods were augmented with the introduction of a twin rig system (prawn), a fork rig arrangement for rough grounds and pair seining.

The thrust of the 1986 shellfish programme was directed at the further commercialisation of non-traditional species for which export markets exist. These were spider crab (Tralee and Brandon Bays), velvet crab (Waterford coast and Bantry), whelk (South-east and North-west), shrimp (Waterford Harbour and L. Foyle), scallop (Glengad with Greencastle boats) and clam (Bertraghboy Bay, Galway).

c)   Marine fish farming

Nine commercial fish farming projects were grant aided under the Mariculture Grant Scheme. Of those, eight were salmon projects, including four smolt rearing schemes, and one was an oyster farming project which had previously been grant-aided at the pilot stage.

A total of thirty-five farm projects were approved for grant aid at pilot scale level of which twenty-nine were new enterprises. The majority (nineteen) were suspended culture mussel operations followed by oyster/clams (nine), salmon smolt (four), bouchot mussels (two) and abalone (one).

In 1986, there were in all 130 marine fish farms in production in Ireland as follows: oysters (48); mussels (41); salmon (30); sea farmed rainbow trout (4); scallops (1) and eels (1). The Irish fish farming industry is expanding into other species, and pilot projects on clam farming and abalone are presently being undertaken numbered 5. In addition, research on the feasibility of farming abalone in Irish waters is continuing at the Shellfish Research Laboratory at Carna, Co. Galway. There were 15 freshwater trout farms in operation.

IV. PROCESSING AND MARKETING

a)   Handling and distribution

The bulk of the pelagic varieties -- mainly herring, mackerel and sprat -- were landed and auctioned at ports in the North-west and South-west and, to a lesser extent, on the West coast. The pelagic varieties are unloaded by trailer into bins which are transported by container load from the port to foreign markets or up into stock. (There is no overall change in distribution patterns on the domestic market). Whitefish are packed in 45 kg boxes and increasing quantities of ice are being used at the ports. During the early part of the year there was a large inventory of mackerel which was cleared in the course of the year mainly to the West African market.

Whitefish was largely utilised in fresh form on the home market with significant quantities being exported to the United Kingdom.

b) Domestic market

The annual consumption survey conducted in the Spring of 1986 on the purchasing behaviour of Irish housewives in relation to fish products once again points to the increasing prominence of fresh fish in the Irish diet. Frozen fish and smoked fish are now frequently purchased because of improved distribution. Purchases of fish fingers are at similar levels to previous years.

c) External trade

Final figures for the year 1986 are not yet available; therefore the commentary on external trade is confined to the industry's performance during the first nine months (Jan./Sept.) in respect of which firm figures are available.

For the period Jan./Sept. 1986, exports of fish and fish products amounted to 109 779 tons valued at Ir£66.3 million. As the Nigerian market was closed for the greater part of this period, exports of mackerel valued at Ir£15.5 million were down on the 1985 level of Ir£28 million for the same period. France, Germany and the Netherlands accounted for a large share of Irish mackerel exports.

Herring exports increased in value in the period by 23 per cent to Ir£9 million. Exports of herring roe contributed to this increase in value.

Shellfish exports valued at Ir£19 million showed a significant increase on the previous year's level mainly as a result of increased exports of Dublin bay prawns.

Imports of fish and fishery products for this period amounted to 20 246 tons valued at Ir£25 million which are broadly in line with last year's level.

d) Transhipments

A total of approx. 26 000 tons of pelagic fish were transhipped in 1986 mainly to East European vessels.

V. OUTLOOK

It is expected that the quotas set for Ireland for the 1987 fishing year under the EEC Common Fisheries Policy will be reached. It is hoped that the relative upturn in the whitefish market will continue in 1987 and that the Nigerian mackerel market will continue to be a stable one for Irish exporters.

ITALY

I. SUMMARY

In 1986, there was a slight improvement in Italian fisheries over the previous year. The market was unable to absorb the entire Mediterranean catch of small pelagic species (bluefish), notably because of distribution problems, but there were fewer withdrawals under EEC regulations. At the same time, withdrawals under EEC regulations declined significantly, especially for the larger amounts intended for processing. Catches of oceanic species increased.

There was probably no substantial change in the size of the fleet, though the only available data are provisional official figures for 1984 from the Central Statistics Institute (ISTAT) in Rome.

Imports and exports declined appreciably. Italy's dependence on imports is still considerable, inter alia, due to difficulties arising from the application of the new Law of the Sea.

II. GOVERNMENT ACTION

A.   *National*

The action envisaged under the Plan for the rationalisation and development of Italian sea fisheries, adopted by Act No. 41 of 17th February 1982, continued during 1986. A co-ordinated research and development programme has been launched following the Plan's guidelines for the rationalisation and balanced development of fishing and aquaculture. A list of research projects was reported in 1984; it includes, in particular, evaluation of the biomass of all fish species and this was pursued during 1986.

a)   *Resource management*

The Italian Government continued its action to improve the management of existing resources by adopting the following measures:

-- Approval of the text of a regulation (D.M. 26/10/85) on dredging for bivalve molluscs;

-- Encouragement of artisanal coastal fishing by the use of improved traditional gear;

-- Development of co-operation between producers (outright grants to co-operative associations);

-- Study of the possibility of awarding, by tender, to co-operatives of exclusive exploitation of specified coastal fishing areas;

-- Reduction of production costs by modernising the fleet and fishing gear;

-- Organisation of exploratory fishing voyages in unexploited or under-exploited fishing zones;

-- Setting up of joint ventures between Italian and foreign fishermen or organisations;

-- Other activities currently in hand.

b) <u>Financial aids</u>

It must be pointed out that as a result of the oil crisis, rising operating costs, the loss of traditional fishing grounds (due to the extension of national jurisdiction, notably in West Africa) and the enforcement of stricter conservation measures in waters frequented by the Italian fleet (North Sea, United States and Canadian zones), fisheries assistance policy has concentrated primarily on the exploitation of the resources available in the Mediterranean since Atlantic fishing cannot be expanded.

Act No. 41 of 17th February 1982, concerning the Plan for the rationalisation and development of Italian sea fisheries, defines the principles -- which are conforming to EEC directives -- applying to the administration and distribution of outright grants. These principles state that priority will be given to applications for grants made by co-operatives and co-operative unions. Applications are for:

-- The modernisation of fishing units still in commission;

-- The encouragement of co-operation in the fishing industry;

-- Land installations for the preparation, canning and marketing of fishery products and especially bluefish, but also aquaculture installations on land or at sea;

-- Financing joint fishing ventures with third countries.

Act No. 41 replaced Act No. 1457 of 27th December 1956 (Working capital fund for the fishing industry) and Act No. 479 of 28th March 1968 (Outright grants for the fishing industry). The payments made since it came into force, and especially those for 1983/84, are in respect of grants awarded in the preceding years.

It is an organic act, which provides either capital grants or low-interest loans. As distinct from the earlier legislation, it does not concentrate on increasing the size of the fishing fleet since there is an excess capacity exploiting the available fishery resources. The Act therefore provides low-interest loans for the construction of new vessels on condition that old vessels, equivalent to 70 per cent of the new tonnage, are scrapped. The principal aids available under this Act and the other legislation on fisheries and the fishing industry are indicated below.

<u>Section 10 of Act 41/82 (Central Fund for loans to the fishing industry)</u> foresees that individual or associated fishermen can obtain low-interest loans (40 per cent, with a reduction of up to 30 per cent for investments in Southern Italy, of the reference rate set each half-year by the

Treasury); such loans may not exceed 70 per cent of the total investment (15 years' maximum amortisation for vessel construction and 20 years for land and aquaculture installations).

Section 20 of Act 41/82 (Outright grants) provides for capital grants of up to 30 per cent for individual or associated producers, half of which are reserved for co-operatives and their unions.

Section 21 of Act 41/82 (Premiums for demolishing or scrapping fishing vessels) provides for a premium of L 400 000 per grt to be paid if the recipient neither builds nor buys another fishing vessel within the next five years, falling to L 200 000 per grt if a new vessel is built. Arrangements for this premium will be implemented in accordance with EEC Directive 515/83.

Section 22 of Act 41/82 (Premiums for free disposal of fishing vessels to scientific institutes) provides for the grant of a premium on condition that the vessel is given to an institute recognised by the Ministry of the Merchant Navy and used for fishery research.

Section 4 of Act 388/75 (Starting-up grants for fishery producers' organisations) provides, in accordance with EEC regulations, for starting-up aids for three consecutive years (and for five years in certain cases) to recognised fishery producers' organisations as from the date of recognition of their status. The amounts, conditions and criteria are as laid down in the relevant EEC regulations.

Section 6 of Act 388/75 (AIMA financial compensation to producers' organisations for intervention on the fishery products market) sets up a State enterprise for intervention in the agricultural market, i.e. AIMA, authorised in accordance with EEC regulations, to intervene on the fishery products market and to pay financial compensation to producers' organisations for withdrawals of fishery products when prices fall below those set by the Community. The amounts, conditions and criteria are as laid down in the relevant EEC regulations.

Section 6 of Act 388/75 (Special carry-over premium for anchovies and Mediterranean sardines) authorises AIMA, in accordance with EEC regulations, to grant the fish processing industries (or producers' organisations processing these products) the premium for withdrawal of anchovies and Mediterranean sardines used for processing and sold by producers' organisations at fixed prices. The amounts, conditions and criteria are as laid down in the relevant EEC regulations.

Section 6 of Act 388/75 (Compensatory allowance for Mediterranean sardines) empowers the AIMA, in accordance with EEC regulations, to grant the fish processing industries the allowance for sardines sold by producers' organisations at a fixed price. The amounts, conditions and criteria are as laid down in the relevant EEC regulations.

Aids are provided to reduce the cost of building vessels, including fishing vessels. Many of these aids have been replaced by similar forms of assistance. For instance:

Act No. 361 of 10th June 1982 (Outright grants for shipbuilding in general) provides for six-monthly interest rebates on the cost of building vessels: 2.75 per cent of the construction contract price.

<u>Act No. 599 of 14th August 1982, as subsequently amended (Outright grants for the shipbuilding industry)</u>, provided various degressive outright grants between 1st January 1981 and 31st December 1986 to reduce the cost of shipbuilding.

<u>Act No. 598 of 14th August 1982, as subsequently amended (Supplementary grants for shipbuilding)</u>, provided degressive outright grants from 1st January 1981 to 30th June 1986 of up to 8 per cent towards the cost of ship repair, modification and conversion.

The amount of subsidies and grants made in 1986 according to the legislation quoted above, compared with 1985, is given in Table 1.

c) <u>Public health regulations</u>

The main regulations referring to fishery products are:

-- <u>Sections 69, 70, 71 and 73 R.D. 3/8/1890 No. 7043 (Special regulations for the inspection of foodstuffs, beverages and objects for domestic use)</u>, which covers supervision of fish markets and retail outlets, the removal and destruction of fish unfit for consumption (including conserves) and the prohibition of harmful dyes.

-- <u>Section 344 T.U. of Public Health Act 27/7/1934 No. 1265</u>, which determines the scope of local health regulations generally and the penalties and fines for their infringement.

-- <u>Section 31 of D.P.R. 10/6/1955 No. 854</u> under which every commune must have regulations for a veterinary service and board, with rules of procedure, penalties and fines for infringements, and rules required by local conditions for veterinary assistance in the application of veterinary inspection regulations and control of animal foodstuffs.

-- <u>Section 1 of Act 30/4/1962 No. 283</u> on food production and marketing controls by inspection, sampling, removal and, where necessary, destruction of merchandise on public health grounds.

-- <u>Section 4 of D.P.R. 11/2/1961 No. 264</u> establishing the veterinary powers and obligations of the commune.

-- <u>Section 11 of Act 25/3/1959 No. 125</u> providing, inter alia, for the creation in fish markets of a service to inspect and control species and categories of merchandise, under the responsibility of an expert veterinarian appointed by the commune, with the necessary staff and facilities: the market director is responsible for implementing instructions from the veterinarian.

-- <u>D.M. 10/6/1959, Section 11</u> which lays down public health service rules for fish markets and stipulates that fish may only be sold after inspection. Fish landed or otherwise arriving from abroad or other communes, and fish sold outside markets or intended for processing and canning is subject to mandatory public health inspection under conditions laid down by the public health authority of the Province concerned. Section 41 of the regulation prohibits the use of unsuitable packaging prejudicial to perfect conservation.

Table 1

(L million)

	1985		1986	
	Number	Amount	Number	Amount
Central Fund for loans to the fishing industry (Act No. 41/82, Section 10)	66	7 519.7	24	1 665.5
Outright grants (Act No. 41/82, Section 20)	74	3 900.8a)	86	7 876.1
Direct grants for sea fishing (Act No. 479/68)	4	297.8	ended, replaced by Act No. 41/82
Outright grants for shipbuilding in general (Act No. 361/82)	2	791.0	2	9 936.0
Supplementary grants for the shipbuilding industry (Act No. 599/82)	5	2 248.0	2	1 114.9
Supplementary grants for the ship repair industry (Act No. 598/82)	separate data not available for the fishing industry
Financial compensation for producers' organisations for withdrawals from the market (Act No. 388/75, Section 6)	–	288.5	–	na
Special carry-over premium (Act No. 388/75, Section 6)	–	2 216.4	–	1 103.4
Compensatory allowance (Act No. 388/75, Section 6)	–	–	–	213.8
Annual start-up grants for fishery producers' organisations (Act No. 388/75, Section 4)	–	–	–	–

a) This also includes grants made under Sections 21 (Scrapping premium) and 22 (Premium for the free disposal of fishing vessels) of Act No. 41/82.

B. *International*

Within the framework of existing agreements between the EEC and certain third countries, Italy continued to obtain quotas and fishing licences subject to payment of fees by the shipowners. There were further talks on improving the running of the joint venture with Tunisia. Negotiations with Yugoslavia are still proceeding.

Under the EEC/Senegal agreement of 13th November 1983, extended on 1st October 1986 until 28th February 1988, Italy received, for 1986, fishing licences for 14 freezer trawlers (as against 17 licences in 1985). The fees payable by shipowners were the same as in 1981, i.e. F.CFA 15 000 (FF 300) per grt.

The EEC/Guinea-Bissau agreement of 27th February 1980 was extended from 16th June 1986 to 15th June 1989. Italian fishermen obtained 15 trawler licences (as in 1985) for a payment of 100-120 ECUs per grt by the shipowners.

In August 1982, a three-year agreement was signed between the EEC and Guinea-Conakry to take effect from 1st January 1983. This was extended from 8th August 1986 to 7th August 1989. Vessels registered in EEC countries are allowed to fish within the 200-mile economic zone of Guinea-Conakry upon payment of 100 ECUs per grt, payable by shipowners. In 1986, Italian fishermen obtained seven licences as against nine in 1985.

Under the EEC/United States fisheries agreement, the Italian quota for the 1986 fishing season was set at 3 500 tons, squid being the most important species. It should be added that in 1983 a joint venture was set up between Italian and American fishermen for squid fishing, with freezing and processing of the product being carried out by two factory vessels which either receive the product from American carriers or direct from Italian fishing vessels. In 1986, 11 large Italian vessels were used for this purpose.

Under the EEC/Canada agreement, Italian fishermen obtained licences to fish 2 000 tons of squid, but this quota could not be filled owing to resource shortages.

III. PRODUCTION

a) <u>Fishing fleet and fishermen</u>

As already mentioned, the 1986 figures on the number of fishing vessels are currently being established by ISTAT. However, the following comments may already be made in respect of the data available for 1984.

There was a slight increase in the number of motorised vessels used for coastal fishing; there was also an increase in tonnage, probably due to the withdrawal or replacement of smaller vessels.

The number of units in the long-distance fleet (in 1986 52, representing 34 159 grt as against 47, representing 33 067 grt) increased slightly, whereas the Mediterranean fleet did not change significantly during 1986. Many vessels were laid up for technical reasons (repairs, modernisation), but also for economic reasons.

The estimated number of fishermen was about 54 789 according to the official figures, available only for 1984; data for later years are still being established.

b)   Results

The preliminary statistics show a small increase in landings to 413 564 tons worth L 1 937 373 million. The increase in value over 1985 is estimated at 15 per cent.

The landed weight of fresh or chilled fish increased from 1985 to 1986. There was a slight increase for all products from Atlantic fisheries and other fresh products.

IV. MARKETING

a)   Utilisation

Only minor changes took place in the utilisation of catches. There was a slight increase for frozen fish and a slight decrease for canned fish, accompanied by a drop from 3 per cent to 2 per cent in the catch used for fish meal, etc.

b)   International trade

There were significant reductions (about 20 per cent) in the overall amount and value of Italian imports of all fish and fish products. Continuing large-scale imports of squid at abnormally low prices were responsible for further problems in marketing Italian production.

The amount and value of exports declined appreciably (about 30 per cent for all products).

c)   Consumption

Italian per capita consumption of fish and fish products has stabilised at around 9.5 kg per year, but it must be recalled that there are large geographical differences in consumption patterns. For the last few years there has been an increasing tendency to eat more frozen fish.

V. OUTLOOK

The profitability of Italian fisheries in the years to come will, to a large extent, depend on the restructuring and rationalisation of the fleet. As already noted, demand for Mediterranean bluefish species is not adequate given the present level of fishing; on the other hand, the special EEC carry-over premium and compensatory allowance for sardines and anchovies has partly eased the current imbalance between supply and demand, while the growing demand for crustaceans, molluscs and deep-sea fish has been responsible for a large trade deficit.

JAPAN

I. SUMMARY

Total landings of fish and fishery products in 1986 were larger than in the previous year, and it is projected that they reached 12.68 million tons, a level second to the all-time high of 12.82 million tons registered in 1984.

The consumption of fishery products remained the same as in 1985 as a whole, although there was a slight drop in that of fresh fishery products.

Imports of fishery products in 1986 at 1 870 000 tons, were up 18 per cent from the previous year. Due to the impact of the rising yen, the value of imports of fishery products in yen terms stood at Y 1 137.7 billion, down 3 per cent from the previous year. Total exports were valued at Y 217.5 billion, 24 per cent lower than in the previous year, mainly because canned fishery products for export were significantly influenced by the Yen appreciation.

The fuel oil prices in 1986 significantly dropped but the prices of other fishery materials remained relatively stable. On the other hand, the landing prices of fishery products were even lower than the level of 1984, a new low in recent years.

The international climate which surrounded Japan's fisheries became all the more severe, and it was an important task to promote fisheries in Japan's coastal and offshore waters even further.

In the face of such severe circumstances prevailing both at home and abroad, the Government continued to take various measures designed to ensure a stable supply of fishery products, maintenance and development of fisheries and a growth in the consumption of fishery products.

II. POLICIES

a)   International

In order to maintain smooth fisheries relations between Japan and foreign countries, Japan has concluded fishery agreements with many coastal nations for fishing operations. Nevertheless, the conditions for fishing operations have, year by year, become more severe, with reduced fishing quotas, strengthened controls on fishing operations and increased fishing fees.

In the United States' 200-mile zone, fishing operations have been carried out in accordance with the Japan-United States Fishery Agreement which came into effect in 1983. However, the greater part of the allocation for 1986 was withheld by the United States as the allocations were tied to the problems raised by the United States relating to Japanese salmon fishery operations, including alleged catches of North American origin salmon by the

Japanese fleet. The linkage of the salmon issue to the allocation was resolved after reaching an agreement between the United States and Japan and finally between the United States, Japan and Canada. However, due to increases in the United States' domestic annual harvest and its over-the-side purchases, the catch quota for Japan in the United States zone in 1986 dropped by 47 per cent, to 475 000 tons.

Between Japan and the Soviet Union, fishing operations have been carried out in accordance with the Japan-USSR Reciprocal Fishery Agreement concluded in 1984. However, talks on conditions for the fishing operations of both countries for 1986 faced difficulties as the Soviet side strongly stressed the necessity of the elimination of the difference in the actual catch between Japan and the Soviet Union. Finally, in April 1986 an agreement was reached incorporating, among others, a fish allocation of 150 000 tons (600 000 tons in the previous year) to each country in the other's fishery zone.

As regards talks on Japanese salmon fishing operations under the Japan-USSR Fisheries Cooperation Agreement concluded in 1985, it was agreed to set the catch quota at 24 500 tons for 1986 (37 600 tons in the previous year).

Between Japan and the Republic of Korea, fishery relations have been maintained under the Japan-Korea Fishery Agreement which came into effect in 1965.

In addition, governmental or industry-oriented fishery agreements have been concluded, inter alia, with China, Australia, New Zealand, Canada and some South Pacific islands, and Japanese fishing boats operated in their waters to catch squid, bottom fish, tuna, etc.

As regards fishery resources in their respective 200-nautical mile fishing zones, coastal nations increasingly attach importance to resources which will contribute to the development of their fishery sector. Hence, coastal States have taken a tougher stance than in the past, towards the operation of foreign fishing vessels. In such an international climate, in 1986 alone, Japan was forced to decrease its fishing fleet by as many as 500 vessels primarily in off-shore trawl and salmon fisheries.

On the other hand, Japan has continued to aid coastal nations with funds and technical cooperation on governmental and non-governmental bases for the mutual development of fisheries in cooperation with the coastal nations.

b) <u>Domestic</u>

Development of infrastructure for fishery production:

In order to work for the systematic and efficient development of the infrastructure for fishery production, efforts have been made to develop fishing port facilities, sea conservation facilities and to improve the coastal environment, to improve the livelihood of fishermen and upgrade their welfare, and to develop their living environment.

Promotion of fisheries in Japan's coastal and offshore waters:

i) In order to maintain and cultivate fishery resources and work for a sophisticated use of fishing grounds in Japan's coastal and offshore waters, the installation of artificial fishing reefs, the development of fishing grounds suitable for propagation and aquaculture, the streamlining of sea farming centres and the development of technology on farming fisheries have been accelerated.

ii) In order to accelerate up the development of fisheries in a more efficient and effective manner than in the past, centering on fisheries of the type "fisheries to breed and grow", it was decided to encourage the development of technology from new perspectives. In this context, the system of Marine-Forum 21, an organisation for the joint development of technology which includes representatives from the industry, the academic community and the government, was developed.

iii) In order to work for the systematic development of Japanese salmon resources, it was decided to carry on projects to increase the number of fry to be hatched and released, and efforts were made to develop high quality resources.

iv) In order to work for the reasonable utilisation of resources with appropriate management of fisheries and fishing grounds, as well as improvement and upgrading of fishery administration and fishery profit yields, reasonable fishery management systems were studied and good examples were surveyed and analysed as in the past.

v) To improve the knowledge of fishery resources, data on mainstay fisheries and fish species were collected and analysed as in the past. In order to use fishery resources in an efficient manner and stabilise the management of fisheries, information services on catches and sea conditions, etc., were carried out and experiments for the practical use of systems for the collection and analysis of information with the use of artificial satellites were also undertaken as in the past.

vi) Studies on the comprehensive counter-measures against fish diseases, measures for fish quarantine at hatcheries were accelerated up and projects for the reduction of damage from fish diseases were undertaken.

Measures for stabilisation of fishery management:

i) In order to strengthen fishery management in ligh of severe internal and external conditions, loans were advanced as in the past from the Agriculture, Forestry and Fishing Finance Corporation and other institutions to smaller fishermen striving to improve the structure of their fishing operations, e.g. by decreasing the number of vessels.

ii) To reconstruct the management of small and medium fishermen facing hard times, low-interest emergency loans were newly advanced.

Measures for stabilisation of the management of the processing industry:

In coping with changes in the conditions of access to raw materials for the processing industry, such as the strengthening of international controls and increases in the production of mass-catch fish, loans were advanced to step up the processing of fishery products, the development of new products and the effective use of fishery products for food.

Increase in the consumption of fishery products:

i) In order to increase the consumption of fishery products in a comprehensive manner, i.a. television programmes were presented, brochures distributed and stations for the propagation of fish diet established.

ii) In order to meet and promote the growing demand of restaurants for fishery products, certain experimental programmes to provide restaurants with stable supplies throughout the year, such as joint purchase of raw material by processing industry organisations, were conducted.

Rationalisation of distribution and processing of fishery products:

Improvements were undertaken to streamline distribution and processing facilities of major landing ports. Efforts were also made to promote the processing of under-utilised fish species for food.

III. PRODUCTION

a) <u>Production</u>

Japan's total fishery production in 1986 increased from the previous year, and it is projected to reach 12.68 million tons, this is next to the all-time high (12 820 000 tons) registered in 1984. The production of distant-water fisheries was lower than in the previous year due to cuts in the catch quotas within the 200-mile exclusive fishing zones of foreign countries and other factors. As regards coastal and off-shore fisheries, judging from the quantities landed at major landing ports, it is projected that production was larger than in the previous year. The catches of sardines, mackerel, skipjack, etc. were up while those of tuna, Pacific saury and squid were down.

b) <u>Prices of fishery products</u>

The average price of all fishery products at 51 major fishing ports in 1986 was down 14 per cent on the previous year. The prices of horse mackerel, Pacific saury and Japanese common squid rose (the catches decreased) whereas the prices of mackerel (the catches of which significantly increased) fell.

The consumer prices of fishery products in 1986 remained low throughout the year and were slightly lower than the consumers' price index.

c)   Prices of fishing materials and equipment

The domestic price of heavy fuel oil A, used primarily for fisheries, dropped significantly in 1986 but that of ropes levelled off and there were signs of slight a rise in prices of fishing nets.

d)   Earnings

A check in 1985 on coastal fisheries suggests that the revenue from fisheries decreased due to lower catches even though the prices of fishery products at port were higher than in the previous year. On the other hand, expenditures remained the same as in the previous year due to increases in costs for fishing vessels and gear, however the costs of manpower and oil dropped due to decreases in employed manpower and fuel prices. Earnings were lower than in the previous year. In marine aquaculture, the increase in the fishery revenue was higher than the increase in the expenditure, with the earnings higher than in the previous year. In small and medium-scale fisheries, profit, albeit slight, was registered for the first time in six years due to rises in the landing prices of fishery products and the drop in fuel prices.

As for the 1986 outlook, the drop in the price of fuel oil, which continued during the year, contributed favourably to the profitability of the fisheries sector. But the appreciated yen and the depreciated dollar resulted in an increase in the imports of fishery products to a new high; as a consequence prices of some fishery products dropped to some extent, and this served as a minus factor in terms of profitability.

e)   Fishing vessels

The number of powered marine fishing vessels was 400 000 units, or 2 744 000 grt, smaller both in number and tonnage than in the previous year. By tonnage group, the number of vessels in the 5 to 9 tons and 100 to 199 tons brackets increased, while those in other brackets decreased. By type of fishery, the number of gill netters and square netters operating primarily along the coast increased, whereas medium trawlers (fishing at West of 130°E) and distant water trawlers significantly decreased.

f)   Number of fishermen

The number of fishermen in 1985 was 432 000, down 1.7 per cent from the previous year. In particular, the number of persons engaged in off-shore and distant-water fisheries decreased by 4 per cent from the previous year to 95 000 as the number of vessels under strengthened international controls decreased due to the rationalisation of management. By age, the number of fishermen dropped in all age brackets other than those of 60 years of age and above, whose number was higher than in the previous year.

IV. DISTRIBUTION AND MARKETING

a) <u>Distribution of catch</u>

The total supply of fish and shellfish in 1985 amounted to 13 710 000 tons, down 2 per cent from the previous year. Of this quantity, domestic production was 11 450 000 tons, down 5 per cent from the previous year.

The supply of fish meal, feed and bait for non-human consumption dropped 6 per cent in 1985 from the previous year due to a drop in sardine production and its share of the total also slightly dropped to 33 per cent. Supplies of products for human consumption dropped 2 per cent from the previous year. Annual per capita consumption of fishery products in 1985 increased 0.3 kg. from the previous year to 35.8 kg. The supply of animal protein from fishery products increased 0.1 gr. to 18.5 gr. per person per day.

In 1986, the supply of fishery products for human consumption appeared to be the same as in the previous year. However, due to a rise in the production of sardines and mackerel, the supply of products for non-human consumption may increase.

b) <u>Consumption</u>

In recent years, changes in consumption patterns of average Japanese households have resulted in a decrease in the purchase of fresh fish and shellfish, while purchases of processed products increased. A similar tendency was also observed in 1986. By species, consumption increases were observed for tuna, sardines, skipjack, salmon and shrimp, whereas consumption fell for jack mackerel, mackerel, Pacific saury, yellowtail and squid.

c) <u>Trade</u>

Imports of fishery products in 1986 at 1 870 000 tons were up 18 per cent from the previous year, while the value at Y 1 137 700 million was down 3 per cent. By quantity imports of shrimp and prawn, tuna, squid and cuttlefish, pollack and octopus increased, whereas salmon decreased.

Exports of fishery products at 760 000 tons were down 3 per cent from the previous year and the value at Y 217 500 million was down 24 per cent. Exports of canned fishery products were, in general, down under the impact of the appreciated yen; exports of canned sardines, mackerel, skipjack and tuna decreased considerably both in quantity and value.

V. OUTLOOK

The Japanese fisheries environment has worsened, i.a. caused by tightened international control, stagnating incomes and prices, congested fish cultivating grounds and the high age of fishermen. As a result, a broad range of problems are having an adverse impact on the regional economies.

Given this situation, efforts are being made to supply high quality fishery products and stablise prices in response to demand for fishery products, with a view to working for a stabilisation of the fisheries incomes and a stable supply of food to consumers.

In addition to the management and reasonable utilisation of fishery resources for long-term stabilisation of the fishing industry, there are calls for a positive adjustment of production structures and prompt establishment of a management infrastructure in order to overcome continued difficulties. These difficulties are discernible as fishery products exports are stagnating due to the ongoing appreciation of the yen and the levelling off of fish prices.

NETHERLANDS

I. SUMMARY

Due to high fish prices and low fuel prices, the financial results of all fishing activities were favourable.

In the trade sector, the quantity imported is still larger than the amount exported, but in money terms exports are still larger than imports. In 1986, imports decreased and exports increased. For imports as well as exports EEC countries are the most important trading partners.

II. GOVERNMENT ACTION

a) <u>Resource management</u>

The general framework of the Dutch resource management was the Common Fisheries Policy of the European Community.

Since 1st February 1985, a licence system has been established for vessels taking catches to which quotas apply. Since October 1986 no more licences have been granted and through these measures the total fishing capacity has been frozen.

As in previous years, the Dutch quotas for plaice and sole were divided into individual vessel quotas. These individual vessel quotas are allotted to 540 fishing vessels (1976 regulation).

Within this regulation it is possible for the producers organisation to form one or several groups of vessel quotas. A group of vessel quotas is the sum of the individual vessel quota brought in by the participants.

The most important new measure in the national fisheries policy, was the obligatory laying-up regulation. It was decided that individual fishermen had to keep their vessels in harbour for 10 weeks per year. Fishermen participating in a group had to keep their vessels in harbour for 8 weeks per year.

It was also decided to decrease the length of the beams of vessels fishing within the 12-mile zone to 6 metres each.

In February 1986, in derogation of the EC regulation, the minimum sizes for plaice, haddock, whiting and cod were raised.

b) <u>Financial support</u>

Under EC regulations, financial aid was granted for 11 projects for the modernisation of fishing vessels, 11 aquaculture projects and 6 projects for the modernisation of the processing industry. Several deepsea trawler owners made use of the EC regulation for withdrawal of vessels. Owing to this, the capacity in the deepsea trawl sector has been considerably diminished.

In 1986, national support was given for an experiment on electric fishing and for research into storage conditions for matjes herring in cooled seawater, and for the installation of filleting machines on two pelagic trawlers (see Table 1).

Table 1

Type of aid	Brief particulars	Amount provided(Gld) 1985	1986
a) Grants to cover experimental voyages	Grants were given to undertake fisheries in American waters	0	5 855 200
	Grants were given within the framework of economising fuel costs by using certain blends instead of diesel	4 800	16 250
b) Grants for experiments with new techniques	Grants were given for an experiment for electric fishery	0	133 675
c) Grants for the modernisation of processing industries (EEC-regulation 355/77)	National contributions were given for the building and modernisation of several private entreprises	0	203 300
d) Grants for new building or modernisation of vessels and aquaculture (EEC-Regulations 1852/78-2908/83)	National contributions for vessels	684 730	282 800
	National contributions for aquaculture projects	0	229 800
e) Grants for improvements onboard vessels	Grants were given for the installation of filleting and gutting machines on board	0	73 350
f) Grants for reduction of capacity (EEC-Regulation 515/83)	Grants were given for withdrawal of vessels	0	2 543 750

c) <u>Sanitary regulations</u>

The basis of sanitary regulations is laid down in the Commodity Act (Warenwet): Fish Decree (Visbesluit) and Shrimps Decree. More specific regulations are laid down in decrees of the Commodity Board and Industry Board for Fish and Fish Products. As from 25th July 1986 a regulation setting microbiological criteria for shrimps and prawns came into force (under Article

5 of the "Decree for Shrimps"). This regulation gives criteria for pathogenic germs, especially staphylococcus aureus and salmonella.

III. PRODUCTION

a) <u>Fleet</u>

In 1986 the number of beam trawlers declined from 620 to 605. Many vessels were sold to other EC countries, in particular the United Kingdom, possibly in anticipation of their replacement by new units.

The trawler fleet consisted of 21 vessels at the end of 1986 (23 in 1985). In the mussel culture and cockle fisheries, the number of vessels remained stable. Technological innovations were realised onboard a few middle-water trawlers, with the construction of filleting equipment. In 1986, a new vessel type -- a cutter with freezing facilities -- joined the fleet.

b) <u>Operations</u>

In 1986, the fishing effort of the cutter fleet decreased due to the declining number of vessels and the enforcement of a reduction on the number of fishing weeks. As a result of the restricted effort, fuel expenditures of the cutter fleet were 5 per cent down in terms of volume.

The trawler fleet developed new activities in connection with the limited catch opportunities on EC fishing grounds. Considerable quantities, especially of mackerel, were obtained in over-the-side sales.

c) <u>Results</u>

In 1986, the financial results of all fishing activities were favourable in connection with substantially lower fuel prices. The prices of white fish remained at a high level, as in 1985.

Against this background, the cutter fleet had good financial results. For the trawler fleet there was an improvement brought about by the increased market opportunities for herring.

The financial results of the mussel culture showed profits in spite of poor yields; the low volume was compensated by a high price level.

The financial results of fish farmers differ somewhat. The expectations of the first eel farms are favourable, but catfish farms suffered from unstable markets.

IV. PROCESSING AND MARKETING

a) <u>Processing</u>

In 1986, there was a stabilisation in investments in the processing industry in the flatfish sector. Owing to the growth in processing capacity, exports of unprocessed flatfish decreased.

In the past five years, substantial investments were made for the expansion and modernisation of the herring processing industry, and the capacity of the marinating industry has grown. Due to these investments, exports of processed products increased. Several pelagic fishing vessels were equipped with filleting machines and horizontal plate freezers.

In the mussel sector investments were made which allow for the production of processed mussel products.

b) Domestic market

Because of the diminished landings of some roundfish, prices increased considerably. In 1986 the flatfish market was stable and the total volume of withdrawn fish decreased considerably. In 1986 the Commodity Board for Fish and Fish Products undertook promotion activities for herring, shrimps (Crangon-crangon) and mussels.

At the end of 1986 a third producer organisation was recognised by the Dutch authorities. In 1986 the two producer organisations applied the official withdrawal prices of the EEC.

c) External trade

i) Imports

The quantity of imports decreased from 525 774 tons in 1985 to 476 357 tons in 1986 (-9.4 per cent). The value decreased from Gld 1 016 million in 1985 to Gld 947 million in 1986 (-6.7 per cent). The most important items imported, by volume, are fishmeal and fish oil -- 59 per cent of total quantity but only 22 per cent of total import value. The decrease in imports was caused by the decline in imports of fishmeal and fish oil.

ii) Exports

The quantity of exports increased from 511 938 tons in 1985 to 559 982 tons in 1986 (+9.4 per cent). The value rose by 4.4 per cent from Gld 1 791 million in 1985 to Gld 1 870 million.

By volume, 51 per cent of total exports in 1986 consisted of frozen fish, especially herring and mackerel, corresponding to a 22 per cent share in total value. Due to the lower dollar rate, average prices decreased. Increased quotas for herring made it possible to expand herring exports and new markets were found.

Exports of fresh fish represented 10 per cent of total export quantity and 20 per cent of total export value. This decrease was due to lower landings and to the growing capacity of the processing industry.

Exports of frozen fillets remained stable and represented 6 per cent of the total export quantity and 14 per cent of the total export value.

NEW ZEALAND

I. SUMMARY

In spite of the NZ$ appreciating again during 1986 against the currencies of two of the three main trading partners, the value of exports rose 20.6 per cent to NZ$657 million over 1985. Orange roughy is now established as the most valuable fish species.

Legislation was enacted giving effect to a new policy to address the problems of the inshore fisheries. This is based on a system of individual transferable quotas (ITQs).

II. GOVERNMENT ACTION

a) Resource management

i) Inshore fisheries

An amendment to the Fisheries Act was enacted on 25 July 1986 giving effect to the individual transferable quota (ITQ) system for shallower water fin fisheries. It is New Zealand's policy to manage all fisheries, where appropriate, through a system of individual transferable quotas (ITQs).

A feature of the new policy is a "buy-back of fishing rights scheme" aimed at reducing pressure on stressed fish stocks. In the latter half of 1986 the Government paid out NZ$45 million for 15 800 tons to fishermen under the scheme.

A proposed policy for the future management of the rock lobster fisheries was presented to the industry and discussions with the industry throughout the country have been held. The Government's preferred option is for the introduction of individual transferable quotas (ITQs). The matter is still under discussion.

ii) Squid

Squid fishing has, and will continue, to operate under an interim policy until 1 October 1987 when individual transferable quotas (ITQs) may be introduced for domestic fishing companies and fishermen. From 1 October 1986 squid jigging joint ventures were replaced by a system in which New Zealand companies chartered foreign vessels.

iii) Tuna

Tuna fisheries in New Zealand waters are managed on an effort control rather than a total allowable catch (TAC) or an individual transferable quota (ITQ) system. This approach will be continued from 1 October 1986 - 30 September 1987.

iv) Deepwater fisheries

These fisheries have to date been managed by a system of individual company allocations. With the introduction of ITQs in the inshore fisheries from 1 October 1986 the deepwater fisheries were incorporated into this system and will be managed from that date through individual transferable quotas. The same monitoring, reporting and enforcement measures will apply to all ITQ fisheries.

v) Resource rentals and licence fees

New Zealand industry

1986/87 year resource rentals payable by the New Zealand industry on quota species are indicated in Table 1.

Table 1

Species or class of fish	Rate per ton of quota if taken domestic vessel ($NZ)	Rate per ton of quota taken by foreign-owned vessel or charter ($NZ)
Hake	22.50	45.00
Hoki	8.25	16.25
Ling	27.50	55.00
Orange roughy	100.00	200.00
Oreo dories	11.25	22.50
Silver warehou	25.00	50.00
Squid		
-- Southern Islands (trawl)	23.75	47.50
-- Elsewhere (all methods)	47.50	95.00
Any other species of class of fish (excluding paua, oysters, scallops and rock lobsters	3.00	6.00

Foreign licensed vessels

The schedule of foreign licensed fishing fees is shown in Table VI of the Statistical Annex.

b) Financial support

Apart from modest regional development grants, mainly to assist aquaculture, all financial assistance from the government to the industry has either been terminated or will do so in a few years.

The Export Performance Taxation Incentive terminates in the year ending 31 March 1987 and the Export Market Development Taxation Incentive terminates on 31 March 1990.

c) Economic efficiency

Marketing

On 1 October 1986 the Trade Commission Service of the Department of Trade and Industry became known as Tradcom which is a commercial organisation providing a wide range of export marketing services on a partial cost recovery basis. Tradcom has 40 posts throughout the world.

d) Other government action

i) Exchange rate

Since March 1985 the NZ$ has been floating. After falling in late 1985 the NZ$ rose again in value, particularly against the currencies of the United States and Australia; which together with Japan are New Zealand's main trading partners in fishery products.

ii) Goods and services tax

This sales tax was introduced on 1 October 1986. The tax is charged at a rate of 10 per cent on virtually all sales of goods and services. Exported fish products are zero rated and the principle effect has been to raise domestic prices.

iii) Barriers to trade

Some progress was made in easing the situation regarding access into Japan of squid caught by New Zealand domestic vessels.

e) Sanitary regulations

There were no changes to sanitary regulations during 1986.

III. PRODUCTION

a) Fleet

See Table I(a) in General Survey. The declining number of smaller vessels and the increasing number of larger vessels reflects the declining productivity of inshore fisheries and the growth of deep water fishing.

b) Operations

i) Costs of fishing

The price of bulk diesel fuel fell 22.3 per cent from October 1985 to October 1986, but then rose in the same month by 10 per cent with the introduction of the goods and services tax.

ii) Aquaculture

Marine farming

1986 saw a further 20 licences and 13 leases issued. Numbers of farms with the following species are:

Mussels	364
Oysters	149
Salmon	9
Scallops	14

There are currently 746 675 hectars of water in active leases and 13 857 992 hectars in active licences.

Freshwater farming

Since 1 January 1986 a further 7 full fish farming licences and 7 provisional fish farming licences have been issued. This brings the total number of active full licences to 20 and provisional licences to 8.

In addition, there is now a total of 24 active freshwater fish processing and dealing licences, 8 of these having been issued since 1 January 1986.

Currently, all licenced freshwater fish farms are located in the South Island. However, as a result of the broadening of the species which may be farmed under the Freshwater Fish Farming Regulations 1983, this situation may change during the coming year with the development of farms raising prawns and other species in the North Island.

Salmon farming

There are now 15 ocean ranching farms, 19 pond rearing and 12 sea cage farms.

Returns of salmon to freshwater on the east coast of the South Island have continued to grow and the future for ocean ranching is promising if the catching at sea of salmon can be adequately controlled.

c) Results[1]

i) Domestic

Total domestic landings were 160 090 tons in 1985, 3.6 per cent down on 1984. Finfish, including tuna, was down 3.4 per cent to 126 442 tons and shellfish fell by 4.8 per cent to 32 145 tons.

Orange roughy continues to assert itself as the most valuable finfish species and landings increased by 26.5 per cent to 26 649 tons. Other significant finfish landings were red cod 13 926 tons, snapper 9 090 tons, barracoutta 8 197 tons and hoki/whiptail 7 262 tons.

The rock lobster catch again demonstrated stability with landings at 5 489 tons, virtually the same as the previous year. Scallop landings fell 31.2 per cent to 3 204 tons while farmed mussel production rose by 13.2 per cent to 10 759 tons. The wild oyster catch fell by 6.4 per cent. Farmed oysters showed a significant fall in production of 24.5 per cent to 795 tons.

ii) Foreign chartered vessels

The total catch by chartered foreign vessels fell by 7.0 per cent in 1985 to 145 251 tons. The drop was due to a reduction in the catch of squid by 21.5 per cent to 43 084 tons due to considerable difficulty encountered in recruiting vessels, as the finfish catch showed a slight increase to 102 161 tons. The orange roughy catch continued to decline to 13 349 tons (17.6 per cent).

iii) Foreign licensed vessels

The foreign licensed vessel catch in 1985 was 88 531 tons considerably down on the previous year (25.2 per cent). Squid catch declined by 29.2 per cent to 42 718 tons and finfish by 23.5 per cent to 41 976 tons.

Country allocations and catches for 1983/84 and 1984/85 seasons are shown in Table V in the Statistical Annex.

IV. PROCESSING AND MARKETING

a) Processing

The first experimental fishing quotas for hoki were granted in 1986 to three New Zealand companies for the at sea production of surimi. Amongst the developments in processing were squid burgers and stir-fry squid.

Investment also began for the on-shore processing of hoki into surimi.

The modernisation of processing technology continued, e.g. laser filleting machines, electronic weighing machines.

The number of employees engaged in processing is estimated to have risen from 3 600 in 1985 to 4 000 in 1986.

b) Promotion activities

The Fishing Industry Board in 1986 undertook a wide range of promotional activities including TV and radio advertising, trade fairs and seminars for dieticians, chefs and school teachers.

Promotion continued aimed at achieving greater acceptability of darker flesh fish.

c)   External trade

The total value of exports increased 20.6 per cent in 1986 to NZ$657 million f.o.b. over 1985, while the volume rose 9 per cent to 158 183 tons. The increase in value is in spite of the appreciation of the NZ$ again in 1986 against the US$ and the Australian dollar.

Rock lobster exports increased in value 17.4 per cent to NZ$104.3 million while the volume increased 17.6 per cent to 3 158 tons.

Orange roughy is now established as the most valuable finfish; exports in 1986 being worth NZ$160 million. Other significant finfish exports were hoki NZ$64.8 million, snapper NZ$41.9 million and warehou NZ$17.2 million.

Squid exports showed a considerable drop of 36.9 per cent to NZ$64.5 million. Other shellfish exports were paua (abalone) NZ$11.9 million, farmed mussels NZ$12 million, scallops NZ$8 million and farmed oysters NZ$5.6 million.

There was an encouraging increase in salmon exports by 100 per cent to NZ$3.1 million. This industry is still based on the caged product. The major markets for New Zealand exports of fish products in 1986 were the United States NZ$262.8 million, Japan NZ$232.2 million and Australia NZ$73.7 million (see Table IV(b) in the Statistical Annex).

Imports of fish products increased 17.4 per cent in value in 1986 to NZ$42.7 million CIF.

V. OUTLOOK

It is intended to bring all commercial fish species under the system of individual transferable quotas unless there are sound reasons for not doing so. Squid will be brought under the ITQ system from 1st November 1987 and consideration is being given to including jack mackerel. Paua, rock lobster and scallops will come under the ITQ system in the future.

NOTES AND REFERENCES

1.   The fishing season is from 1 October to 30 September. Calendar year catch statistics thus cover part of two seasons.

NORWAY

I. SUMMARY

Preliminary figures show that the total Norwegian catch in 1986 decreased to the lowest quantity since 1964. However, the total first-hand value of the catch increased due to a relative shift from less valuable species in 1985 to more valuable species in 1986.

There was an increase in total export value compared to 1985. The most important products were frozen fillets and fresh farmed salmon.

The export of farmed salmon continued to show a considerable increase.

The stock situation for cod, haddock and herring improved in 1986. The capelin stock in the Barents Sea is, on the other hand, at an extremely low level, and the stocks of shrimp and saithe have decreased from 1985 to 1986.

II. GOVERNMENT ACTION

a)   Resource management

The catches of Arcto-Norwegian cod have been relatively low over the last few years. Expectations for the future are, however, more optimistic, as research indicates increased catches. Although the total quota was raised from 1985 to 1986, the stock situation was still vulnerable and necessitated the continuing of approximately the same regulatory measures north of 62°N as in 1985.

Due to small catches the implementation of catch regulations for cod were partly omitted during Autumn 1986. Thus, the vessel quotas for trawlers were increased and an announced fishing ban at Christmas time was not accomplished.

The stock situation for capelin in the Barents Sea has worsened dramatically since Autumn 1985. Due to this development only a very small quota was fixed for the winter season 1986, and the capelin fishery was completely closed during the summer season. The capelin quota for the winter season was divided between the purse seiners and the trawlers, and the vessel group quotas were again divided into individual vessel quotas. The total Norwegian capelin quota in the Jan Mayen fishery zone increased from 191 750 tons in 1985 to 198 230 tons in 1986. About 50 000 tons of the quota was by an agreement with Iceland allocated to be fished in Icelandic waters during the winter season. Altogether 67 purse seiners participated in this winter fishery, and they were allowed to make one trip each.

Contrary to 1985 EEC and Norway managed to reach a quota agreement in the fisheries for herring in the North Sea for 1986. The Norwegian fishery was regulated partly by individual vessel quotas and partly by trip quotas. In addition a part of the total quota was reserved the inshore coastal fishery.

Norway, Sweden and EEC did not manage to reach a quota agreement for the fishery for herring in the Skagerrak for 1986. The Norwegian fishery was, however, regulated within a total quantity corresponding to the traditional Norwegian share of TAC (which the three parties during consultations did not formally agree upon but had found reasonable 147 000 tons).

The improved stock situation for Atlanto-Scandian herring north of 62°N permitted an increase of the total quota from 60 450 tons in 1985 to 162 750 tons in 1986. As in 1985, the fishery was regulated by vessel group quotas, which for the purse seiners again were divided into individual vessel quotas. The trawlers were regulated by trip-quotas and the coastal fishing vessels were regulated by quotas allocated to groups of vessels by size.

Norway and the EEC reached agreement on a total quota of 55 000 tons of mackerel in the North Sea. The Norwegian quota was divided between purse seiners, trawlers and coastal vessels. The share allocated to the purse seiners was divided into individual vessel quotas.

In addition Norway achieved in exchange of quotas in the Norwegian Zone a mackerel quota in Faroese waters and west of 4°W in the EEC zone. Norwegian vessels could also fish mackerel north of 62°N in the Norwegian Economic zone and in international waters. The fishery west of 4°W was regulated by an equal quota per vessel, whereas the fishery north of 62° was unregulated except for a total quota.

The total Norwegian catch of blue whiting in 1986 (285 300 tons) was higher than in any previous year. In EEC and Faroe Islands waters, Norway had quotas of 250 000 tons and 50 000 tons respectively. Except for the total quotas the blue whiting fishery was unregulated.

Due to the poor stock situation for saithe north of 62°N the catches dropped from 108 000 tons in 1985 to 62 000 tons in 1986. Some fishing grounds had to be temporarily closed due to a relatively large share of undersized fish. The stock situation was unsatisfactory also for saithe in the North Sea, and the Norwegian catches dropped from 94 000 tons in 1985 to 60 000 tons in 1986. Except for a total quota there were no quota regulations in the Norwegian saithe fisheries in the North Sea and in coastal waters. The coastal purse seine fishery was, however, restrained due to relatively large quantities of undersized fish.

The stock of shrimps in the Barents Sea declined during 1986. Accordingly the Norwegian catches dropped from 90 000 tons in 1985 to 56 500 tons in 1986. Except for the closing of some fishing grounds due to high by-catches of juvenile cod and haddock there were no regulations in the Norwegian shrimp fisheries in the Barents Sea.

b)   <u>Financial support</u>

A main objective of the Government has been to secure fishermen's incomes at about the same level as that of industrial workers. To secure a normal activity within the fishing industry, and to maintain settlements in sparsely populated areas in the northern and western parts of Norway, it has been necessary to grant financial support to the fishing industry. Another objective has been to increase the profitability (excluding Government support) in the Norwegian fisheries.

The financial support package, excluding loans made avilable through the State Fishery Bank, is governed by an agreement between the Norwegian Fishermen's Association and the Norwegian Government. In 1986 a total amount of NKr 1 330 million was made available to the fishing industry. In 1983 and 1984, the total financial support each year was NKr 1 100 million and in 1985 the support was NKr 1 375 million.

Existing support schemes were transferred to 1986, roughly along the same guidelines as in 1985. Table 1 presents the total value of the annual financial support package and presents figures that show the relative share of the various support schemes.

Table 1

Val. : NKr million

	1983	1984	1985	1986
Total value	1 100	1 100	1 375	1 330
Price support to first-hand sales of fish	48.7%	54.2%	52.9%	47.9%
Support to reduce operational costs	26.9%	21.8%	17.9%	17.3%
Social schemes	12.9%	11.5%	13.7%	16.0%
Structure programmes	10.5%	7.7%	12.5%	16.2%
Other programmes	1.0%	4.9%	3.0%	2.6%

A main objective has been to increase the relative share of support given to promote structural reorganisation and innovations in the fishing industry. Here, there has been a gradual increase over the years 1983-1986. Support to promote structural reorganisation comprises scrapping schemes for capacity reductions, readaptation schemes and schemes to improve the efficiency of the fishing fleet and the fish processing industry.

The purpose of the readaptation schemes is to support investments to enhance rationalisation and reorganisation in the fishing fleet and the fish processing industry. Two readaptation schemes which were introduced in 1984, one for the fleet and one for the industry, were implemented on a broader basis in 1985 and 1986. Further details about readaptation schemes for the fleet are given in Chapter IIIa) Fleet.

Information about scrapping schemes for capacity reduction and schemes to improve the efficiency of the fishing industry is given in c) Economic efficiency, below.

In 1986 the new fishing vessel financing scheme, endorsed by the Parliament on 22 November 1985, was elaborated and effected. The financing scheme is similar to the financing scheme permitted by the OECD terms on export financing of vessels. The new financing scheme, which was administered by the State Fishery Bank, was in 1986 bound by an upper limit of a total sum of contracts of NKr 640 million. The new financing scheme is not expected to increase the volume of new fishing vessels to be built. The State Fishery Bank also applied a grant scheme of NKr 21 million to ease financing of coastal fishing vessels in accordance with given regional priorities.

Fish processing companies located in certain remote areas can, as most manufacturing industries in these areas, benefit from special grants and loans from the Regional Development Fund.

c)   Economic efficiency

In 1986, NKr 123 million were made available for capacity reductions in the fishing fleet and in the fish processing industry.

The Government continued the scrapping schemes for fishing vessels above 10.67 metres, which was introduced in November 1984, and for purse seiners which was introduced in 1985. A total of NKr 125 million was granted and about 194 vessels were scrapped in 1986. The scrapping scheme will be continued in 1987.

The Government also continued the special schemes giving grants to processing plants as an incentive to close down. This action is considered necessary in order to both reduce the number of plants and the capacity in the processing industry, thereby securing an adaptation of the industry to the fishing resources available. A total of NKr 37.2 million for this purpose was granted to five fish processing plants, and to five other plants previously active in the production of fish meal.

In 1986, NKr 18.7 million was paid out to improve the efficiency of the fishing fleet, and NKr 12.9 million for improvements in the fish processing industry and in the marketing of fish and fish products. About NKr 7.4 million out of the NKr 12.9 million was paid out to the fishing industry as well as to sales and export organisations, in order to increase the efficiency of different marketing activities.

d)   Bilateral arrangements

In Norway, fisheries management decisions result from an elaborate process of consultations between a multitude of affected parties, both at the domestic and international levels. The scientific advice provided by ICES in relation to total allowable catches (TACs) is fundamental to any subsequent management decisions.

In 1985, for 1986, Norway had consultations regarding fishing arrangements with the USSR, the EEC, the Faroe Islands, Iceland, Sweden, Poland, the German Democratic Republic, Portugal and Spain. In addition, Norway participated in multilateral consultations on regulatory measures in the North East Atlantic Fisheries Commission (NEAFC) and in the Fisheries Commission for the North West Atlantic (NAFO).

For 1986, Norway agreed on total quotas in the North Sea and the Barents Sea with the EEC and the USSR, respectively. During these consultations, the allocation of total quotas of joint stocks between the two parties was established. The yearly reciprocal arrangements give each of the two parties the right to fish parts of the quotas of the joint stocks in the zone of the other party. The agreements with the Soviet Union provide for a reciprocal capelin fishery of 120 000 tons, as well as flexible arrangements in relation to cod and haddock fishery on a transboundary basis.

In addition to the reciprocal fishing rights for the joint capelin stock, agreed upon with the USSR, 711 030 tons were allocated to foreign countries (including USSR quotas) in the Norwegian Economic Zone north of 62°N (in the Barents Sea and the Norwegian Sea). This total allocation consisted of 402 500 tons of blue whiting (including a quota to the USSR of 385 000 tons which may be fished in the Norwegian Economic Zone and in the fishery zone around the island of Jan Mayen); 78 550 tons of redfish; 99 130 tons of cod; 23 800 tons of haddock; 7 800 tons of saithe; 7 250 tons of Greenland halibut; 48 000 tons of capelin (a genuine USSR quota); 3 500 tons of catfish and 40 500 tons of other species.

In addition to the above-mentioned quota of blue whiting, the USSR was allocated a total of 235 500 tons including 65 000 tons of redfish. EEC fishermen were allowed to catch 43 750 tons, which included 11 000 tons of cod. The Faroe Islands received a total quota of 20 600 tons. The German Democratic Republic, Poland and Portugal were allowed to catch a total of 26 100 tons of which 12 500 tons was blue whiting.

Regarding joint stocks in the North Sea, EEC fishermen were permitted to catch 557 250 tons in the Norwegian Economic Zone south of 62°N, of which 230 000 tons was Norway pout and sandeel. The Faroe Islands were allocated a quota of 23 200 tons. A total of 2 000 tons was allocated to Poland and 5 875 tons to Sweden.

In the fishery zone around Jan Mayen and in addition to the above-mentioned quota of blue whiting allocated to the USSR, foreign countries were allocated quotas of altogether 168 275 tons, mainly capelin (to Iceland) and blue whiting (to Poland, the EEC, the German Democratic Republic and the Faroe Islands).

e) Other government action

By regulation of 28 July 1986 a quota of 18 600 tons of cod was introduced for the fishery of non-coastal states to the stock in 1986 in the Fishery Protection Zone around Svalbard.

By revision of regulations of 6 April 1982 relating to minimum mesh size, minimum size of fish and by-catches, a minimum mesh size of 100 mm is introduced in the trawl fishery for demersal species in the Norwegian Economic Zone south of 64°N. The regulation takes effect as from 1 July 1987.

The Norwegian Government 3 July 1983 decided a temporary ban on commercial whaling as from 1988. By Royal Decree of 8 August 1986 the Government has appointed a group of scientists to evaluate the present situation of the minke whale stock in the North-East Atlantic.

f) Structural adjustments

The different schemes for readaptation and scrapping of fishing vessels and processing plants are considered to have had a positive effect on economic efficiency and overall performance of the fishing industry, as mentioned in section c) Economic efficiency.

g) Sanitary regulations

There have been no significant changes in national legislation on this matter.

   i) Agencies having jurisdiction or related functions

   -- Ministry of Fisheries, Box 8118 Dep, 0032 Oslo 1;

   -- Directorate of Fisheries, Box 188, 5001 Bergen;

   -- Canning Industries Institute for Control, Box 329, 4001 Stavanger;

   -- Ministry of Health and Social Affairs, Box 8011 Dep, 0030 Oslo 1;

   -- Health Directorate, Box 8128 Dep 0032 Oslo 1.

   ii) Laws and regulations

   -- Act of 28 May 1959 on quality control of fish and fishery products;

   -- Ministry Decree of 1 July 1986 on general quality regulation of fish and fishery products, which will come into force from 1 January 1987;

   -- Ministry Decree of 12 October 1972 on general quality regulation of canned fish;

   -- Royal Decree of 8 July 1983 on general food regulations;

   -- Royal Decree of 8 July 1983 on hygiene regulations.

   iii) Labelling requirements

   -- General regulation of 25 September 1986 on labelling of prepacked food;

   -- Ministry Decree of 11 November 1987 on general regulation of labelling of prepacked seafood for domestic trade;

   -- Ministry Decree of 23 January 1967 on labelling of canned fish and fishery products.

   iv) Sanitary control

   The Ministry Decree of 1 July 1986 has general provisions concerning fishing, handling, processing, storing, packing, transportation, etc.

of all fishery products except canned fish. The Directorate of Fisheries shall examine all fishery products before sale/distribution, and also seafood of foreign origin. Imported seafood cannot be delivered from Customs Services without approval from the Directorate. The Directorate decides when the goods shall be controlled before delivering and whether samples must be taken.

v) <u>Canned products</u>

According to Ministry Decree of 12 October 1972 all control of production of canned products is carried out by the Canning Industries Institute of Control.

III. PRODUCTION

a) <u>Fleet</u>

Due to the general problem of increasing costs and, for some important species, declining resources, the number of new fishing vessels above 25 grt was extremely low in 1983 and 1984 (11 and 9 respectively). In 1985, the building activity increased due to expected improvement in the stock situation for cod and herring. The positive trend accelerated in 1986 as 32 new fishing vessels above 25 grt were built for Norwegian fishermen.

Twenty of these 32 vessels were multi-purpose fishing vessels for gill-netting, shrimp trawling and Danish seining, while 2 of the vessels were large purse-seiners, 2 were stern factory trawlers, 2 shrimp freezing trawlers, 1 large stern trawler and 4 new-developed scallop factory trawlers. In addition, 2 former supply vessels were rebuilt and converted into scallop factory trawlers. One former coastal cargo vessel was rebuilt as a coastal shrimp trawler and 1 coastal fishing vessel was reconstructed and re-registered as a fishing vessel. In 1986 no fishing vessels above 25 grt were imported.

A total of 36 vessels above 25 grt entered the Norwegian fishing fleet, while 51 vessels were withdrawn from fishing. Seven of the withdrawn vessels were sold abroad (2 purse seiners, 1 shrimp trawler, 1 saithe purse seiner, 1 stern factory trawler, 1 longliner and 1 multi-purpose coastal fishing vessel), 3 were sunk and 41 were scrapped. The reason for the high number of scrapped vessels is the general scrapping schemes for coastal fishing vessels and for purse seiners which were introduced by the Government in 1985 and continued in 1986. A relatively high number of new fishing vessels and a high number of scrapped vessels have led to a lower average age in the fishing fleet.

As mentioned in the section on Financial support, the Government has also supported different readaptation and cost reducing schemes in 1986. The schemes consist of supporting further introduction of energy saving equipment, investment for improving the working conditions onboard fishing vessels, and the establishment of joint management services.

As in the previous year, technological innovations for fishing vessels were focused on new hull designs for reduced friction and larger propeller dimensions, estimated to reduce the fuel costs by up to 20-30 per cent.

In 1986 fillet machinery was installed onboard purse seiners and longliners, changing them from purse fishing vessels to combined freezer/factory/fishing vessels.

A new design for stern factory vessels with a more compact hull design was introduced in 1986. A number of contracts were signed for renewal of both former stern factory trawlers and former salt fish trawlers.

In 1986, the first 4 scallop factory trawlers of a new design were introduced and further contracts were signed.

Design of large scale surimi producing vessels was also introduced in 1986. The first 3 allowances were granted by the Norwegian Government, for building vessels for this purpose. The vessels will not be ready before 1988.

b) Operations

In 1986, 8 of the 9 former salt fish trawlers were in operation as stern factory trawlers. Most of them will, however, be replaced by newbuilding during 1987 and 1988.

Scallop trawling combined with handling and processing onboard was introduced to the Norwegian fisheries in 1985. In 1986, there has been a substantial increase in this fleet as 4 new and 2 rebuilt factory trawlers entered the fishing fleet, and 4 or 5 fishing vessels were partly rebuilt for this activity.

c) Results

Preliminary records indicate that the total catch decreased from 2.2 million tons in 1984 to 2.0 million tons in 1985, which is the lowest quantity since 1964. The total first-hand value however showed an increase in nominal terms from NKr 4.5 billion to NKr 4.9 billion.

There were, however, considerable variations amongst the main species. The low quantity in 1986 was mainly due to a further sharp decline in the catches of capelin (from 641 000 tons in 1985 to 272 000 tons in 1986).

Also the catches of Norway pout decreased sharply. The catches of herring, sandeel, mackerel and blue whiting increased both in quantity and by value. All in all, for the pelagic fish group there was a decrease both in total catch and in total first hand value.

The gadoid fish group, however, showed an increase in the total first-hand value (20 per cent) and a decrease in the total catch (5 per cent). The main reason was increased prices for the main species in this group (cod, saithe, haddock, etc.). The catches of saithe decreased sharply from 1985 to 1986.

The crustaceans (mainly shrimps) showed a modest increase in total first-hand value and a sharp decrease in total catch. The scallops catches increased from almost zero in 1985 to 15 400 tons and NKr 100 million in 1986.

Table 2 indicates the percentage distribution of the total value by pelagic fish, gadoids, shellfish and others for the period 1983-1986.

Table 2

	1983	1984	1985	1986
Pelagic fish	31	29	24	20
Gadoids, etc.	50	50	53	58
Shellfish	16	18	19	18
Others	3	3	4a)	4a)

a) Includes scallops.

In the annual costs and earnings studies for the harvesting sector in Norway, the wage-earning-ability (defined as gross income minus total costs, excluding wages per man/year) is the most commonly used measure of profitability. On average, this increased from NKr 101 000 in 1984 to NKr 106 000 in 1985 for fishing units above 13 metres overall length, and smaller fishing units (8-13 metres overall length) showed an increase in the wage paying ability per man/year from NKr 42 000 in 1984 to NKr 52 000 in 1985. The average wage per man/year may be different from the wage earning ability per man/year.

Figures for 1986 are not yet available. However, profitability in the gadoids fisheries is expected to have increased in 1986, while in the pelagic fisheries and shrimp fisheries the profitability is expected to have decreased in 1986.

d) **Fish farming**

The growth in the seafarming of salmon was in 1986 even larger than in 1985. The total live weight of rainbow trout in 1986 was about the same as in 1985.

In Table 3, figures are presented for live weight and first-hand value of farmed fish. This table also shows the increasing number of sea farms. Preliminary figures for 1986 indicate that the total quantity of farmed salmon and rainbow trout will amount to about 50 000 tons, a 47 per cent increase from 1985.

Table 3

	1983	1984	1985	1986a)
Salmon (tons)	17 298	21 881	29 473	45 494
Rainbow trout (tons)	5 405	3 569	5 141	4 248
Total (tons)	22 703	25 806	33 796	49 742
First-hand value (NKr million)	776	960	1 434	1 717
Number of delivering sea farms	362	410	487	500
Total licences for salmon and trout farming per 31/12	483	586	607	689

a) Preliminary figures.

The total first-hand value shows an increase of about 20 per cent, i.e. a lower increase than the increase in the total weight delivered. This is due to an average first-hand value about 17 per cent lower than in 1985. Production is expected to increase also in the future.

The major part of the fish farms are sea water fish farms with floating net cages situated in the western and northern parts of Norway. Very few of them are situated in the inner parts of fjords or near big rivers. Pollution from the environment has therefore not been a serious problem for Norwegian fish farming in 1986.

IV. PROCESSING AND MARKETING

a)   Handling and distribution

As regards handling and distribution, there has not been any major development or change in 1986.

b)   Processing

In 1986, a relatively higher share of the catches was utilised to fresh and frozen products than in 1985, while fish meal and fish oil production declined. About the same share of the catches as in 1985 was converted into dried and salted products. There have been no major developments or changes in the technology of processing in 1986. The lack of capelin catches from the Barents Sea has caused serious problems for the oil/meal processing plants in North of Norway. In Norway as a whole only 18 plants were producing at the end of 1986, compared with 28 plants two years earlier.

c)   Domestic market

In 1980, annual consumption of fish was estimated to be just over 30 kg per capita. By 1983, consumption was estimated to be 36 kg per capita. Major uncertainties are, however, attached to these estimates. The Government's aim is to increase the consumption to at least 40 kg per capita per year by 1990.

In order to achieve this, a programme has been launched to assist private companies and organisations of fishermen in their efforts to promote domestic consumption of fish.

d)   External trade

The total imports of fish and fish products in 1986 amounted to NKr 790 million, an increase of about 27.3 per cent. The two most important trading partners were Finland (NKr 194 million) and Denmark (NKr 136 million).

The total export value increased by about 8.8 per cent to NKr 8 911 million. The most important products were frozen fillets (including breaded fillets) (NKr 165 million) and fresh reared salmon

(NKr 1 459 million). Excluding exports of hardened fat, the three most important outlets were the United States (NKr 1 456 million), the United Kingdom (NKr 1 320 million) and France (NKr 855 million).

Exports of some fish and fish products to EEC countries are still subject to considerable tariff barriers, leaving significant products outside the "Fishery Letter" attached to the free Trade Agreement between the EEC and Norway, and thus given tariff reduction.

There have been no exports of stockfish to Nigeria in 1986 because of the non-issuing of import licences. This has led to structural changes in production, especially for the small processing plants. Important export products, such as fresh salmon, enjoy low tariffs bound in GATT.

V. OUTLOOK

Within the cod fisheries, research indicates increased catches of cod and haddock and reduced catches of deep sea prawns. Within the pelagic fish group, the herring quota is expected to be at a moderate level while the stock situation of capelin indicates low quotas. In 1987, there will be no capelin quota.

The support schemes given to promote structural reorganisation with capacity reductions will be continued and given equal importance as last year.

PORTUGAL

I. SUMMARY

1986 was an important year, being marked by a recovery in the fisheries sector and by Portugal's integration into the Common Fisheries Policy of the EEC.

A start was made in applying Community regulations to fish resources and, with the cooperation of the economic partners, national laws governing fisheries and aquaculture, as well as regulations affecting fishing in the Atlantic, inshore waters and fishing vessels, were prepared. However, these laws and regulations could not be approved and promulgated until 1987. A system of control to qualify for Community aid was studied and the results have now been published.

Application to the Community regulations regarding structures to the fishing fleet, aquaculture and the fish processing industry began, and these regulations have been widely observed by the economic partners. 64.2 per cent of the applications submitted have been approved by the EEC, i.e. 10.4 per cent of approvals relating to 11 countries. Government aid for the construction and modernisation of vessels amounted to Esc 221 million.

Ten producers' organisations which received official recognition in 1986 were granted financial support amounting to Esc 24 812 million. The system of the common market policy began to be applied, especially within the sardine sector.

The decentralised vocational training system was fully reintroduced and its legal and constitutional framework defined. Construction and modernisation of suitable premises were started in 8 ports and short-term training courses introduced for the principal fishing communities. The number of enrolments in Lisbon, where the headquarters are situated, has tripled, and programmes and teaching methods have been reorganissed.

1986 was a very important year for the restoration and modernisation of auction halls and for the storage and refrigeration equipment of fishing ports. The setting-up of 9 new auction halls and the reorganisation of 5 existing infrastructures were pursued and almost completed.

II. GOVERNMENT ACTION

a)   Resource management

Due to the presence of territorial waters and an exclusive economic zone which cover a total ocean area of over 500 000 square nautical miles, and as fishing activities will tend to become more diversified in the near future and because of the responsibilities deriving from membership of the European Communities, the need to develop the fisheries surveillance system in

Portuguese waters and to adapt it as soon as possible to the new situation was recognised.

Consequently, a joint technical committee consisting of representatives of the Ministry of National Defence and of the fisheries sector has been created by a combined decision of the Ministers of Defence and of Food and Fisheries dated 24 April 1986. Its task is:

-- To submit proposals for improvement of the existing surveillance system by, inter alia, creating a liaison between the maritime control service and the aerial detection service and linking them to the National Fisheries Data Bank.

-- To design the future fishing operations inspection and detection system and prepare a definitive plan.

A report has been produced by this committee and submitted to the EEC Commission with a view to joint financing.

The possibilities offered by the Ria Formosa Nature Reserve, which is internationally recognised as a lagoon area of great importance for the preservation of natural ecosystems, bird nesting, fish spawning and mollusc reproduction, have led to the launching of an integrated, coordinated programme. This is the most effective method of preserving and exploiting the possibilities in question while harmonising social and economic interests.

Consequently, in accordance with Resolution n° 63/86 of the Council of Ministers issued on 5 August, the Office for the Coordination of the Ria Formosa Integrated Development Programme was created with the following objectives:

-- To assist and develop food production through the implementation of aquaculture, pisciculture and mollusc farming schemes.

-- To expand and reorganise the infrastructures and equipment which serve as a support for productive activities.

-- To conduct surveys and research aimed at ensuring better use of existing resources and the development and improvement of the Ria Formosa.

-- To preserve and improve the environment and natural resources and to combat pollution by controlling discharges of pollutants and solving essential decontamination and drainage problems.

Mention must also be made of:

-- The establishment of the National Fisheries Data Bank.

-- The development of experimental fishing.

-- The implementation of Community laws making it obligatory to keep fishing logs and to inspect landings, especially of fish species subject to TAC and to Community quota restrictions.

-- The redrafting and harmonisation of existing national laws in conformity with Community laws.

b) Financial assistance

In accordance with EEC Regulation 2908/83, pluriannual guidance programmes for the fishing fleet and for aquaculture were approved by Decisions 86/350/EEC and 86/351/EEC, respectively, of 14 July. In accordance with EEC Regulation 355/77, a model programme was prepared to serve as a basis for the selection of projects covered by this Regulation.

Financial assistance amounting to Esc 2 367 978 000, comprising Esc 1 732 157 000 from the FEOGA and Esc 635 821 000 from the Portuguese Government, was allocated as indicated in Table 1.

Table 1

('000 Esc)

	FEOGA	Portuguese Government
Fishing fleet Construction of new vessels and modernisation of existing ones	1 187 180	427 016
Aquaculture Conversion of salt-beds, improvement of water quality, installation of new fish-breeding stations, etc.	64 347	43 504
Structures and marketing Olive oil and sauces, frozen fish, algae, etc.	480 630	165 301

The amount of assistance granted, strictly on a national basis, was Esc 564 275 000. This consisted of Esc 346 012 000 in non-repayable grants to producers' organisations for experimental fishing, renewal of the fleet, and salt production, and Esc 218 263 000 to be spent by the State Secretariat for Fisheries for research and technical assistance, computerisation, new office accommodation, vocational training and the Data Bank.

-- A method was devised (Decree n° 341-A/86 of 8 October) for the application of the norms laid down by EEC Regulation n° 2908/83 concerning investment projects for submission by individual fishermen, fishermen's cooperatives or fishing companies established in or having their head offices in Portugal, and projects for the restructuring, modernisation and development of the fishing fleet and of aquaculture.

-- Rates of Portuguese financial participation in projects co-financed by the Community, as well as in projects receiving only national assistance, were also defined.

-- In accordance with EEC Directive n° 83/515, a system of financial aids was created (Decree n° 341-B/86 of 8 October) for initiatives aimed at achieving a temporary or permanent reduction in the number of fishing vessels.

-- Decree n° 422-D/86 of 24 December defined the modes of application to Portugal of the norms laid down by EEC Regulation n° 2909/83 regarding the granting of financial encouragement to experimental fishing operations.

In the framework of EEC Directive n° 83/515, criteria have been defined for the selection of projects which would qualify for scrapping subsidies:

-- Gillnetters, salters only, which operated in the Canadian EEZ in 1985.

-- Side trawlers, salters only, more than 30 years old.

-- Gillnetters, partial freezers but over 30 years old.

-- Vessels over 30 years old which have not been extensively converted within the past ten years and are no longer fit for use in prevailing conditions or vessels whose conversion would involve a high level of investment.

-- Other vessels which engaged in other types of fishing in 1985 or 1986 with obvious difficulties.

In all cases the vessels will have to comply with the following conditions:

-- Their conversion must be technically or economically impracticable, so that their scrapping does not impair the continued fishing capacity of Portugal's distant-waters fishing fleet.

-- Scrapping must not have serious effects on employment in the company concerned.

The application of these criteria led to the authorised scrapping of 8 vessels totalling 9 640 grt with grants totalling Esc 791 862 000, to a proposal for temporary laying-up (Esc 4 391 000) and to an experimental fishing campaign (Esc 12 000 000).

c) Profitability

The aim of the programmes mentioned above is to improve profitability through better use of available capacity, either by building more modern vessels to replace obsolete ones or by modernising existing ones, as well as by reducing, through scrapping, the number of distint-water fishing vessels whose operations, because of their characteristics, are of less than average profitability. Due to the exploitation of the Economic Zone and to the increased value of the product, the reorientation of fishing operations towards alternative fishing-grounds and the improvement in conservation conditions on-board have contributed radically to an increase in profitability.

The following measures have been adopted in the areas of processing and marketing:

-- A special plan for the fish processing industry has been prepared and submitted to the Commission for approval. Two sets of investment projects covering the year 1986 for the fish processing industry (frozen, and canned in olive oil and sauce) have been submitted to the EEC Commission. Nine projects have already been approved. Projects for 1987 have also been submitted.

-- Ten producers' organisations were recognised and are now in operation. Eight have received financial assistance to enable them to implement the functions assigned within the framework of the Market Organisation for Fish and Fishery Products. Implementation of the pre-membership financial assistance programme for producers' organisations is now in its final stage. Most of the investment funds have been channelled into fishing equipment.

-- The Community's first-hand sale intervention system has been introduced by grading fish products and fixing prices for fish withdrawn. Compensatory payments are now being made to producers' organisations.

-- The restitution system for olive oil used in fish canning, as provided for under Community Regulations, has been brought into operation. This enables the canning industry to obtain olive oil at international market prices and thus operate under more competitive conditions.

Esc 350 million were allocated to cover olive oil restitution and compensatory payments in respect of the period from 1/1/1986 to 31/12/1986. Esc 315 007 000 were made available to the INGA (National Institute for Agricultural Guarantees). Of this, Esc 310 002 000 were spent by the former Portuguese Fish Canners' Institute and the Portuguese Institute for Fish Preserves and Products. This sum was allocated as follows:

	('000 Escudos)
Restitution to producers in respect of olive oil - EEC Regulation n° 136/66, art. 20bis	253 714
Compensatory payments - EEC Regulation n° 3796/81, art. 13	56 288

The Portuguese Institute for Fish Preserves and Products has assumed responsibility for controlling the quality of cod -- a role which was formerly held by the Cod Trade Control Commission, now disolved.

In collaboration with the National Institute for Fishery Research and the Ricardo Jorge National Institute for Public Health, the Institute for Fish Preserves and Products has set up a system for the sanitary inspection and certification of bivalve molluscs intended for export to Spain.

Together with the traditional methods of quality control in the fields of bromatology, microbiology and packaging, the histamine control method has now been introduced.

d) <u>Bilateral/Community Agreements</u>

Agreements to amend the fishing agreements between the EEC and the Governments of Guinea Bissau (16 June), Equatorial Guinea (27 June) and Guinea Conakry (8 August) were concluded.

The purpose of these agreements is to define the principles and rules which should govern fishing by vessels flying the flags of Community Member States in waters under the jurisdiction of those republics.

A fishing agreement between the EEC and Mozambique was signed on 11 December and is being applied on a temporary basis since 1 January 1987.

The new protocol to the EEC/Senegal agreement came into force on 1 January.

The Fishing Cooperation Agreement between the Governments of the Portuguese Republic and the Islamic Republic of Mauretania, which was signed at Nouakchott on 15 January 1984, was approved by Government Decree n° 13/86 on 10 November. It defines the general conditions and basic principles of cooperation between the two States in this domain. The agreement was applied up to 31 March 1987.

Government decree n° 14/86 of 13 November approved the Protocol modifying the Sea Fishing Agreement between the Governments of the Portuguese Republic and the Kingdom of Morocco, signed in Lisbon on 18 October 1985. Under this Protocol, the two parties shall develop technical cooperation in sea fisheries and related activities and shall consult each other on matters of world fishing policy within the framework of the competent international organisations and in the light of recent developments in the Law of the Sea, in order to harmonise their respective positions on questions of common interest.

e) <u>Other measures</u>

In view of the need to ensure supplies in optimum conditions of certain fish products intended for canning in sauce, Decree n° 230/86 of 14 August has fully adopted the tariff suspension applied to tuna by the Community of Ten.

So far as dried and salted cod and cod in brine, and fresh cod imported from third countries, are concerned, the tariff suspension can be fully applied within the limits of the quotas to be fixed by the Community. If they are imported from other Member States, the existing residual duties will be abolished.

The duties on frozen sardine, bonito and mackerel and on anchovy imported from other Member States are fully suspended within the limits of the quotas to be fixed annually by a joint decision of the Minister of Finance, the Minister for Agriculture, Food and Fisheries, and the Minister for Industry and Commerce.

A vocational training course for candidates for the Seamen's Register who have not obtained the compulsory school attendance certificate, under the supervision of the Lisbon Fishery School, was created by order n° 577/86 of 6 October.

The Vocational Training Centre for Fishermen -- the FORPESCAS Centre -- was created under Order n° 596/86 of 11 October under the aegis of the Ministry for Labour and Social Security. Under an agreement concluded between the Institute for Employment and Vocational Training and the Lisbon Fishery School, its functions will be to contribute to the technological development of the fishing sector, establish contacts, sign agreements and prepare and implement the domestic and international cooperation programmes which may prove necessary for this purpose.

Under this agreement, the Institute has placed its training centres throughout the country, especially those along the coast, at the disposal of the FORPESCAS Centre, for the provision of the training which is required in the fishery sector.

Vocational training courses for the fish canning industry have also been organised with the collaboration of the Lisbon Fishery School to provide training for persons already in employment as well as for new candidates for employment in the sector.

Under the terms of Decree n° 40/86 of 12 September, registered seamen employed in the fishing industry who are eligible for social security benefits will henceforth qualify for old age pensions from the age of 55, provided they have completed 30 years service.

A special pricing system for salted and dried cod and similar varieties has been created by order n° 77/86 of 30 December. The prices are fixed by agreement between the Administration (represented by the General Directorate for Competition and Prices), the importers (represented by the Food Distributors Association -- ADIPA), the producers (represented by the Association of Owners of Vessels operating in Distant Waters -- ADALPA) and the representatives of industry (ADAPI). The duration of the agreement shall never be less than 3 months and it shall be deemed to have been tacitly and repeatedly reenewed as long as it has not been denounced by one of the signatories. In the absence of an agreement, prices shall be fixed by the Secretary of State for Foreign Trade, shall remain in force for a period of three months from the expiry date of the previous agreement and shall be repeatedly renewed for a similar period provided they have not been denounced by one of the parties concerned, in accordance with the terms of the Order in question.

The Cod Trade Control Commission was dissolved by Decree n° 224/86 of 12 August and on the same date Decree n° 225/86 created the Associated Frozen Food and Cod Companies -- a limited liability company whose function is to trade in, import and export fishery products (frozen and cod) and other activities in the context of the support for the production and processing of these products.

Under Decree n° 266/86 of 3 September, the Portuguese Fish Canners Institute was dissolved and replaced by the Portuguese Fish Preserves and Products Institute which, in accordance with the principles of the Treaty of Rome, will be responsible for the control and regulation of the market for fish products both before and after their processing. Its functions will include, inter alia, supporting measures connected with the preservation, consumption and sanitary and quality control of these products, controlling manufacturing operations, issuing analysis reports and certificates of origin, promoting technological studies and manufacturing processes, etc., ensuring the proper functioning of the market network and following the operation of the international market in these products.

Under Decree n° 294/86 of 19 September, the holding of State capital shares in "PESCRUL -- Sociedade de Pesca de Crustaceos, SARL" was transferred to "IPE -- Investimentos e Participaçoes do Estado, SARL";

III. PRODUCTION

a) <u>Fishing fleet</u>

Statistics concerning the fishing fleet are not available. 72 projects received Community aid in 1986, of which 41 related to the construction of new vessels and the others the modernisation of existing units. The distribution was as follows:

--For the mainland:

<u>Construction of new vessels</u>
Deep-sea trawlers	4
Coastal trawlers	3
Boats for crustacean fishing	3
Seiners	7
PINE	5
Owner-operated boats	8
Tuna boats	1

<u>Conversions</u>
Deep-sea trawlers	2
Coastal trawlers	4
Boats for crustacean fishing	1
Seiners	16
PINE	3
Owner-operated boats	4
Tuna boats	1

Ten construction projects for the Autonomous Regions were approved -- 9 for the Azores and 1 for Madeira.

b) <u>Operations</u>

Applied research: This is administered by the INIP. Resource prospection operations were conducted in the Portuguese EEZ either by the "N/E Noruega" or by the "N/E Mestre Costeiro".

Aquaculture: The INIP maintained two continuous production units in operation -- one for micro-algae and the other for planktonic herbivora.

Fish breeding: Laboratory experiments were conducted in the reproduction and growth of larvae of different varieties including bass, sole and sea bream.

Eel farming: An experiment was conducted in producing fodder for eels based on ensilaged blue whiting and trumpet fish.

Studies were conducted in the rearing of shellfish (young seedlings of giant Algarve shrimps and Japanese shrimps) and oysters. Experiments were made in rearing fish in floating cages (trout, rainbow trout, carp, perch).

Aquacultural pathology: The INIP has been following the evolution of the natural oyster-beds in the Prado estuary, has lent its support to the rearing of trout and other varieties, and has continued to carry out periodic samplings (Ria de Faro-Olhao in the framework of the NATO-PO-FISHES Programme and the Franco-Portuguese Cooperation Programme).

Thirteen experimental fishing operations were carried out -- four for tuna and other migratory varieties, one in pelagic trawling and eight for scabbardfish.

c) Results

There was a 2 per cent growth in production in 1986 compared with the previous year.

d) Aquaculture

Twenty-three of the projects submitted to Brussels were approved. Most of them were in the following domains:

Alevins breeding farms	1
Water quality improvement	2
New fish farms	2
Conversion of salt beds	16
Improvement of operations	2

At present there are 56 fish breeding stations licensed by the Secretariat for Fisheries. Most of them are situated on the Sado and in the Algarve. Their total area is 560 hectares. If the salt beds and adjoining lands which have been brought into use for this purpose are added, the total area assigned to fish farming amounts to about 1 690 hectares.

The total estimated output of these farms is nearly 11 000 tons. It is made up as follows and -- apart from the mullet and trout -- most of it is exported (see Table 2).

Most of the cages for bivalve molluscs are located on the Algarve coast. They occupy a total area of 1 030 hectares, corresponding to some 1 600 breeding stations. Their average output is estimated at 8 200 tons, over half of which is exported.

Table 2

	(tons)
Trout	1 165
Mullet	420
Sar	262
Eel	590
Bass	52
Sole	37
Bream	67
Cockle	7 000
Bivalves	1 220

IV. PROCESSING AND MARKETING

Most of the actions undertaken in this field have been mentioned under the heading of profitability.

In addition, nine projects have been approved by the EEC Commission and have received financial support amounting to Esc 645 936 000, including Esc 165 306 000 provided by the Portuguese Government. These projects involve a total investment of Esc 1 051 177 000. As indicated in Table 3, they deal with the processing and marketing of fishery products.

Table 3

('000 Esc)

	Total Investment	FEOGA	Portuguese Government
Frozen products	781 977	360 739	123 329
Olive oil and sauces	72 970	21 776	8 465
Auction halls (Lagos et Matosinhos)	196 230	98 115	33 512
Total	1 051 177	480 630	165 306

Apart from the projects mentioned above, the construction and equipment of nine auction halls was pursued, and all of them are now completed. Six were financed through the Directorate General for Ports and three by the Auctions and Sales Service (SLV). In addition, five old auction halls were renovated, being self-financed by the SLV.

As far as ports are concerned, the work on improving protection and mooring facilities and the supply of services for the fleet in fishing ports was pursued.

This work entails a series of projects ranging from the erection of shelters and port improvements to infrastructures, including support equipment for auction halls and associated services, in order to ensure better working conditions in the handling and canning of fishery products. A start has also been made in the computerisation of auctions. Computer equipment has already been installed in the auction hall at Portimao, and it is planned to install similar equipment at Matosinhos, Docapesca and Peniche.

As regards market organisation, at the end of the year there were ten producers' organisations. They were given Government assistance to purchase computers and refrigeration equipment and also in regard to operations and management.

V. OUTLOOK

The work to be undertaken in 1987 forms part of the Multi-Annual Guidance Programme for the period 1987-1991.

The aims of this Programme are:

-- To stabilise the medium-term overall capacity of the fleet by developing new sectors that would enable fishing operations to be re-directed to deeper, more distant waters where fish stocks are under-exploited.

-- To modernise the fleet by encouraging the replacement of the oldest vessels of unsuitable size.

-- To modernise certain parts of the fleet by introducing new fishing techniques, thus ensuring energy saving and reducing overall operating costs.

-- To improve working conditions and safety for crews.

-- To improve the preservation and processing of catches on board.

-- To improve possibilities of access to alternative fishing grounds, either in the national EEZ or in other zones.

-- To promote the development of on-shore logistical support infrastructures.

-- To valorise fishery products and non-traditional varieties by diversifying the processing plants.

-- To develop marine and inland water fish farming, with special attention to varieties whose breeding offers good prospects, to breeding methods and to market prospects.

In the light of the resource conservation policy, it is planned to rationalise the operations of the local fleet (which receives financial aid only from national sources) through the use of more selective fishing gear (lines), the construction of bigger vessels -- generally to replace more than one vessel -- capable of operating further from the coast, and the modernisation of existing vessels, priority being given to those which offer better storage and safety conditions.

It is planned to stabilise the overall capacity of the inshore fleet through structural and maintenance measures.

The replacement of oversized trawlers will be encouraged, priority being given to replace vessels which are proposed to operate with gillnets or longlines, and which have refrigeration or freezing equipment.

Existing purse-seiners will be replaced by units of similar size but with better storage and conservation systems and better crew accomodation.

Multi-purpose vessels will be replaced with units of a more suitable size for operating in distant waters.

It is intended to convert viable units of the deep-sea fleet by modernising their conservation and processing equipment. It is being recommended that the fleet be modernised by replacing the oldest units with more modern vessels. Temporary lay-offs are also agreed to whenever unexpected difficulties are encountered in securing access to fishing grounds.

It is also planned to create a system of financial incentives for the re-direction of fishing activities into such branches as experimental fishing, or for the permanent or temporary reduction of a vessel's operations, or for the scrapping of the oldest units as soon as they are obviously no longer suitable for use and incapable of being converted.

Certain forms of cooperation with third countries will also be developed through, for instance, agreements within the Community framework to extend fishing possibilities.

The chief aim in the domain of aquaculture is to promote integrated exploitation of the areas feeding the salt-beds and which have potentialities for intensive and extensive fish farming. Fish farming enterprises will be encouraged to transfer their operations to inland waters, thus reducing the catching of alevins in nursery zones.

The application of technical measures to conserve resources and of the control measures called for by Community regulations depends upon the development of a taxation and control system involving Community participation.

As far as the processing and marketing of fish products is concerned, it is intended to restrict recourse to imported raw materials (sardine, tuna, razorfish) for canning in sauces, by increasing domestic production.

It is planned to introduce general carry-over arrangements for sardines withdrawn from auction when they do not fetch the common withdrawal price fixed by the producers' organisation.

Quality controls will be intensified in order to ensure that the rules concerning raw materials and finished products are observed.

The establishment of industrial units will be encouraged in order to improve quality and raise productivity.

Consumption of new products will be promoted through smoking, freeze-drying, the manufacture of pre-cooked dishes, the use of varieties that are plentiful along the Portuguese coast, consumption of products canned in olive oil or sauce, and of products whose sales are declining.

The regulation of markets will be encouraged by supporting the creation and operation of producers' organisations which will continue to be provided with appropriate infrastructures, and by rationalising the unloading and transport of fishery products.

SPAIN

I. SUMMARY

In 1986, Spain joined the European Communities. This most important event largely influenced the activity and development of the Spanish fishing which was singled out for special treatment in the Act of Accession.

Spain's joining the EEC involves full application of the common fisheries policy, except in areas subject to a transitional period.

In some cases, this means radical change in certain Spanish government measures described in previous Reviews.

In other cases, the common fisheries policy must be adapted for application to Spain because of the country's particular administrative structure and the need to continue certain pre-existing policies.

II. GOVERNMENT ACTION

a) Resource management

The following measures were introduced in 1986 as regards resource management:

-- Resolution of 19th February 1986 of the General Secretariat for Fisheries publishing the official registers of bottom trawling, "rasco" (bottom gillnet), bottom longline and "volanta" (bottom gillnet) fishing vessels operating in the national waters of the Cantabrian Sea (BOE -- Official State Gazette -- of 7/3/1986).

-- Resolution of 19th February 1986 of the General Secretariat for Fisheries publishing the official registers of "arte claro" (type of purse seine) fishing vessels, purse seiners and surface longliners operating in national waters (BOE of 8/3/1986).

-- Ministerial Order of 6th March 1986 concerning change of port of registry and type of fishing, amending certain provisions of the Ministerial Order of 28th July 1981 regulating fishing by Spanish vessels operating in Moroccan waters (BOE of 12/3/1986).

-- Resolution of 11th March 1986 of the General Directorate for Foreign Trade laying down the 1986 import quotas for fisheries products subject to the supplementary trade mechanism (BOE of 19/3/1986).

-- Resolution of 19th March 1986 of the General Secretariat for Fisheries providing for the annual revision of the register on 1st January, pursuant to the Ministerial Order of 8th June 1981 governing the fishing operations of the cod fleet.

-- Resolution of 26th March 1986 of the General Directorate for Foreign Trade publishing the total quantities of fisheries products subject to the supplementary trade mechanism expected to be imported from the EEC in April, May and June (BOE of 9/4/1986).

-- Resolution of 26th March 1986 of the General Directorate for Foreign Trade laying down an import quota for April, May and June for fisheries products subject to quantitative restrictions (BOE of 2/4/1986).

-- Ministerial Order of 31st March 1986 concerning surface longliners operating in Portuguese waters (BOE of 9/4/1986).

-- Ministerial Order of 4th April 1986 establishing a fishing reserve in the island of Tabarca (BOE of 10/5/1986).

-- Ministerial Order of 10th April 1986 providing grants for exploratory fishing in international waters (BOE of 9/5/1986).

-- Ministerial Order of 21st April 1986 regulating catches of pelagic species in the Cantabrian Sea and the North West (BOE of 29/4/1986).

-- Ministerial Order of 5th May 1986 setting up the register of organisations of fish producers (BOE of 10/5/1986).

-- Ministerial Order of 13th June 1986 regulating the marketing of fisheries, seafood and fish farm products.

-- Royal Decree 2133/86 of 19th September regulating amateur fishing in Spanish territorial waters around the Canary Islands (BOE of 17/10/1986).

-- Royal Decree 2200/1986 of 19th September regulating fishing gear and methods in Canary Island waters.

-- Royal Decree 2571/86 of 5th December amending Royal Decree 2349/84 of 28th November regulating purse seine fishing in national waters.

b) Financial support

Following Spain's entry into the EEC, the fisheries administration and the industry itself have been obliged to make a considerable effort to adapt to the vast area covered by the common fisheries policy (productive investment on behalf of the fishing fleet, fish farming, industrial and commercial infrastructure, etc.).

The following aids were granted by the government to the private sector in 1986 under the Social and Economic Plan for the fishing industry:

-- Construction and improvement of infrastructure at ports;

-- Construction and improvement of the infrastructure of fish farming installations;

-- Restructuring of the fishing fleet.

Government financial support for the fishing industry was channelled through the General Secretariat for Fisheries, the FROM and the Banco de Credito Industrial.

i) <u>General Secretariat for Fisheries</u>

The General Secretariat for Fisheries has provided assistance in the following areas:

Fishing fleet

— Ministerial Order of 30th January 1986 providing economic incentives to change the type of fishing vessel operated (BOE of 5/2/1986).

— Ministerial Order of 11th April 1986 providing economic incentives for temporary or permanent cessation of fishing activity.

— Ministerial Order of 11th April 1986 extending the period of validity of the Ministerial Order of 15/7/1985 (Economic incentives to encourage certain investment operations by enterprises in the fishery sector).

— Ministerial Order of 11th April 1986 extending the period of validity of Ministerial Order No. 16088 of 15th July 1985 (BOE of 9/5/1986) (Financial aid for the gradual reduction of fishing capacity).

Conservation of stocks

— Ministerial Order of 11th April 1986 extending the period of validity of the Ministerial Order of 21st March 1985 (BOE of 7/5/1986) (Subsidies to offset the effects of the ban on nonnat fishing).

— Royal Decree 2134/86 of 19th September laying down minimum sizes for catches of certain species in Canary Island waters (BOE of 17/10/1986).

Tax exemption

— Ministerial Order of 16th April 1986 laying down the procedure for reimbursement of tax, or exemption, on purchases of "B" diesel oil for coastal fishing vessels (BOE of 17/4/1986).

Subsidies for the building, modernisation and conversion of the fishing fleet

— Royal Decree 519/86 of 7th March on aids for the modernisation and conversion of fishing vessels with a length between verticals of between 9 m and 12 m (BOE of 15/3/1986).

— Ministerial Order of 17th March 1986 specifying the procedure to be followed with regard to applications for aid for the building, modernisation and conversion of the fishing fleet made under the terms of Royal Decrees 2339/85 of 4th December and 519/86 of 7th March concerning fishing vessels with a length between verticals of between 9 m and 33 m (BOE of 22/3/1986).

-- Ministerial Order of 12th May 1986 concerning the time limit for applications for the economic incentives granted for changing the type of fishing vessel operated (BOE of 14/5/1986).

Quantitative aspects

The total amount of aid disbursed under the terms of the above measures was Ptas 2 787.2 million, broken down as follows:

	(Ptas million)
-- Subsidies to fishermen's associations and their federations	70.4
-- Restructuring of the fishing fleet	2 101.8
-- Fish farming and artificial reefs	162.1
-- Surveys of new fishing grounds	361.3
-- Subsidies for suspension of fishing operations	91.6
Total	2 787.2

ii) Banco de Credito Industrial

In compliance with Royal Decree 1609/1985 of 17th July, the Banco de Credito Industrial has taken over from the independent Credito Social Pesquero (Social Fund for Fisheries). In 1986, loans granted by the Banco de Credito Industrial amounted to Ptas 13 728 million, allocated to the construction of vessels. Total loans pending at 31st December 1986 amounted to Ptas 5 118 million.

iii) FROM

In 1986, this Organisation granted aids to Fish Producer Organisations and to Autonomous Communities under Community rules.

The aids transferred to the Autonomous Communities amounted to Ptas 310.5 million, and the aids for storage of fisheries products amounted to Ptas 84.6 million.

Improvement of marketing operations

A total of Ptas 351 million was allocated in the form of loans to improve the marketing of fisheries products.

Marine fish farming

In 1986, the co-ordinating role assigned by Act 23/1984 on marine fish farming to the National Advisory Council for Marine Fish Farming was strengthened. The Council's main tasks are as follows:

-- To study control standards for species imported for fish farming.

-- To co-ordinate the teaching of fish farming.

-- To work out and study the national fish farming Plans provided for by Act 23/1984 on marine fish farming.

c)   Economic efficiency

In order to adjust the fishing effort to resources, limits have been set per vessel and per day on catches of pelagic species in the Cantabrian Sea and the North West. At the same time, the conservation policy for certain specified fishing zones has been continued (zones where fishing is banned, fishing reserves). Action has also continued on the basis of exploratory fishing plans for the conservation of zones whose balance is jeopardised by overfishing.

Lastly, co-ordination with the Autonomous Communities and the monitoring of fishing operations have been stepped up.

As regards licensing, Community and Portuguese waters are now also included in the fishing plans. Licences have also been granted for fishing in Moroccan waters.

The efficiency of distribution has been improved by the introduction in Spain of the producers' organisation structure. In just six months, 31 fish producers' organisations have been set up, covering nearly 90 per cent of all Spanish fishing operations. At the same time, full application of Regulation (EEC) No. 3796/81 on the common market policy has prompted the introduction in Spain of all the Community market regulation mechanisms.

d)   Bilateral agreements

The following adjustments were made in 1986 to the bilateral agreements listed in the 1985 Review:

Norway: This agreement expired on 31st December 1986, when Spain became a party to the agreement between the EEC and Norway.

United States: The Spanish fleet has been experiencing increasing difficulties in obtaining access to resources. Although the quota allotted for 1986 was 3 821 tonnes, Spain only landed 1 803 tonnes, mainly of squid and other similar species. Between 8 and 15 freezer ships were normally operating. This agreement expired on 30th June 1987, when Spain became a party to the EEC/United States agreement.

Canada: The bilateral agreement signed by the two countries on 10th June 1976 expired on 9th June 1986.

South Africa: In 1986, the quota granted was 2 000 tonnes for five specified vessels, each of which had to pay a fee of 2 500 Rands for the licence to fish and 45 Rands per tonne of hake caught.

Angola: This agreement, which expired on 2nd November 1985, was extended until 2nd May 1987.

Guinea Bissau: In May 1986, a new fisheries protocol to the Fisheries Agreement between the EEC and Guinea Bissau was initialled, covering the conditions regulating Spanish fishing under the bilateral agreement between Spain and Guinea Bissau, itself incorporated in the Community Agreement on 16th June, when the latter provisionally entered into force.

Guinea Conakry: With the entry into force of a new protocol to the Fisheries Agreement between the EEC and Guinea Conakry on 8th August 1986, the bilateral agreement between Spain and this country was incorporated in the Community Agreement.

Equatorial Guinea: The new protocol to the Agreement between the EEC and Equatorial Guinea came into force on 27th June 1986, the current agreement between Spain and Equatorial Guinea being incorporated therein.

Mauritania: Notice of termination of the framework agreement with Mauritania was given on 6th December 1986 and it expired on 5th April 1987.

Morocco: This agreement expired in July 1987. It will be renegotiated under the future agreement between the EEC and Morocco.

Mozambique: The framework agreement with Mozambique expired in December 1986. The quotas allotted in 1986 were 1 800 tonnes for surface shrimps and 900 tonnes for deep-water shrimps.

Sao Tome: The procedures regarding the entry into force of the Fisheries Agreement between Spain and Sao Tomé initialled in November 1984 have been halted. The first Community negotiations took place in Sao Tomé in November 1986.

Senegal: The new protocol between the EEC and Senegal came into force in November 1986. Spain became a party to it on 1st March 1987, when the current bilateral agreement expired.

Seychelles: A new protocol to the agreement between the EEC and the Seychelles was signed in December 1986. It came into force on 18th January 1987, when the current bilateral agreement between Spain and the Seychelles was incorporated in the Community Agreement.

Gambia: A fisheries agreement between the EEC and the Gambia, covering a period of three years, was initialled in November 1986.

Cape Verde: The agreement with this country signed on 25th September 1981 expired on 24th September 1986, new negotiations on an agreement between the Community and Cape Verde are underway.

Madagascar: The agreement between the Community and Madagascar has been modified to take into account the substantial increase in the Community's tuna fleet following the adhesion of Spain.

Other agreements

Treaty of Paris: Spain has also had access to resources in the Svalbard zone since 1925, when it signed the Treaty of Paris. Cod catches in this zone amounted to 3 737 tonnes in 1986.

NAFO: Since 31st December 1986, Spain is not a contracting party to this Organisation but has been represented since that date by the EEC. The quotas allocated to Spain since its adhesion to NAFO in 1983 have been extremely low, not taking into account Spanish requests in comparison with their earlier activities in this zone and the socio-economic needs of its fleet. For this reason, objections, on a number of occasions, and in particular, in 1986, have been put forward by Spain in conformity with the procedure foreseen in the Convention.

ICSEAF: The quota allocated by this Organisation in 1986 to Spain amounted to 106 437 tons of hake.

e) Other government action

In 1986, the General Secretariat for Fisheries made a considerable effort to encourage the creation of fish producers' organisations. Thus, Royal Decree 337/86 of 10th February laid down the procedure for government recognition of these organisations, which were assigned two major objectives:

-- To rationalse fishing operations.

-- To improve the marketing of their products.

The Spanish Oceanographic Institute, which is ancillary to the General Secretariat for Fisheries, undertook research in the following areas:

i) Research in national and Community fishing zones

-- Survey of fish resources in the ICES zone.

-- Survey of fish resources in the Mediterranean.

-- Evaluation of fish resources in the Canaries.

ii) Research in international fishing zones

-- Evaluation of resources in the central East Atlantic.

-- Evaluation of tuna resources.

-- Evaluation of fish resources in the Antarctic.

iii) Fish farming research

-- Industrial production of molluscs.

-- Industrial production of algae.

-- Fish farming research.

iv) Oceanographic research on the marine environment

-- Monitoring network on toxic dinoflagellates.

- Oceanographic research in Galicia.

- Multidisciplinary oceanographic survey of the Strait of Gibraltar and the adjacent seas.

- Studies on marine pollution in the Cantabrian and adjacent seas.

- Studies on marine pollution in the Mediterranean.

- Studies on marine pollution in the Canaries.

- Research on the marine environment in the Antarctic.

- Fisheries sampling and information network.

- Study on tides.

f) <u>Structural adjustment</u>

As from 1986, the Community legislation on structural policy for fisheries applies to Spain (Regulations 2908/1983, 2909/1983 and 355/1977 and Directive 83/515). Spanish legislation has been fully adjusted to comply with these regulations.

Guidance programmes have been drawn up for 1986 for the construction and modernisation of shipping vessels and for marine fish farming.

Investment projects related to the above matters have been submitted to the EEC Commission, together with plans for exploratory fishing and the adjustment of fishing capacity.

g) <u>Sanitary regulations</u>

The regulations for 1985 remain in force.

III. PRODUCTION

a) <u>Fishing fleet</u>

The figures have not yet been updated for 1986.

b) <u>Operations</u>

The prices of fisheries products rose by about 8 per cent in 1986.

c) Results

In 1986 total landings were 1 742 722 tons, which represents a decrease of 2.31 per cent compared with 1985. Overall results are as follows: decreases in landings of fresh fish (-3.69 per cent) and frozen fish (-9.37 per cent) and an increase in aquaculture production.

d) Fish farming

In 1985, production in the fish farming sector was as follows:

-- Fish and crustaceans: 473 tonnes of fish and crustaceans in intensive production (pots and tanks) and semi-intensive production (lagoons, saltwater and brackish water ponds). For fish, the highest yields were of salmon (150 tonnes) and bream (127 tonnes) and for crustaceans, shrimps (40 tonnes) and giant prawns.

-- Molluscs: 249 671 tonnes of molluscs (deepwater and surface). The highest yields were: mussels (245 645 tonnes) and oysters (3 164 tonnes). The total number of installations on an area of 21 786 ha was 4 787 (30 under construction). Floating mollusc tanks account for 70 per cent of all installations.

IV. PROCESSING AND MARKETING

a)b)c)

The sardine-anchovy group (with 54 per cent of the total) came first for canned fish and molluscs and semi-preserved and salted anchovy, 87 per cent of the raw material processed being of national origin. The group consisting of hake and similar fish made up nearly a third (28.6 per cent of the total) of all semi-preserved smoked and salted and frozen products, 76 per cent of the raw material being also of national origin.

As regards canned fish and molluscs and semi-preserved anchovies the trend is generally towards automation and mechanisation, except for anchovy preserves, where the entrenched cottage industry is slowing down the introduction of new methods.

In the semi-preserved smoked products, salting and freezing subsector the trend is towards new methods of thawing, cooking, washing, cutting and filleting, besides deep-freezing and icing. Improvements have also been made in the storage, canning, equipment and transport processes, both in factories and elsewhere (isothermic and refrigeration equipment). All these plans for new methods have been incorporated in the programme submitted by Spain to the EEC Commission (under Regulation 355/77 of 15th February) with a view to obtaining aid from the Guidance Section of the EAGGF (European Agricultural Guidance and Guarantee Fund).

No new forms of production have been introduced in the canned and semi-preserved anchovy subsector. However, in the semi-preserved salted and frozen products subsector as a whole, there is a marked tendency to diversify products that is due to functional catering and the vast and varied field of prepared, pre-cooked and fish-based dishes. In addition, developments in vacuum-packing of frozen products might lead to the industrialisation of surplus species or those under-exploited for lack of a market.

In 1986, investment in the two above-mentioned processing subsectors amounted to Ptas 6 708.9 million, while government grants for new plant and improvements totalled Ptas 1 237.1 million.

The breakdown of employment in the processing industry was as follows:

	Workforce
-- Canned products and semi-preserves	38 986
-- New processing industries	7 300a)

a) Excluding casual workers.

The coefficient of utilisation of plant by the above industries is about 65 per cent for canned products and semi-preserves and 95 per cent for new processing plant (whose rate of return is very high owing to the sophisticated technology used).

d) External trade

In 1986, Spain imported 351 141 tonnes of fisheries products, totalling over Ptas 100 000 million, which represents a rise of 8.6 per cent in the volume of imports and 44 per cent in their value over the previous year.

Exports of Spanish fisheries products amounted in 1986 to 220 000 tonnes, worth Ptas 55 000 million, a drop of 4.6 per cent in volume and 8.3 per cent in value from 1985.

These figures show a sharp fall in the rate of coverage of the fisheries trade balance: from 86 per cent in 1985 to 54.9 per cent in 1986.

Imports rose substantially for frozen fish (34 per cent), dried and salted fish (35 per cent) and crustaceans and molluscs (92 per cent); imports of fresh and chilled fish, on the other hand, were down (-25 per cent).

Exports of fresh and chilled fish doubled, and frozen fish exports increased by 14 per cent. However, those of other products were down: dried and salted fish (-27 per cent), frozen crustaceans and molluscs (-26 per cent), frozen cephalopods (-21 per cent). Exports of canned products also declined.

One year after its accesstion to full membership of the EEC, Spain imported 34 per cent by volume and 42.8 per cent by value of fisheries products from the Community and exported 65 per cent by volume and 47.7 per cent by value of its own fisheries products to the Community.

V. OUTLOOK

In the short and medium term, the future activity and development of the Spanish fishing industry depends on its progressive involvement in Community fisheries policy.

Thus, as regards fish resources, the activity of the Spanish fishing industry will depend on the Community renewing Spanish bilateral fishing agreements and any participation Spain obtains under EEC fishing agreements.

The future of the fisheries structures will depend on acceptance of the multiannual guidance programmes that Spain has submitted to the EEC Commission for approval.

Finally, gradual participation in the Community quotas and transitional factors, such as adjustment to the CCT and absorption of the Community's preferential agreements, will have a considerable impact on trade policy.

SWEDEN

I. SUMMARY

No major changes were made during 1986 regarding systems for price regulation, fishery management or support to the fishing industry, including the processing industry.

According to preliminary figures, the total domestic landings fell by roughly 25 000 tons, or 11 per cent. However, due to increasing prices for groundfish species, mainly cod, the total catch value decreased only slightly.

Exports fell by 17 000 tons or 13 per cent in terms of quantity but remained principally unchanged by value due to increasing prices. Imports remained unchanged by quantity but increased moderately in terms of value.

II. GOVERNMENT ACTION

a) <u>Resource management</u>

Catch quotas in the Swedish fishing area are normally determined by negotiations with neighbouring countries. However, for cod and salmon in the Baltic, it has not been possible to determine a common TAC for 1986 and, consequently, there have been no internationally determined catch quotas in this fishing area. The Swedish Government therefore unilaterally fixed the catch quotas for cod.

In order to protect the stock of wild salmon, continuous efforts have been made to restrict the fishing of salmon in the Baltic. Furthermore, rigorous restrictions have been introduced to protect the wild salmon in the coastal and river-areas.

b) <u>Financial support</u>

The main support within the price regulation system consists of price supplements, based on norm-prices for different species. Norm-prices are determined by the government and are normally unchanged during each financial year (July-June). When the average market price falls below the norm-price, price supplement is paid out to the fishermen.

The norm-price support is of essential importance for the herring fleet. On average, price supplements account for approximately 20 per cent of the total receipts from fishing for vessels involved in fishing herring.

On average, the norm-prices were raised by 4.5 per cent for the financial year 1985/86 and by 5.8 per cent for 1986/87. The norm quantity for herring, i.e. the ceiling for which the price supplements will be paid, was decreased by 4 000 tons for 1985/86 and by a further 3 000 tons for 1986/87. For the financial year 1986/87, the norm quantity for herring was fixed at

73 000 tons. However, as a result of decreasing catches of herring, almost all landings in domestic ports could receive price supplements despite a reduced norm quantity.

Other measures within the price regulation scheme during 1985/86 and 1986/87 are listed below:

-- Fixed-price supplements for mackerel, ling, prawns and some fresh-water species;

-- Subsidies for surpluses in connection with purchases by the Swedish Fish Association;

-- Freight support to fishermen in the northern parts of the country;

-- Freight support for industrial fish;

-- Fishermen's security insurance system.

Total costs of the price regulation scheme for the financial year 1985/86 amounted to SKr 65 million, of which SKr 49 million refers to price supplements for herring including costs for surpluses. For 1984/85 the corresponding figures were SKr 72 million and SKr 54 million. According to provisional figures, total costs for the financial year 1986/87 will be SKr 60 million.

c)   Economic efficiency

Higher efficiency in the fishing industry, as well as in the processing industry and in the distribution channels for fish, is considered to be an important prerequisite of achieving the primary objectives of the fishery policy.

The following amounts were set aside for the financial year 1985/86 (the amounts for 1984/85 are given in brackets) for rationalisation measures in the fishing industry:

-- Low-interest loans for financing investments in vessels, including technical equipment, gears and safety-related equipment, etc.: SKr 25 million (SKr 28 million);

-- Loans with Government warrant for investments in vessels and technical equipment: SKr 5 million (SKr 10 million);

-- Subsidies to promote purchase of new vessels: SKr 3 million (SKr 3 million);

-- Subsidies to promote transition to a more profitable fishing direction: SKr 1 million (SKr 1 million);

-- Subsidies for fuel-saving measures and safety-related measures: SKr 2.5 million (SKr 2 million);

-- Scrapping premiums to promote a quicker renewal of the fishing fleet: SKr 1.1 million (SKr 1.6 million).

For the processing industry, low-interest loans (interest 5 per cent) have been granted in order to increase the efficiency of the production. These loans amounted to SKr 4.5 million for the financial year 1985/86 and to SKr 10.6 million for 1984/85.

Further in order to increase the efficiency of the first-hand purchase and the distribution of fish, the Government has decided that price supplements will only be granted to fishermen who deliver their catches to buyers cooperating in a special association. This system came into force on December 1, 1986, and as a consequence thereof, price support will not be paid to those fishermen who deliver their catches to a great number of small purchasers (restaurants, retail shops, etc.)

d) Bilateral arrangements

Sweden has concluded bilateral quota agreements with the EEC, Finland, the German Democratic Republic, Norway, Poland and the USSR. Details of these agreements are contained in the Statistical Annex.

e) Other Government action

In order to break the trend of continuing rising prices and to secure the deliveries of cod to the domestic filleting industry, an export levy was imposed on all exports of cod. This system was introduced on March 1, 1986 and the levy amounted to SKr 0.50 per kilogram.

f) Environmental problems

In the field of environmental problems there are several things that should be observed. As regards coastal and sea-water areas, the most important problem is the eutrophication of water in certain areas. A government programme of measures has recently been laid down in order to reduce these problems. Within the framework of this programme there will also be contacts with the neighbouring countries concerned. Another problem recently discovered in sea-water on the west-coast is a high percentage of dioxine. Further studies in this field are to be carried out in the near future.

For several years the most important environmental problem for the inland-lakes has been the problem of acidification. Measures financed by government grants have continuously been taken to restore the affected lakes by liming. There have also been actions, financed by public funds, to restore the fish stocks in such lakes. A newly-discovered problem concerning a great number of lakes in northern Sweden is the consequence of the Chernobyl disaster, which has led to an unacceptably high percentage of cesium in certain districts.

g)   Structural adjustments

In an attempt to make the fishing fleet less dependent on fishing herring, several measures have been taken in order to modernise and better equip the vessels. There is also an incentive to encourage the replacement of old fishing vessels by more effective new ones. In this way, it is also possible to attain a most desirable improvement of the fishermen's working conditions.

If a newly-built vessel is intended to be used in herring fishing, the owner can only count upon government loans provided that at least one old vessel is taken away from the fleet.

h)   Sanitary regulations

No fundamental changes concerning sanitary regulations in the field of fisheries have been effectuated during 1986.

III. PRODUCTION

a)   Fleet

Recent figures on the composition of the fleet are available concerning so-called "team" fisheries only. These figures are based on selling notes for full-time fishing vessels, i.e. vessels which have landed fish for at least 25 weeks during a year.

According to the above-mentioned definition, the full-time fleet consisted of 547 vessels during the financial year 1984/85. This figure increased by 42 units to a total of 589 vessels in the ensuing 1985/86 financial year. The total tonnage of the fishing fleet increased slightly to approximately 32 000 grt.

b)   Operations

The total cost of fishing decreased by 8 per cent between 1985 and 1986 mostly owing to a considerable decline in fuel prices. The cost of fuel decreased by 32 per cent.

c)   Results

Owing to decreasing catches of, mainly, herring and cod, the total catch fell by some 25 000 tons between 1985 and 1986. Landings abroad by Swedish vessels remained principally unchanged from 1985 to 1986. However, there was an evident change in the composition of the foreign landings towards a greater share of industrial fish than in the previous year.

Total catch, including landings abroad, amounted to 200 000 tons in 1986. The value of landings was about the same as 1985 (Skr 700 million).

Prices of herring remained on a non-profitable level. For cod, on the other hand, prices continued to rise throughout the year, thus causing problems for the cod processing industry.

Fleet profitability, as a whole, seems to have been satisfactory during 1986. However, for parts of the herring fleet, the profitability is totally dependent on price supplements.

The yield of Swedish aquaculture in 1985 was 2 260 metric tons of fish for consumption, which when converted to round fresh weight is the equivalent of 2 665 tons. The dominating species was rainbow trout (2 532 tons). Furthermore, there were 415 tons of cultivated blue mussels harvested. The total value of the aquaculture production amounted to SKr 73.7 million. Estimated figures for 1986 indicate a production of roughly 3 100 tons of rainbow trout.

The number of enterprises engaged in aquaculture was 675 in the year 1985, of which 196 produced rainbow trout for human consumption and 6 produced blue mussels. 163 establishments cultivated fry for restocking purposes. For compensatory purposes 2.4 million fry of salmon and trout were released, mainly in rivers running to the Baltic.

In the course of 1986 the market situation for aquaculture production has grown considerably worse. Falling prices in addition to increasing tendencies of over-production is causing severe problems to the aquaculture industry. So far, no Government subsidies have been considered necessary, but if the present situation continues, there may be a clear risk that such measures will be unavoidable in the future.

IV. PROCESSING AND MARKETING

a) Handling and distribution

Unfortunately, continuous efforts to improve methods of storing fresh fish in retail shops have not yet given the desirable results.

b) Processing

The herring processing industry is partially dependent on exporting herring flaps. Unfavourable price development in connection with increasing supplies of herring fillets on the European market has led to serious problems for the processing industry. For some companies, a shift towards more production of cod fillets has shown to be advantageous. However, because of decreasing catches accompanied by higher prices, not even cod production has been able to avoid problems.

c) Domestic market

Official statistics do not reveal any particular short-term trends in the consumption of fish. However, efforts are made to increase the consumption of mainly domestically produced fish products.

d) External trade

Swedish exports mainly consist of fresh herring, fresh cod, some fresh water fish, preserved herring and shellfish products.

Decreasing catches of herring and cod implied diminished exports by roughly 17 000 tons. By value, exports remained unchanged on the whole. Export prices for herring increased moderately whilst the prices for cod increased considerably from 1985 to 1986.

Total imports increased by roughly SKr 260 million but remained relatively unchanged by quantity. Different kinds of prepared fish products and shellfish products account for the main part of the increase in value.

V. OUTLOOK

A continuous decline of the catches of herring is likely to occur during 1987. The development of the catches of cod is more difficult to predict, but no major changes are to be expected.

The development towards more prepared imported fish products is also likely to continue. The composition of exports will probably depend upon the price relation between herring and cod, which is difficult to predict.

TURKEY

I. SUMMARY

During 1985 total harvest increased to 578 073 tons mainly due to the development of the Turkish fishing fleet. In comparison the 1984 harvest was 566 900 tons. No important changes are expected for 1986.

In 1986, more credit schemes to the fishing industry were made available at lower rates of interest and measures taken for the protection of stocks came into effect. Important steps were made regarding the development of new fishery products.

II. GOVERNMENT ACTION

During 1986, under the fifth five-year Development Programme (1985-1989), there was a 7.8 per cent increase in production, a 7.6 per cent increase in domestic consumption and a 16.6 per cent increase in exports. In this context, the main objectives of the programme are to determine the stocks, identify the new production areas, carry out studies on offshore fishing and improve the production of fishery products.

Other aims are to:

-- Put emphasis on training (in the processing industry);

-- Improve the fishermen's conditions and to improve the storage and marketing of fish;

-- Take the necessary measures against water pollution and to improve water protection and control services (as stipulated in Fishery Products Law 1380).

The Government underlined the importance of stock control particularly in the Black Sea, and quotas were set for the stocks of migratory fish i.e. anchovy and mackerel. Quotas were also fixed for red mullet, sardine, coral, etc.

The Ministry of Agriculture and Forestry together with the Black Sea University decided on a project to set up a control on dolphin stocks and to determine the biological characteristics of dolphins in the Black Sea. This project will begin in 1987/88.

The improvement of six lagoons was completed in 1986 and a further three lagoons will be improved.

The fixing of sponge stocks began on the Aegean coast and will be completed in 1989, as this is an important export product.

In the 1986 Circular No. 20, which specifies the protection of present stock and also prevents overfishing, prohibitions and limitations were even greater than in previous years.

The Government announced the implementation of a 200-mile EEZ in the Black Sea on 5th December 1986, with a view to carrying out research on the resources and to maintain and manage these resources to the benefit of the country.

In 1986 the Government encouraged investment in the fishery sector and thus reduced interest rates for loans, tax relief and customs exemption.

From 7th June 1986 a premium was paid from the "source supporting fund" amounting to 30 per cent in regions considered as "development priority regions" and 25 per cent to other regions. Credit interest by the Ziraat Bank was reduced from 34 per cent to 22 per cent in 1986 and the sum of the credit available increased to TL 20 million; in 1985 it was TL 12 million. Most of the credits were for the purchasing of equipment in order to develop the fishing fleet and aquaculture.

Around TL 9.5 million was invested by the Government in the fishing industry in 1986. The construction of 11 fishermen's shelters was completed and a further 31 are under construction.

No real developments were observed in the processing and marketing sectors. Final data are not yet available but anchovy production increased by 5-7 per cent while crawfish production decreased by 90 per cent due to disease.

Two private sector aquaculture projects were initiated in the production of sea fish and the total capacity is about 500 tons (seabream production).

In 1986, a treaty was signed with the Italian Government for the purpose of improving lagoon fishing on the Turkish coast. Within the framework of this treaty exploitation will be carried out at the Buyuk Menderes River estuary and at Bodrum in the Aegean Sea and a processing unit and hatchery will be designed in Tuzla Lagoon. Under the MEDRAP (Mediterranean Regional Aquaculture Project) work has begun on production facilities on sea fish in Beymelek Lagoon in Antalya Province.

A change was made to Fishery Product Law 1380 and became valid from 28th May 1986; the disciplinary clauses were enlarged, directed towards prohibited fishing and protection, and the right to hire out fishermens' lodgings was given to the cooperatives.

III. PRODUCTION

The fishing fleet increased by 26 per cent to 8 604 vessels. Of this figure 768 vessels had more than 100 hp and 308 were more than 20 metres in length. There were 2 671 in the East Black Sea region, 688 in the West Black Sea region, 3 020 in the Marmara region, 1 337 in the Aegean Sea and 888 in the Mediterranean region. The number of trawl boats increased from 403 to 422 and purse seiners increased from 373 to 611. This increase in the fishing fleet did not directly effect fish prices and fishing methods did not change.

Total production in 1985 was 578 073 tons. This was made up as follows: sea fish (519 911 tons); shellfish and molluscs (12 691 tons) and freshwater fish (45 471 tons). Eighty-seven per cent of the total production came from the Black Sea area and consisted of 55 per cent anchovy and 22 per cent small mackerel. Other important species are sardine, bonito, red mullet and seabream. Fifty-six per cent of the shellfish and mollusc production came from the Black Sea region and 34.2 per cent of freshwater production came from the Eastern Anatolia Region.

IV. PROCESSING AND MARKETING

Of the total 1985 production, 78 per cent was for direct human consumption and 22 per cent was used as fishmeal. However, 45 per cent of the 1984 production was used in the fishmeal sector and 55 per cent for direct human consumption. Although the wholesale share increased from 49 per cent to 69.2 per cent, the cooperatives' share did not change (2.6 per cent). The share of the canning industry increased from 0.09 per cent to 1.52 per cent. In 1985 the amount exported was 11 444 tons; in the first ten months of 1986 it was 8 397 tons. Imports were at 366 tons in 1985 and 1 798 tons in 1986. The main import items were crustaceans and molluscs, e.g. common octopus, lobsters.

V. OUTLOOK

In 1985 production was at 578 073 tons but it has not been possible to appraise the 1986 production yet. At the end of 1986, 117 fishermen's shelters were completed which should contribute to an increase in production. However the storage, classification and marketing of fishery products is still inadequate.

Improvements in protection and sanitary regulations, processing and breeding of fish have been made. Further studies will be continued in 1987 and the following measures will be taken.

i) Protection and control of all species, water pollution and sanitary conditions will be monitored carefully so as to increase production.

ii) The determining of new fishing areas.

iii) Breeding will be examined.

iv) Improvements to fishermen's shelters, sea ports and slips will be made.

In 1987, a 7.2 per cent increase in production, a 7 per cent increase in domestic demand and an 18.2 per cent increase in exports are expected.

UNITED KINGDOM

I. SUMMARY

For British fishermen, 1986 saw some improvement compared with 1985. Although the quantity of fish landed by British vessels was down on 1985, this was compensated for by a rise in the overall value of fish caught.

Demersal species account for almost three-quarters of the value of the catch. Haddock replaced mackerel as the leading species caught, in terms of quantity, going for human consumption landed in the United Kingdom. Various management schemes were operated in order to achieve continuity of fishing and to ensure that UK quotas under the Common Fisheries Policy (CFP) were not exceeded.

Imports decreased in quantity but increased in value, whilst exports increased in terms of both quantity and value.

In 1986, the Seafish Industry Development Programme continued to progress, helping to tackle a range of problems facing the fishing industry, in the fields of marketing and training.

II. GOVERNMENT ACTION

a) <u>Resource management</u>

Under the revised Common Fisheries Policy (CFP), proposals for Total Allowable Catches (TACs) and quotas for 1986 were agreed at the Council of Fisheries Ministers held in December 1985 and are set out in EEC Regulations 3721/85 and 3777/85. This was the second year for which the TACs and quotas had been agreed before the start of the fishing year and on this occasion the two Regulations cover both the stocks included in the 1985 Regulation and the additions and amendments made necessary of the Accession of Spain and Portugal to the European Community. The Regulations also laid down certain conditions under which fishing operations may be conducted, relating to by-catch, closed areas and closed periods. These conditions were similar to those in force at the end of 1985, except for some minor changes to the seasonal closures for herring fishing in the Irish Sea and sprat fishing in the Skagerrak and Kattegat.

In order to effect greater quota management and in line with EC legislation, certain changes were made to three Sea Fish Licensing Orders as from 21st August 1986:

    a) Following the inclusion of Norway lobster and pollack in the EEC quota system, the species were added to the list for which it is prohibited to fish without a licence.

b) Similarly, because of the accession of Spain and Portugal certain ICES sub-areas were included for quota species as per the amended EEC quota regulation.

c) For the purposes of UK quota management control, hake, megrim, Norway lobster and pollack were excluded from the by-catch provision which exists in ICES sub-area VII and VIII because it is more effective to manage these fisheries on a quantitative basis. In addition licensing for <u>all</u> vessels, irrespective of length, was introduced for the Mourne herring fishery which, for the first time in 1986 was one amalgamated quota consisting of the Mourne and the Isle of Man stocks. In general, the wording of the exemption from the requirement to obtain a licence to fish quota species for small vessels was clarified from "under 10 metres" to "10 metres and under".

Following consultation with the industry, North Sea sole, Area VII megrim, Area VII and VIII haddock and Area VII (excluding VIIa) and VIII cod were designated as "special pressure stocks" in order to conserve the particular stocks in those areas. As a result, the four additional pressure stocks became subject to restrictive licensing policy.

Despite the existing UK restrictive licensing scheme there is increasing pressure on a number of white fish stocks subject to quotas in Area VII which is exacerbated as a result of the transfer of pressure stock licences from smaller vessels to considerably larger ones. The restrictive licensing scheme is currently undergoing a major review, in consultation with the industry but while this problem is examined more thoroughly a moratorium on the transfer of Area VII white fish pressure stock licences from vessels below 80 feet in length to vessels over 80 feet in length was introduced on 21st November 1986 until further notice.

An order was made to enable proceedings to be taken against the owner or charterer, as well as the master, of a vessel which fails to comply with the regulations on conservation and control measures or co-operate with the Fisheries Inspectorate. It also requires vessels receiving transhipped fish to keep and submit records of transhipments.

During 1986, agreement was reached on a Council Regulation designed to provide for a general strengthening of the arrangements for inspection monitoring and keeping of records, as well as requiring Member States to provide the Commission Inspector with such access and means as they need to oversee national inspection and monitoring arrangements in order to ensure that enforcement in all Member States is carried out on an effective but even-handed basis. The Regulation also empowers the Commission to close a fishery, after informing the Member States concerned, when the total allowable catch is exhausted and provides for Member States which overfish their quotas to provide compensation to the other Member States concerned. This is the first time that agreement has been reached to penalise overfishing and it will strengthen the incentive to observe the quotas.

Further measures to tackle illegal fishing for salmon (and migratory trout) are contained in the Salmon Act 1986 which received Royal Assent on 7th November 1986 to come into force two months later. The Act makes it an offence for a person to handle salmon where he believes, or it would be reasonable for him to suspect, that the fish had been illegally taken. It also provides powers for the establishment of salmon dealer licensing schemes. Among its other measures the Act updates the arrangements for the administration of salmon fisheries in Scotland.

The Importation of Live Fish of the Salmon Family Order 1986 came into operation on 18th February 1986. This order permits imports (under licence and with appropriate fish health certification) of live salmonids from Northern Ireland. This is the only exception to the prohibition on imports into Great Britain of live fish of the salmon family under the Diseases of Fish Act 1937, as amended.

An order has been made under the Animal Health Act 1981 banning the import of dead ungutted salmon and trout into Great Britain other than by licence. The measure is aimed at preventing the introduction of the serious fish diseases, Viral Haemorrhagic Septicaemia (VHS) and Infectious Haematopoietic Necrosis (IHN), via the viscera of dead salmonids and complements the existing prohibition on the import of live salmonids. Licences to permit individual countries to continue exports to Great Britain of salmon and trout "in the round" will be considered where a country can satisfactorily demonstrate the absence of VHS and IHN from its waters and also operate its own controls on imports of uneviscerated salmonids, so as to prevent fish from an unacceptable source entering Great Britain through the "back door".

b)   Financial support

National grants and loans towards the construction and modernisation of fishing vessels during 1986, administered by the Sea Fish Industry Authority, amounted to £9.7 million and £3.1 million respectively, compared with £7.6 million and £3.2 million in 1985.

During the year, £3.4 million was paid in grants to owners who had permanently removed their vessels from the fishing fleet. This represents the withdrawal of approximately 8 500 tonnes.

European Community grants of about £10.5 million were approved for the United Kingdom, for the construction and modernisation of inshore fishing vessels and aquaculture.

The Highlands and Islands Development Board also administered grant and loan schemes during the year. The financial aid given was £1 352 555 for grants and £2 407 987 for loans.

Improvement grants continued to be given for harbours used for fishing vessels. In Scotland loans are also available for other than local authority harbours. During 1986 loan expenditure totalled £396 218, and United Kingdom grant-aid expenditure amounted to £3 599 702.

c)   Economic efficiency

The Seafish Industry Development Programme has continued to progress, enabling the Sea Fish Industry Authority to tackle the range of problems facing the fishing industry in the fields of marketing, promotion and training. Since commencement of this £15 million programme about £7 million of the total UK Goverment support of up to £7.9 million has been made available to the Authority.

d) <u>Other Government action</u>

The Sea Fish (Marketing Standards) Regulations 1986 introduced financial penalties for breaches of the European community common marketing standards for size and freshness for certain fresh or chilled fish.

The Food Advisory Committee has recently completed a review of coated and ice-glazed fish products. Its report was published on 10th March 1987.

e) <u>Sanitary regulations</u>

Microbiological standards for fish and fishery products are not laid down by statute. Good manufacturing practice with proper controls throughout every stage of processing, storage, distribution and sale is considered more important than end product testing. This is governed in the United Kingdom by statutory food hygiene regulations, i.e. The Food Hygiene (General) Regulations 1970 (No. 1172).

The composition and labelling provisions of the Food Act 1984 would also apply to fish and fish products. In addition, specific levels of lead in canned fish have been prescribed under the Lead in Food Regulations. The latest of these regulations (Lead in Food Regulations 1985) permits a lead level of 3.0 mg/kg in fish canned before 31st December 1985; fish canned later than that date, or sold after 31st December 1987, must comply with the general limit of lead in food, i.e. 1 mg/kg.

The following non-statutory, advisory guidelines have been issued:

i) "The Food Hygiene Code of Practice No. 10. -- The Canning of Low Acid Food." This is a guide to good manufacturing practice in the canning of low acid foods, including fish.

ii) The "Advisory Memorandum on Hygiene Production of Low Acid Canned Food." This guide is intended for overseas producers and summarises those points considered essential to prevent a hazard to the consumer.

iii) "Guidelines for microbiological quality of imported frozen cooked prawns" (1975). This paper describes suggested sampling techniques, methods of microbiological examination and microbiological guidelines.

All foodstuffs imported for human consumption are subject to the Imported Food Regulations 1984 (No. 1918), which broadly require food to be sound, wholesome and fit for human consumption.

The main regulations concerning sanitary regulations in Scotland are the Food Hygiene (Scotland) Regulations 1959 (SI numbers 1959/413, 1959/1153,

1961/622, 1966/967 and 1978/173 refer). These regulations secure the hygienic handling of food by any person, who, or whose clothing, is liable to come into contact with food in the course of a food business and to regulate the construction equipment and maintenance of such things as premises, vehicles and stalls in which food is handled. Also included are requirements for the notification of diseases likely to cause food poisoning, the treatment of certain kinds of food, including fish, the disposal of refuse, drainage sanitation and lighting of food premises. The provisions do not apply to food in impervious containers or to docks or buildings in docks used in connection with unloading fish or to warehouses. Subsequent amending regulations mainly updated the principal regulations in respect of the cleaning of utensils and, apply to stalls and vehicles, whereas certain provisions previously applied solely to food premises.

III. PRODUCTION

a) Fleet

Details of the fishing fleet and of the number of fishermen are given in Table 1 to the General Survey.

b) Operations

1986 saw no significant changes in this field.

c) Results

Table I(a) in the Statistical Annex shows the quantities and values of United Kingdom landings by U.K. vessels in 1986. Compared to 1985, total landings for all purposes fell by 6 per cent to 716 583 tons, while values rose by 11.5 per cent to £361 million.

Again in 1986, mackerel was the most important U.K. species in terms of quantity caught for human consumption but landings fell by 26 999 tons to 132 053 tons. Landings of haddock, however, increased by 5.3 per cent to 130 509 tons and haddock replaced cod as the most important in terms of the total value caught (up 17.1 per cent to £79 million). Whiting landings for human consumption fell by 17.6 per cent to 40 984 tons and herring and saithe rose by 33 per cent and 27.5 per cent respectively. Landings of sprats for human consumption fell by 51.2 per cent to 3 777 tons.

As in 1985, prices at first sale generally increased.

Shellfish landings rose by 16.7 per cent to 87 270 tons and the value rose by 18.9 per cent to £77.2 million.

Table I(b) in the Statistical Annex shows the provisional quantities and values of landings by foreign vessels in the United Kingdom. Compared to 1985, landings rose by 0.7 per cent to 53 119 tons, and the total value fell 10.9 per cent to £26.3 million. Cod accounted for much of the decrease with landings down 20.3 per cent to 17 850 tons. Herring landings increased by 7 311 tons to 8 644 tons.

Table I(c) in the Statistical Annex shows the provisional landings by UK vessels in foreign ports. Compared to 1985, these have fallen by 23.2 per cent to 35 684 tons, with mackerel contributing 47.7 per cent to the total tonnage.

Over-the-side sales in 1986 amounted to 171 229 tons, a decrease of 11.4 per cent over 1985. Mackerel provided 54.4 per cent of the total tonnage involved and herring 45.5 per cent.

Provisional withdrawals from the market under EC support arrangements totalled 10 547 tons, of which 9 455 tons was mackerel, and 573 tons was haddock.

Production of fishmeal fell from 50 187 tons in 1985 to 44 181 tons in 1986, a reduction of 12 per cent. Fish oil production fell by 22.2 per cent, from 9 644 tons in 1985 to 7 500 tons in 1986.

d)   Fish farming

Production statistics in relation to fish farming are now being collected for England and Wales, as they have been in Scotland and Northern Ireland, from an approximate total of 800 fish farms, with an annual production ranging from one to several hundred tons. (Final figures for England and Wales in 1986 are not yet available).

Estimated total production, by species, in 1986 is given in Table 1.

Table 1

	1986	Utilisation
Rainbow trout	12 500	84 per cent table production; 16 per cent restocking angling waters
Brown Trout	200	Mostly restocking angling waters
Salmon	10 000	All table production
Carp	50	Mostly table production
Eels	30	All table production
Turbot	70	All table production

IV. PROCESSING AND MARKETING

a)   Handling and distribution

Figures on utilisation of catch for 1986 (see Table 3 to General Survey) show a slight increase in fish utilised fresh/chilled, while the amount reduced to meal and oil decreased slightly.

b) Processing

Estimated figures on utilisation of the 1986 catch (see Table 3 to General Survey) show a slight decrease in fish utilised fresh/chilled, while the amount reduced to meal and oil rose slightly.

c) Domestic demand

Consumption of fish per head of population remained static compared with 1985.

d) External trade

Total imports fell in quantity (by 7.4 per cent) but rose in value (by 14.7 per cent) compared to 1985. Exports rose in both value and quantity. The value of imports rose to £834 million, while exports rose to £329 million resulting in a net trade deficit of £505 million, 8.8 per cent more than the £464 million recorded in 1985.

i) Imports

The quantity of fresh and chilled imports increased by 1 000 tons in 1986, and the value increased by 19.3 per cent to £99 million. Cod, with an increase of 4 000 tons to 50 000 tons, continued to be the major species imported. 50.4 per cent of all fresh and chilled imports originated in Iceland. Imports of frozen fish showed a 7.8 per cent decrease, while frozen fillet imports rose by 9 000 tons to 96 000 tons, 77.1 per cent of which was cod.

Canned fish imports rose 22 per cent to 105 000 tons; with a value of £220 million. The largest increases involved canned tunny and bonito (54.2 per cent) and canned salmon (47.4 per cent).

Imports of shellfish rose by 8.7 per cent to 50 000 tons, while the value of these imports rose 37.9 per cent to £200 million. Imports of prawns and shrimps increased by 6 000 tons to 26 000 tons and the value by 47.5 per cent to £90 million.

Overall fish meal imports were as 1985 (236 000 tons). Imports from West Germany increased by 36.5 per cent to 117 000 tons while those from Chile fell 36.5 per cent to 33 000 tons. Compared to 1985, imports of fish oil in quantity terms fell 35.8 per cent to 170 000 tons and in value terms by £37 million to £28 million. Imports from Norway fell by 68 per cent to 23 000 tons, while those from Iceland fell by 40.9 per cent to 39 000 tons. Imports of fish oil from Japan increased by 41.9 per cent to 44 000 tons.

ii) Exports

Quantities exported in 1986 increased by 46 000 tons to 350 000 tons, while the value rose by 25 per cent to £329 million. Exports of fresh and chilled fish remained by far the most important category in terms of both quantity and value, increasing in quantity by 29 000 tons to 241 000 tons and by value from £85 million to £105 million. Fresh, chilled and frozen mackerel exports (the main species and inclusive of transhipments) rose 4.6 per cent to 137 000 tons and in value by £2 million to £19 million. In terms of quantity, exports of frozen fillets increased by 30 per cent to 13 000 tons, while canned products decreased by 15.4 per cent to 11 000 tons. The valuable trade in shellfish rose 6 000 tons to 44 000 tons, and increased in value by 32.6 per cent to £134 million.

V. OUTLOOK

On the basis of the catch quotas allocated, the United Kingdom will be faced with reduced landings in 1987 for the important species of cod, haddock and saithe. It is therefore expected that as landings decline, imports will continue to play an important role in satisfying demand.

It is estimated that Atlantic salmon production in Scotland in 1987 will be around 16 000 tons and in 1988 in excess of 25 000 tons.

UNITED STATES

I. SUMMARY

A total of 4.3 million tons of fish, shellfish and other aquatic products valued at $3.1 billion, was harvested by U.S. fishermen in 1986 from international and domestic waters. In 1985, 4.0 million tons was harvested, valued at $2.6 billion.

New fisheries management plans were implemented for Western Pacific bottomfish and seamount groundfish, Gulf of Mexico red drum, and Northeast (New England) groundfish.

The U.S. domestic tuna industry continues to decline. Investments in the U.S. freezer/trawler fleet and surimi processing are expanding. Joint ventures for under-utilised species, continue to reach new record levels.

In 1986, total imports ($7.6 billion) and exports ($1.4 billion) of edible and non-edible fishery products, reached record levels.

II. GOVERNMENT ACTION

a) <u>Resource management</u>

i) <u>Major legislation</u>

The Magnuson Fishery Conservation and Management Act establishes U.S. jurisdiction over the fishery resources within 200-miles of the U.S. coastline. Regional fishery management councils (Councils) composed of government and industry representatives develop fishery management plans (FMPs). These plans are reviewed by the public, and implemented by the federal government in accordance with national standards and time schedules stipulated by the Magnuson Act. Principal changes to the Magnuson Act in 1986 were:

a) Council nominees must be knowledgeable, experienced and dependent for their livelihood upon the fisheries represented in the council region.

b) Council members and officers must disclose any financial interest in the fisheries within their jurisdiction.

c) The schedule of the federal government review of FMPs was shortened.

d) The regulations concerning foreign fishing permits were revised -- permits are to be issued annually; permit applications can be disapproved entirely or in part; permits may be modified, suspended, revoked or denied; minimum health/safety standards are to be maintained for U.S. observers, permit applications must specify safety standards and certification of compliance.

e) Councils may comment on any federal or state activity that may affect habitat of a fishery resource under their jurisdiction and federal agencies must provide detailed written response to councils within 45 days.

f) The performance of foreign fishing nations is to be reviewed annually with respect to contributions to the development of U.S. fisheries and harvest of U.S.-origin anadromous salmon. Nations may be assessed the higher foreign fishing fees if their performance is deemed unsatisfactory by the Secretary of Commerce, after consultation with the Secretary of State.

The Fish and Seafood Promotion Act of 1986 provides for the establishment of a National Fish and Seafood Promotional Council and Product-Specific Councils. The National Council's objectives are to develop annual plans and budgets to generically market and promote fisheries products, including consumer education, research and other activities of the Council, such as funding referenda to establish any Product-Specific Councils formed under the act and coordinating their activities. Legislative authority for the National Council expires on 1st October 1990. Funding for this Council will be primarily through monies transferred from the Saltonstall-Kennedy Fund ($750 000 in financial year 1987, $3 million each in financial years 1988 and 1989 and $2 million in financial year 1990). The Council will consist of the Secretary of Commerce or his designee, who shall be a non-voting member, and 15 voting members appointed from the seafood industry. The Product-Specific Seafood Councils will conduct product specific promotion, including consumer education and research, and develop seafood quality standards for fish or fish products. The Product-Specific Councils will be funded through self imposed industry.

ii) <u>Fishery management plans</u>

Three new fishery management plans (FMPs) were implemented in 1986.

-- <u>The Bottomfish and Seamount Groundfishery of the Western Pacific Region</u> establishes a fishery monitoring system for use in determining future management actions, a moratorium on fishing on the Hancock Seamount, permit requirements for bottomfish fishing in the Northwestern Hawaiian Islands, prohibition of the use of poisons or explosives and use of bottom trawls and bottom-set gillnets in harvesting bottomfish.

-- <u>The Red Drum FMP</u> for the Gulf of Mexico closes the fishery to directed commercial fishing in the U.S. fishery conservation zone to allow for stock rebuilding.

-- <u>The Northeast Multispecies FMP</u>, which completely replaces the previous FMP, establishes minimum sizes for commercial and recreational harvests of cod, haddock, yellowtail flounder, pollock, American plaice, winter flounder and witch flounder. Area closures were established <u>for</u> the Southern New England stocks of yellowtail flounder, winter flounder and haddock<u> to protect them</u> during spawning.

Five FMPs were amended in 1986. Uniform gear marking requirements for the offshore American lobster fishery and prohibitions against taking lobsters with red crab fishing gear in waters deeper than 200 metres were instituted. The meat count regulations of the original FMP for the Atlantic sea scallop fishery were continued while allowing exemptions to the regulations for the conduct of experimental fishing operations beneficial to the sea scallop resource. A prohibition on the taking of berried female Gulf of Mexico stone crabs and requirements that harvested stone crabs be held in shaded containers were instituted. Measurements that define legal sized Western Pacific lobsters were adjusted. Regulations for the Alaskan tanner crab fishery were temporarily suspended pending study of alternative long term management measures.

iii) <u>Allocations</u>

No changes were made in 1986 to the basic management mechanisms that allocate amongst domestic and foreign harvesters. However, a greater proportion of the fishery resources were reserved for domestic use because of the growth of U.S. harvests through over-the-side sales to foreign processing vessels (joint ventures) and the growth in the U.S. factory trawler fleet.

Table 1

DIVISION OF OPTIMUM YIELDS INTO DOMESTIC & FOREIGN QUOTAS
FOR SPECIES WHICH ARE SURPLUS TO DOMESTIC NEEDS

(million tons)

	Optimum Yield	Domestic Annual Harvest	Total Allowable Level of Foreign Fishing
1977	2.6	0.4	2.1
1978	2.7	0.6	2.1
1979	2.7	0.6	2.1
1980	2.9	0.7	2.2
1981	2.8	0.7	2.1
1982	2.7	0.7	1.9
1983	3.0	1.0	1.9
1984	3.3	1.3	1.9
1985	3.3	1.8	1.4
1986	3.2	2.3	0.8

<u>Foreign</u>: In 1986, 0.6 million tons of fish were allocated to seven countries from the 0.8 million tons total allowable level of foreign fishing (TALFF). Japan received the largest allocation -- 0.5 million tons. The People's Republic of China received allocations in U.S. waters off Alaska for the first time ever this year. Much of the unallocated TALFF was due to lack of foreign interest in Atlantic mackerel.

<u>Conditions of access</u>: Foreign access to the U.S. fishery resources is predicated on the review of the foreign nation's contribution to U.S. fishery development. As specified in the Magnuson Act allocations of fish to foreign

nations are dependent upon: the extent to which the foreign nation restricts market access to U.S. fishery products through tariff and non-tariff trade barriers; that nation's purchases of U.S. fish products from U.S. harvesters and processors; to what extent the nation has cooperated in the enforcement of U.S. fishing regulations; to what extent the nation utilises the fish harvested for domestic consumption; to what extent the foreign nation contributes to the growth of the U.S. fishery by minimising gear conflicts or through the transfer of technology; to what extent the nation has traditionally fished in the U.S. waters; that nation's contribution to U.S. fishery research; and other matters that are deemed important, such as when the United States denied the Soviet Union access to U.S. stocks because of the Soviet invasion of Afghanistan. Through public notice and comment, policies and procedures concerning the evaluation of foreign performance under these criteria were established and implemented. New conditions were also added to the permits for foreign fishing, such as restrictions governing area and seasonal closures of the Gulf of Alaska pollock fishery.

    iv)   <u>Enforcement and surveillance</u>

The United States strictly enforces its 200-mile EEZ. In 1986, there were numerous cases of unauthorised entry and fishing within the U.S. zone. Several of these resulted in payment of fines; $271 164 was collected in fines during the year.

    v)   <u>Fees and permits</u>

Fees for foreign fishing in 1986 were set to achieve a target amount of $49.7 million, of which $49.5 million were to be collected from tonnage fees for foreign harvests and the balance from fishing permit fees. The target reflects a prorated portion of the fiscal year 1985 cost of administering the Magnuson Act due to foreign fishing. The calculation of these costs was revised to include indirect costs for at-sea enforcement and from other increases in programme costs. Since these costs have been increasing during a period of declining foreign effort, the rate per ton of foreign harvest continues to increase. In 1986 this rate was 35.6 per cent of the ex-vessel value of the fish to the foreign fleets where ex-vessel value is based on the prices received in foreign markets.

The Magnuson Act was amended to allow the Secretary of Commerce to establish higher fees during a fiscal year for nations found to be harvesting anadromous species of U.S. origin at unacceptable levels or to be failing to take sufficient action to benefit the conservation and development of U.S. fisheries. The fee schedule was revised for FY87 to require any fishing nation so found to pay an additional incremental amount of 78.6 per cent of its total tonnage fees. No foreign nation paid such fees during 1986.

Over 800 foreign vessels from 11 different countries were permitted to either fish, receive fish at-sea from U.S. vessels, transport fish or to scout for fish in U.S. waters in 1986.

b)   <u>Financial support</u>

During 1986, five financial support programmes were in force, virtually unchanged from 1985:

i) The Fisheries Obligation Guarantee Programme provides long term financing or refinancing of the construction cost of fishing vessels and shoreside facilities. Credit standards are high; the only benefits are extended terms and interest at the lowest rate for federally-guaranteed obligations. The programme is self-supporting and receives no government funding.

ii) The Fishermen's Guarantee Fund indemnifies participating commercial fishermen against certain costs of seizure and detention by foreign nations made on the basis of oceanic rights and claims not recognised by the United States. Covered costs include direct costs such as damage to the vessel and its equipment, loss or confiscation of vessels or equipment, dockage or utilities fees and up to 50 per cent of any gross income loss due to seizure and detention. The current fee for voluntary participation is $30 per gross registered ton. This programme is largely funded by participant fees. Effective 1 July 1986, the responsibility for administering these programmes was transferred to the Department of State from the Department of Commerce.

iii) The Fishing Vessel and Gear Damage Compensation Fund reimburses U.S. commercial fishermen for vessel loss or damage caused by foreign vessels or their crews and for the gear loss or damage caused by another vessel. Casualties must have occurred within the U.S. waters. This programme is funded by a surcharge applied to fees charged foreign countries for fishing in the U.S. waters.

iv) The Fishermen's Contingency Fund reimburses U.S. commercial fishermen for vessel or gear casualties caused by oil and gas related activities on the outer Continental Shelf (OCS). This programme is funded by assessment of oil and gas companies operating on the OCS.

v) The Capital Construction Fund provisions of the Merchant Marine Act presently allow certain U.S. merchant and fishing vessels to defer Federal taxation on the income from the operation of those vessels.

vi) The Saltonstall-Kennedy Act makes available to the Secretary of Commerce up to 30 per cent of the gross receipts collected under the customs laws from duties on fishery products. The Secretary must use a portion of these funds each year to make available grants to assist persons in carrying out research and development projects which address aspects of United States fisheries, including, but not limited to, harvesting, processing and associated infrastructures. For financial year 1987 $7.4 million was appropriated for the programme.

c)   Economic efficiency

i) Limited entry

No limited entry systems were implemented in 1986; although several federally funded limited entry studies are being developed. Regulations that limit individual vessel effort are being considered for the Atlantic surf clam fishery which is currently regulated through a moritorium on new vessels, while the possibility of a limited entry system for the South Atlantic and Gulf of Mexico spiny lobster fishery is being explored. A major study that addresses the problem of matching capital to resources in the fish harvesting industry has been initiated and will form the basis of two regional and one national limited entry conferences.

ii) <u>Disincentives</u>

No changes have occurred in the rules regarding the assessment of taxes or fees as a disincentive to the over-capitalisation of the fleets. The Magnuson Act continues to restrict the amount of fees charged domestic entities for the administrative costs associated with issuing permits. Federal policy continues to restrict fees assessed on foreign activities to the cost of management rather than the utilisation of fees as a management tool.

iii) <u>Improvements in gear and vessel efficiency</u>

There are no dedicated programmes for the development or improvement of vessel efficiency. The use of Trawl Efficiency Devices (TED) for shrimp vessels, however, are currently being proposed through regulation for certain areas of the Gulf of Mexico in an effort to protect endangered species of turtles. A coincident benefit of the use of TEDs is that they are able to separate and eliminate a high percentage of unwanted by-catch, resulting in a marked improvement of operating efficiency.

d) <u>Bilateral arrangements</u>

The United States agreed on 8 October 1986 to extend its Governing International Fishery Agreement (GIFA) with the USSR through 31 December 1987. Such agreements provide for fisheries access and other related matters.

e) <u>Other government action</u>

i) <u>Decision-making process in management</u>

As discussed previously, several changes to the fishery management schedule for federal review of FMPs were made in order to implement regulations in a more expedient manner. The federal government is to provide an immediate evaluation of proposed fishery management plans or amendments submitted by the fishery management councils. If the proposed regulations are consistent with the national standards of the Magnuson Act and if the standards are met, the document is to be published for public comment within 15 days of receipt, while the public comment period for such documents was shortened to 60 from 90 days.

ii) <u>Processing, marketing and distribution of fish and fish products</u>

To strengthen and develop the U.S. fishing industry, $7.9 million worth of research and development projects were federally funded in 1986 through the Salstonstall-Kennedy Act. This grant programme stressed harvesting, processing and marketing projects. Of the 54 projects funded, over

$2.4 million was for the development of Alaska groundfish with $854 000 specifically designated for the development of Alaska pollock surimi.

The U.S. Department of Agriculture began a test programme to purchase up to 6 500 tons of frozen fried Alaska fish nuggets for use in school lunch programmes. These nuggets are a new product made from Alaska pollock modelled after the popular chicken nugget (breaded bite-sized portions of fish).

iii) <u>International questions, negotiations</u>

The U.S. concluded negotiations with member governments of the Forum Fisheries Agency for access for U.S. tuna vessels to a broad ocean area including the waters of the Pacific Island States.

f) <u>Structural adjustment</u>

<u>Industry changes</u>: The trend toward "Americanization" of U.S. fisheries continues, as U.S. producers gear up to utilise an increasing share of the resources available in U.S. waters. The term "Americanization" refers to U.S. policy to encourage greater U.S. exploitation of coastal resources. The total catch by all vessels in U.S. waters -- U.S. and foreign -- has been relatively stable. U.S. effort, however, is rapidly replacing foreign operations, and the U.S. fleets now harvest more than three-fourths of the total catch. The expansion of U.S. fishing operations reflects important changes in fleet structure and strategies. Joint ventures (JVs) have become a major factor along with significant investment in U.S. processing vessels. Increasing numbers of U.S. catcher vessels are fishing almost exclusively under contract to foreign operators. Catches are off-loaded at sea onto foreign processing vessels. Also, new U.S. factory vessels are coming on line, on all coasts. Advanced technology and a more favourable investment climate have significantly increased prospects for successful U.S. factory ship operations. These operations are growing so rapidly that in the future the processing of many species will shift from being predominately on-shore to being predominately at-sea.

Fish processors will be increasingly subjected to new environmental regulations. For example, several New England fish meal and oil processing plants were closed in 1986 for environmental reasons. Modernisation costs to meet the environmental guidelines were determined to be prohibitive.

Fish processors in the Gulf of Mexico are benefiting from the decrease in world-wide oil prices which has caused high unemployment within the Gulf of Mexico regional economy. The newly created labour pools of unemployed oil industry workers have turned to fish processing companies for work. Fish processors have traditionally been plagued by a lack of capable employees and high employee turnover rates.

g) <u>Sanitary regulations</u>

During 1986, there were no significant changes in U.S. Sanitary Regulations. The U.S. Food and Drug Administration (FDA) is the regulatory agency responsible for enforcing the provisions of the Federal Food Drug and Cosmetic Act. The FDA uses from the Federal Code of Regulations Title 21:

Part 110 -- "Current Good Manufacturing Practice in Manufacturing, Processing, Packing, or Holding Human Food"; Part 113 -- "Thermally Processed Low-Acid Foods Packaged in Hermetically Sealed Containers"; and Part 123 -- "Frozen Raw Breaded Shrimp", as a basis for sanitation compliance. These regulations, which the FDA promulgates and enforces, govern food handling and processing practices which apply to all foods (except meat, poultry, and dairy products) $\underline{*}$ produced or sold in the United States.

III. PRODUCTION

a) <u>Fleet</u>

The U.S. fishing fleet, comprising vessels of 5 grt and larger, in 1985 had an estimated 24 300 vessels. In addition, 104 000 motor boats of less than 5 grt and 1 500 small non-motorised boats less than 5 grt operated in 1985.

b) <u>Operations</u>

i) <u>Changes in the cost of fishing</u>

Vessel operating costs remained at about the same level as those in 1985, with the exception of a slight increase in insurance costs. However, low fuel costs have made many fishing ventures more profitable. This is particularly true in the shrimp industry where it takes an estimated one gallon of fuel to catch one pound of shrimp. Ice and other fuel dependent costs were also lower.

ii) <u>Changes in fishing methods, gear, equipment</u>

The U.S. fishing fleet operating off Alaska increased in 1986 to 15 factory trawlers and three factory motherships. The factory trawlers represented an increase of three vessels over those operating in 1985. All of the domestic motherships were initially deployed in 1986. One mothership was designed with state-of-the arts equipment for processing Pacific cod and pollock fillets, as well as headed and gutted cod, pollock, rockfish, and sablefish. Almost 70 per cent of the Alaska groundfish processed by U.S. firms was processed at sea. Providing headed and gutted fish to the Japanese market was a significant breakthrough in 1986. Construction of 11 new factory trawlers was underway in 1986 and these are all expected to enter the groundfish fishery in 1987.

High prices attracted large numbers of small and medium size vessels, 40-75 feet long, to the fisheries for sablefish and Pacific cod.

The first floating surimi processing ship, a 130 foot vessel, entered the Alaska pollock fishery in 1986. Several are under construction and should enter the fleet in 1987.

The first pair trawling vessel was introduced into the Alaska pollock fishery in 1986.

In 1986, seven U.S. factory trawlers were operating in the northeast (New England) fisheries. These vessels were harvesting squid, mackerel, and butterfish, with one vessel harvesting shrimp as well. One of these vessels is being transferred to Alaska, while there is another factory trawler which has been in dock for the past few years. There is also at least one new vessel being built to enter the fishery in 1987.

The Portland Fish Exchange in Portland, Maine was established in 1986 and is the first "display auction" used in the United States. This auction allows lot bidding where buyers view the refrigerated product available for sale.

The U.S. tuna fleet declined 20 per cent in 1986 to 89 vessels with a total capacity of about 42 000 tons. The high cost of financing and insurance have been a primary cause for this decline. Only one major cannery still exists in the Continental United States.

iii) <u>Exploitation of under-utilised species</u>

Domestic catches of Alaska pollock increased over 40 per cent in 1986. U.S. capacity replaced all directed foreign fisheries in the Gulf of Alaska for pollock in 1986 (Alaska waters are divided into two management regions, the Bering Sea and Aleutian Islands and the Gulf of Alaska). Directed U.S. fisheries for Rock sole and Greenland turbot became significant in 1986, with approximately 5 000 tons and 3 000 tons being taken respectively.

As a result of the continuing short supply of many traditional Atlantic groundfish species, there is building interest in harvesting and processing ocean pout (Macrozoarces americanus) to substitute for Atlantic whitefish.

Continued improvement in the size of the Northern shrimp (Pandalus borealis) resource has offered fishermen a viable alternative to offset the decline in New England groundfish.

Gulf and South Atlantic shrimp fishermen have renewed their interest in culling out saleable fish caught incidental to shrimping. Fishermen now are more cognizant about the profitability of flounder, croaker, snapper, grouper, and crab that are normally discarded after the shrimp are removed.

For the first time, bluefin tuna caught in waters off Los Angeles, California may find a durable niche in the Japanese market. Air-freighted fresh bluefin tuna from Los Angeles began to arrive in Japan in mid-November 1985, and in just two months the shipments totalled over 2 300 fish. The Japanese buyers gave the fish a highly favourable rating for their oil-rich meat and excellent flavour, and above all for their size and flavour which resembled those of the Japanese domestic bluefin tuna. Another appealing factor for the Los Angeles bluefin tuna is the timeliness of their supply. The three-month off-season for Japanese domestic bluefin from November through January is the catch season in waters off Los Angeles.

iv) <u>Preservation, processing and handling onboard ship</u>

Two factory ships began surimi operations in 1986.

As Alaska pollock utilisation increased in 1986, the use of the Baader 182 filleting machine has become standard for processing volume quantities. Headed and gutted cod and rockfish were also important product forms for the U.S. industry.

c) Results

i) Commercial catch rates and trends

Total U.S. harvest: Total commercial harvests (edible and industrial) including joint ventures and landings at non-U.S. ports were 4.3 million tons in 1986, valued at $3.1 billion, up from 1985 total harvests of 3.9 million tons, valued at $2.6 billion.

Joint venture harvest: Joint venture catches by U.S. fishermen, unloaded onto foreign vessels were 1.3 million tons valued at $154.9 million. This was a 44 per cent increase over 1985, when 911 000 tons were caught, valued at $104.3 million. The major species were flounders, Pacific hake and Alaska pollock.

Foreign harvest: The foreign catch of fish (excluding tunas) and shellfish in the U.S. EEZ was 588 000 tons in 1986, a 49 per cent decrease compared with 1985 and 53 per cent below the average for the preceding 5 years. As in other years, the U.S. EEZ off Alaska supplied the largest share of the foreign catch (84 per cent) followed by California, Oregon and Washington (12 per cent), and the North Atlantic (4 per cent). Alaska pollock comprised 60 per cent of the foreign catch; Pacific flounders, 13 per cent; Pacific hake, 12 per cent; Pacific cod, 9 per cent; and other fish and shellfish the remainder. Japan continued as the leading nation fishing in the U.S. EEZ with a catch of 385 700 tons, 66 per cent of the total foreign catch. Catches by vessels of the Republic of Korea, the second leading nation were 97 300 tons representing 16 per cent of the catch in 1986.

Table 2

DOMESTIC AND FOREIGN HARVESTS FROM THE U.S. EXCLUSIVE ECONOMIC ZONE (EEZ)

('000 tons)

	Landings	U.S. Harvest Joint Venture	Total	Foreign Harvest	Total EEZ Harvest
1977	706	0	706	1 699	2 405
1978	664	0	664	1 754	2 418
1979	805	11	816	1 650	2 466
1980	858	62	920	1 628	2 548
1981	924	140	1 064	1 655	2 719
1982	802	255	1 057	1 415	2 472
1983	721	435	1 156	1 312	2 468
1984	683	665	1 348	1 315	2 663
1985	761	911	1 672	1 164	2 836
1986	1 117	1 310	2 427	588	3 015

U.S. EEZ: The total U.S. and foreign catch from the U.S. EEZ (3 to 200 miles offshore of the U.S. coast) was 3.0 million tons in 1986 (up 6 per cent) compared with 1985. However, the U.S. share was 80 per cent of the total, up 21 per cent from 1985.

ii) Trends in domestic U.S. landings of main species

Total commercial landings: Commercial landings (edible and industrial) by U.S. fishermen at ports in the United States were 2.7 million tons valued at $2.8 billion in 1986 -- a decrease of 103 000 tons in quantity, but an increase of $437 000 in value compared with 1985. Increased landings of shellfish such as crabs, shrimp and squid helped offset declines in major finfish species such as flounders, menhaden and salmon. Landings of sea herring, Alaska pollock and tuna increased.

North Atlantic trawl fish: North Atlantic landings of butterfish, Atlantic cod, cusk, flounders, haddock, red and white hake, ocean perch, pollock and whiting (silver hake) were 140 300 tons valued at $190.5 million -- a decrease of 13 per cent in quantity but an increase of 4 per cent in value compared with 1985. Of these species, flounders led in value, accounting for 55 per cent of the total; followed by Atlantic cod, 19 per cent; and pollock, 7 per cent. The 1986 catch of cod (27 700 tons) was the lowest since 1976, while landings of yellowtail flounder (10 400 tons) and haddock (5 000 tons) also reflected continuing downward trends in U.S. harvests because of resource problems. The North Atlantic pollock landings (24 700 tons) were the highest ever recorded despite the fact that the abundance of pollock has declined substantially in recent years from record high levels in the late 1970s and early 1980s.

Alaska pollock and other Pacific trawl fish: U.S. landings of Pacific trawl fish (Pacific cod, flounders, hake (Pacific whiting), Pacific ocean perch, Alaska pollock and rockfishes) were 195 800 tons, valued at $70.1 million -- an increase of 11 per cent in quantity compared with 1985. Landings of Alaska pollock increased 40 per cent to 59 100 tons while Pacific cod landings decreased 13 per cent to 47 400 tons. Pacific whiting landings increased by 57 per cent in quantity in 1986 to 11 600 tons valued at $1.2 million.

Anchovies: U.S. landings of anchovies were 6 100 tons -- a decrease of 8 per cent from the previous year. Almost all of these landings were used for bait with only 500 tons converted to fish meal and oil. A factor contributing to the lower landings was the stiff competition from substitute products in the fish meal and oil market.

Halibut: U.S. landings of Atlantic and Pacific halibut were 35 200 tons in the round valued at $82.9 million -- an increase of 27 per cent in quantity and 116 per cent in value compared with 1985. (The 1986 ex-vessel price was $2 360 per ton as compared with the 1985 ex-vessel price of $1 390 per ton). The Pacific fishery accounted for almost all of the halibut landed.

Sea herring: U.S. commercial landings of sea herring (Atlantic and Pacific) were 95 300 tons valued at $48.9 million -- an increase of 5 per cent in quantity, and a decline of 2 per cent in value compared with 1985. Landings of Atlantic herring (36 000 tons) remain low compared with historical levels because biological recruitment to the fishery was relatively poor throughout the 1980s. Landings of Pacific herring were 59 000 tons, valued at $44.6 million. Although Alaskan landings of Pacific herring of 51 000 tons decreased 9 per cent compared with 1985, the 1986 value was about $400 000 greater than in 1985. Increased spawn-on-kelp harvest accounted for most of the increased value of the fishery, rather than an increase in price paid per ton of sac roe or food/bait herring.

Mackerel: Landings of jack mackerel in 1986 were 10 800 tons, valued at $1.8 million. U.S. landings of Atlantic mackerel were 4 300 tons valued at $1.2 million -- a 43 per cent increase in quantity and 21 per cent increase in value as compared with 1985. The Northwest Atlantic mackerel stock is continuing to expand due to above average strength in recent year classes. Landings of Pacific mackerel were 38 700 tons valued at $6.4 million -- an increase of 13 per cent in quantity and 1 per cent in value compared with 1985.

Menhaden: The U.S. menhaden landings were 1.1 million tons valued at $98.3 million -- a decrease of 13 per cent in quantity and 7 per cent in value. Atlantic landings (256 000 tons) decreased significantly (29 per cent) while the Gulf landings (816 000 tons) decreased slightly (6 per cent) from 1985. Atlantic landings over the short term will probably not reach higher levels and Gulf landings are slightly below the record catches that occurred during 1983-85.

Salmon: The 1986 Alaskan commercial harvest was 127.8 million fish weighing almost 270 000 tons. This is the fourth largest harvest in history and the seventh year in a row that the Alaska salmon harvest exceeded 100 million fish. Alaska salmon accounted for over 90 per cent of the total U.S. landings of salmon which were 298 700 tons valued at $493.9 million.

Sablefish: U.S. commercial landings of sablefish were 38 500 tons valued at $45.9 million -- an increase of 34 per cent in quantity and 60 per cent in value compared with 1985.

Tuna: Landings of tuna by U.S. fishermen at ports in the 50 states, Puerto Rico, American Samoa, other U.S. territories, and foreign ports were 252 000 tons valued at $217 million -- an increase of 8 per cent in quantity and 3 per cent in value compared with 1985. Sixteen per cent of the tuna landings were at ports in the continental United States, principally California with 77 per cent of the continental landings.

Clams: Landings of all species yielded 66 000 tons of meats valued at $135 million -- a decrease of 3 per cent in quantity, but an increase of 5 per cent in value compared with 1985. Surf clams yielded 36 000 tons; ocean quahogs 21 000 tons; hard clams 5 400 tons and soft clams 2 700 tons of meats in 1986.

Crabs: Landings of all species of crabs were 161 000 tons, valued at $270 million -- an increase of 5 per cent in quantity and 33 per cent in value compared with 1985. Hard blue crab landings were 84 000 tons; dungeness crab landings were 10 000 tons; and king crab landings were 12 000 tons. Snow (tanner) crab landings were 50 000 tons valued at $83.4 million -- a substantial increase of 28 per cent in quantity and 62 per cent in value compared with 1985.

Lobsters, oysters, and shrimp: American lobster landings were 21 000 tons valued at $121 million -- a 1 per cent decrease in quantity, but a 5 per cent increase in value compared with 1985. U.S. landings of American lobster have increased steadily during the last decade. This increase has largely been attributable to increases in the inshore trap fishery. (The offshore fishery typically accounts for 20 per cent of total landings.) A continued increase in fishing effort is responsible in part for higher landings in coastal waters. It is currently estimated that in excess of 3 million traps are set annually. Exploitation rates for lobsters in coastal waters are extremely high and there is concern for the long-term viability of the resource. Fishery-independent measures of abundance, however, indicate that lobster population levels are currently stable. U.S. oyster landings yielded 18 400 tons of meats valued at $78.1 million -- an 8 per cent decrease in quantity, but an 11 per cent increase in value compared with 1985. U.S. landings of shrimp were 181 500 tons valued at $662.7 million -- a 20 per cent increase in quantity and 40 per cent increase in value compared with 1985.

Scallops: U.S. landings of all species of scallops were 10 100 tons of meats valued at $107 million -- a decrease of 25 per cent in quantity, but an increase of 15 per cent in value compared with 1985. U.S. bay scallops were 33 tons; sea scallop landings were 9 000 tons and calico scallops were 700 tons in 1986. The 26 per cent increase in sea scallop landings from 1985, the first increase since 1978, was due to marked improvement in the abundance of harvestable sized scallops on Georges Bank and in the Middle Atlantic region from outstanding recruitment of the 1982 year class. Abundance is expected to increase further in 1987 due to the additional strong recruitment from the 1983 year class.

Squid: U.S. commercial landings of squid were 34 000 tons valued at $14.6 million -- an increase of 54 per cent in quantity and 30 per cent in value compared with 1985. Squid landings of the Pacific Coast in 1986 (19 000 tons) continued the strong recovery being made from the climatic effects of El Nino which depressed past landings.

iii) U.S. Recreational catch

U.S. recreational fishermen catch a significant quantity of marine fish (see Table 3). The majority of their catch takes place in inland waters (182 000 individual fish) or in the ocean within 3 miles (197 000 individual fish). The chief mode of catch is the use of private and rental boats (vessels that typically hold 6 or less fishermen); over 34 000 fishing trips were taken by fishermen using these vessels. Off the Atlantic and Gulf Coasts the principal species taken in terms of number were black sea bass, bluefish, scup and Atlantic croaker. Off the coast of Washington, Oregon and California, the principle species taken were surf smelt, kelp bass, rockfish and Pacific mackerel.

iv) Fish farming

Preliminary estimates of U.S. private aquaculture production for 1985 (the last year surveyed), was some 242 000 tons with a value to the producers of $423 million. Estimates for 1986 are unavailable, but it is expected that production in 1986 will increase over 1985. Pacific salmon returns in Alaska from public and private non-profit hatcheries showed significant increases over 1984, while catfish comprises 51 per cent of total U.S. production.

(Data shown -- Table 4 -- are live weight for consumption, except for oysters, clams and mussels which are meat weight. Excluded are eggs, fingerlings, etc., which are an intermediate product level. The "other species" include species such as sturgeon, carp, buffalofish, tilapia, mullet, etc. (Some figures for earlier years have been revised).

Table 3

MARINE RECREATIONAL CATCH

(millions of fish)
(no. of trips)

	Atlantic and Gulf Coast		Pacific Coast		Grand Total	
	fish	trips	fish	trips	fish	trips
1979	439	63	49	8	488	71
1980	463	74	84	15	547	89
1981	331	52	51	11	382	63
1982	371	61	53	11	424	72
1983	398	69	45	11	443	80
1984	356	62	47	10	403	72
1985	382	61	43	10	425	71
1986						

Table 4

	Production ('000 tons)					Value ($ million)				
	1980	1982	1983	1984	1985	1980	1982	1983	1984	1985
Baitfish	10.0	10.0	10.0	11.0	11.0	44	44	44	47	51
Catfish	35.0	91.0	100.0	109.0	123.0	54	120	132	192	189
Clams	0.3	0.3	0.8	0.8	0.7	2	3	10	4	5
Crawfish	11.0	25.0	27.0	27.0	30.0	13	27	30	30	33
Freshwater prawns	0.1	0.2	0.1	0.1	0.1	1	2	2	2	2
Mussels	-	0.2	0.4	0.4	0.4	-	2	2	2	1
Oysters	11.0	10.0	11.0	11.0	10.0	37	34	32	39	40
Pacific salmon	4.0	12.0	9.0	21.0	38.0	3	4	7	17	25
Shrimp	-	-	0.1	0.2	0.2	-	-	1	2	2
Trout	22.0	22.0	22.0	23.0	23.0	37	48	50	54	55
Other species	-	2.0	3.0	5.0	6.0	-	5	7	10	20
Total	93.4	172.7	180.4	208.4	242.0	191	289	317	399	423

IV. PROCESSING AND MARKETING

a) Handling

In 1986 the National Seafood Inspection Programme inspected 11.6 per cent of the seafood products consumed in the U.S. Approximately 6.3 per cent or 121 processing facilities in the U.S. participated in the voluntary programme in addition to 400 lot inspection users. State-federal cooperation agreements for cross-licencing of inspectors exist between the Department of Commerce and the states of Alaska, Alabama, Oregon, Minnesota, Tennessee, Arkansas, Florida, New Jersey, New York, Louisiana, Mississippi and Maine. Cross utilisation agreements for use of inspections staff, also exist between the U.S. Department of Commerce and the U.S. Department of Agriculture. The programme operates on a fee-for-service basis and provides services for domestic as well as foreign exporters.

In the New England region, efforts are underway to improve the quality of New England fish. Processing plants in the State of Maine are certified by state officials if the plants observe strict quality control guidelines. Efforts are also devoted to projects that will improve the quality of at-sea handling of fish products.

There appears to be a new level of sophistication in the area of air freight shipping with new customised containers made to order for shippers. Many airlines have reduced rates to be more competitive.

Cold storage holdings: In 1986 stocks of frozen fishery products were at a low of 117 500 tons on 31 March and a high of 165 000 tons on 30 November. Cold storage holdings were down every month in 1986 compared with 1985, resulting in a 7 per cent annual decline. This decline in holdings of frozen fishery products wa a direct result of a decline in domestic landings, at a time when consumer demand for fishery products was at its highest level ever.

Over-production coupled with lower demand has created a large build-up of catfish that are ready for processing. Producer prices are lower than any time since 1984, at $1.34 per kilo. Institutional demand for catfish products used has been lower than expected.

b) Processing

Major processed products

Total production: The 1986 estimated value of edible and non-edible fishery products was $5.2 billion, $263.6 million more than the $4.9 billion in 1985. The value of edible products was $4.9 billion and the value of industrial products was $276 million.

Canned fishery products: The pack of canned fishery products in the 50 states, American Samoa and Puerto Rico was 590 000 tons (50.1 million standard cases) valued at $1.5 billion.

<u>Canned salmon</u>: The 1986 U.S. pack of natural Pacific salmon was 2.9 million standard cases (64 400 tons) valued at $265.5 million, compared with the 3.3 million standard cases packed a year earlier. Alaskan plants accounted for 92 per cent of the salmon packed.

<u>Canned tuna</u>: The U.S. pack of tuna was 32.7 million standard cases (290 000 tons) valued at $881.5 million -- an increase of 4.7 million standard cases in quantity compared with the 1985 pack. Plants in the United States packed 6 per cent of the total and plants in American Samoa and Puerto Rico packed the remainder. About 25 per cent of the U.S. supply of canned tuna was packed from U.S. caught fish and 48 per cent from imported fish. Imports of canned tuna made up the remaining 27 per cent.

<u>Fish fillets and steaks</u>: In 1986 the U.S. production of raw (uncooked) fish fillets and steaks was 118 000 tons valued at $539 million -- $98.5 million more than the record set in 1985. Cod fillets led all species and comprised 24 per cent of the total.

<u>Fish sticks and portions</u>: The combined production of fish sticks and portions was 194 000 tons valued at $491.9 million compared with a 1985 production of 193 500 tons valued at $478.9 million.

<u>Non-edible fish products</u>: The value of the domestic production of industrial fishery products was $174.5 million -- a decrease of 4 per cent from 1985. The domestic production of fish meal and scrap was 319 000 tons valued at $83.4 million -- a decrease of 3 per cent in quantity and 1 per cent in value from 1985. Menhaden meal production was 269 000 tons. Domestic production of fish solubles was 88 500 tons, 40 per cent less than in 1985. Fish oil production increased 18 per cent to 152 900 tons valued at $43.7 million.

Product development

U.S. processors continued their expansion into Alaska pollock surimi and surimi analogue products. At year end more than 15 plants were in operation or planned. These plants use surimi blocks imported from Japan or the Republic of Korea to produce analogue products, mostly imitation crab, scallops and shrimp. Two shoreside plants in Dutch Harbor, Alaska initiated surimi production while the plant in Kodiak, Alaska continued producing surimi. Two factory ships also began surimi production in 1986. Much of the U.S. surimi was utilised by the analogue plants while small quantities were exported to Japan.

A two year research and development project was initiated in January 1986 through government contract with a U.S. corporation to determine the technical and economic feasibility of producing surimi from menhaden. During the first year a pilot demonstration plant was constructed with the latest surimi processing equipment. By the close of the menhaden season in October 1986, an initial experimental production lot of one ton of surimi from menhaden was produced. The project is required to produce 40 tons of surimi from menhaden during the 1987 menhaden harvesting season from April through October 1987. The surimi will be distributed within the industry to those interested in evaluating the functional and performance characteristics of the material.

The Alaska Seafood Marketing Institute, in 1986, spent $400 000 on the problem of excess supplies of canned pink salmon by developing radio commercials in seventeen U.S. cities. This was done to remove excess inventories of the 1985 pack to make way for the 1986 pack later in the summer. Total pink salmon harvests in 1986 were about 75 million fish with the total pack of pink salmon being over 2.9 million cases.

The State of Alaska in partnership with the industry, contributed $4 million to the promotion of Alaska fisheries in 1986.

In 1986, the world-wide scarcity of whitefish products caused wholesale prices of Alaska pollock to climb rapidly. IQF fillets sold for up to $3.52 per kilo with shatter pack fillets around $4.30 per kilo and blocks reached well over $2.20 per kilo. The U.S. Department of Agriculture made a test purchase of Alaska pollock fish nuggets (0.8-0.9 oz breaded pieces) made either from blocks or by forming equipment for use in schools.

Three firms continue to test market canned skinless boneless chunk pink salmon packed in water. The U.S. salmon industry is still working towards new product forms in addition to the traditional method of packing canned salmon.

c) <u>Domestic market</u>

U.S. per capita consumption of fish and shellfish was a record 6.67 kg. U.S. citizens consumed 4.09 kg. of fresh and frozen fish, 2.45 kg. of canned fishery products and approximately 1.59 kg. of recreationally caught fish. Increased imports and a reduction of stocks led to this record.

The annual index for edible fish and shellfish increased 12 per cent while the annual index for industrial fish increased 6 per cent in comparison with 1985.

d) <u>External trade</u>

The total value of edible and non-edible products resulted in a record import value of $7.6 billion in 1986 -- $947.7 million more than the previous record in 1985, when $6.7 billion of fishery products were imported. The total value of edible and non-edible U.S. fishery products exported in 1986 was $1.4 billion -- an increase of $272.0 million compared with 1985. (Fishery export figures do not include the $104.3 million of at-sea transfers of U.S. flag catches to foreign vessels in joint ventures.)

United States imports of edible fishery products in 1986 were valued at a record $4.8 billion, $771.8 million higher than the previous record for value established in 1985. The quantity of edible imports was a record 1.36 million tons, 102 000 tons more than the previous record established in 1985.

Shrimp and tuna were the major species imported. The quantity of shrimp imported in 1986 established a record with 182 000 tons, 18 700 tons more than the previous record quantity imported in 1985. Valued at $1.4 billion, $281.4 million more than the 1985 value, shrimp imports accounted for 30 per cent of the value of total edible imports.

Imports of fresh and frozen tuna were 256 000 tons, an increase of 39 000 tons from 1985. Canned tuna imports were a record 107 500 tons, worth $228.6 million in 1986. In 1985, canned tuna imports were 97 220 tons worth $228.6 million. Canned tuna packed in water continues to be the preferred style with record amounts coming in from Thailand, Taiwan and the Philippines.

Imports of fresh and frozen fillets and steaks amounted to a record 244 100 tons, an increase of 8 000 tons over 1985. Regular and minced block imports were 165 000 tons, an increase of 13 500 tons from 1985.

Imports of analogue products (surimi) in 1986 amounted to 15 600 tons valued at $58.5 million in 1986, compared to the 15 300 tons and $48.2 million imported in 1985.

Imports of non-edible fishery products were valued at a record $2.8 billion in 1986 -- $198.6 million more than the $2.6 billion imported one year earlier.

U.S. exports of edible fishery products of domestic origin were a record 334 000 tons, valued at a record $1.3 billion, compared with 295 000 tons valued at $1 billion exported in 1985.

Fresh and frozen edible exports were 279 000 tons, valued at $1 billion, increases of 32 300 tons and $260 200 million compared with 1985. These imports were composed principally of salmon (133 000 tons, $552 300 million) and herring (41 100 tons, $70.2 million).

Canned exports of edible U.S. fishery products in 1986 were 36 200 tons valued at $132 200 million, up 9 600 tons and $35.8 million from 1985 levels. Salmon (27 000 tons, $101.2 million) was the major canned export in 1986.

Cured items, consisting mainly of salmon and herring roe, amounted to 13 500 tons and $91.2 million in 1986.

Exports of non-edible products in 1986 were valued at $66.3 million, a decline of $7.6 million from 1985. Menhaden oil exports (85 900 tons, $19.8 million) comprised 30 per cent of the total non-edible exports and declined from 1985 levels (129 400 tons, $36 million).

A partial reason for increased U.S. exports of fishery products was the decline of the value of the U.S. dollar relative to most other currencies. For example, the value of the U.S. dollar dropped 30 per cent against the Japanese Yen between 1985 and 1986, increasing the competitiveness of U.S. fishery products in the Japanese market.

In May 1986, the United States imposed a 5.82 per cent countervailing duty on Canadian fresh groundfish, excluding fillets. This duty, which was applied following an inquiry, is to compensate for the subsidisation Canada provides to its groundfish industry. During the year, investigations of other unfair trade practices regarding fishery products were initiated, but they were not completed by year end.

The duties collected on fishery imports were $188 million at an average ad valorem equivalent of 2.5 per cent compared with an average ad valorem rate of 3.6 per cent on all U.S. imports of fishery and non-fishery items.

V. OUTLOOK

Demand for seafood products in the U.S. is expected to grow as consumer's disposable income increases and consumers realise the nutritional benefits of seafood. This should happen in spite of the forecast of higher prices of seafood items. Consumption in 1987 is expected to climb over 6.8 kilos per person.

Farm raised catfish production, which is now the second leading finfish in the U.S. (salmon is first) is expected to increase slightly in 1987, up from a record setting 1986 level of 95 000 tons.

Surimi production in the U.S. will continue to increase as new processing plants come on line. New analogue products will expand the market for surimi based products. Alaska pollock, with an estimated sustainable yield of over 5.5 million tons is still the preferred species used in surimi production although menhaden and other species also offer significant potential for future development.

Foreign fishing will continue to decline as domestic harvests increase. Alaska pollock will be fully utilised by domestic fishermen. Atlantic mackerel will be the major under-utilised species in U.S. waters.

YUGOSLAVIA

I. FISHING

Yugoslavia has not introduced a quota system for fishing in the Adriatic, although some scientific research is being carried out to see whether quotas are essential for certain stocks of white fish. The main reason for not having quotas is that the demand for white fish and shellfish exceeds the country's catch. This is why the licensing system has not be introduced either in the organisation of fishing.

However, in order to protect certain regions, areas have been defined where some kinds of fishing are banned. For example, trawling is banned in all channels and small gulfs.

With regard to the right to fish, a permit system exists and citizens must comply with the requirements laid down to obtain the commercial fishing permit. The number of permits is not limited, as any citizen can obtain one if he meets the requirements. Fishing equipment (kinds and number of nets, vessel engine power) is subject to restrictions, which already limits the catch, but it is not limited in the case of fishing enterprises.

Several provisions concerning quality (sanitary conditions, packaging, storage) are in force. The most important is the regulation on the quality of fish, crustaceans, shellfish, sea urchins, frogs, turtles, snails and their products (Official Gazette of the SFR of Yugoslavia, No. 65/79). This regulation is of a technical kind, i.e. it sets compulsory standards for production and trade concerning the above species and processing operations (processing method, authorised additives, quality and quantity of fish in canned products, transport conditions, information to be given on the declaration, etc.). Since it is a provision concerning processing techniques, the revision of this regulation will soon be completed.

II. INTERNATIONAL TRADE

a)   Exports

Since the output of bluefish products exceeds domestic market needs in Yugoslavia, 40 to 50 per cent of this production is exported. Exports vary from one year to another, depending on the price trend on the national and world markets and the protective measures taken by the countries importing these products.

Exports (quantities)

Year	1982	1983	1984	1985	1986
Tonnes	16 600	18 000	14 116	17 050	15 900

Value of exports ($000)

Year	1982	1983	1984	1985	1986
$US	18 054	27 153	14 658	16 757	17 244

The government stimulates activity by refunding some of the customs duty (about 24 per cent) to the exporting bodies in the case of fish and fish product exports (Official Gazette of the SFR of Yugoslavia, No. 73/85).

b) Imports

Imports of fish for processing

Year	1982	1983	1984	1985	1986
Tonnes	7 800	4 500	6 630	6 000	5 900

These tonnages refer to imports of types of fish supplementing the range of species for the processing industry as the quantities fished in the Adriatic of these fish are insufficient. In addition, about 15 000 tonnes of white fish are imported for current consumption (for tourism needs), as well as 100 tonnes of fish meal.

III. MEASURES AND PROGRAMMES FOR THE FUTURE DEVELOPMENT OF FISHING

Since the fishing fleet is still too small (compared with processing capacity), the plan provides for the construction of 26 new fishing vessels by 1990 (surface trawling) and the modernisation of the existing fleet. The loans provided for construction are more advantageous since they are paid out of a special fund for the development of agriculture (loans over 15 years, interest 6 per cent). Such a fleet would account for about 300 000 tonnes of small pelagic fish a year and the production of about 800 tonnes of fish meal.

The government grants certain reductions on the price of fuel: fishing enterprises and co-operating commercial fishermen pay a lower price for fuel since they are exempt from the tax on industrial and commercial profits.

Since processing industry capacity is too high compared with the country's catches, the construction of new plant is not planned. However, as fishing is a seasonal activity, cold rooms will be built; this construction operation will be financed partly by the government and partly by the districts in which these cold rooms are built.

Aquaculture

Aquaculture which is one of the traditional fishery activities on the Adriatic coast has developed greatly in the last five years. This is mainly attributable to scientific progress that has resulted in improved production techniques. Natural conditions are extremely favourable for farming: in Istria, the Lim and Rasa canals, islands in the Mid Adriatic and southern Dalmatia, the Gulf of Maloston, the Neretva delta and the mouths of the Kotor. There are many aquaculture centres -- at Rovinj, Zadar, Dubrovnik, and Kardeljevo. In the case of fish, aquaculture is directed at sea bass (and alevins), mullet and eel, and in the case of sea food, at mussels and oysters.

In 1985 total production amounted to 60 tonnes of fish, 1 000 tonnes of mussels and 5 million oysters.

The development plan up to 1990 provides for the production of 1 600 tonnes of quality fish and 9 000 tonnes of shellfish. In addition to those already in existence, two new spawning grounds are being built (at Rovinj and Metkovic) so that production will run smoothly. Some of the capital required has been obtained from the International Bank for Restruction and Development (IBRD) and some from government funds for the development of agriculture.

EUROPEAN ECONOMIC COMMUNITY

I. MANAGEMENT OF RESOURCES

a) Total allowable catches (TAC) and quotas

1986 TACs and quotas (1) reflect the Community's ongoing policy of ensuring as far as possible the stability of fishing activities in compliance with the imperatives of stock conservation which are based on analyses of the most relevant scientific data. Set prior to the start of 1986, these TACs and quotas were amended four times (2) in the light of negotiations with third countries and further advice from scientists during the year.

Table III in the Statistical Annex shows the total quotas for all stocks (main species) allocated to each Member State in 1986 in the zones covered by Community regulations. The figures in brackets refer to the corresponding quantities for 1985.

b) Technical conservation measures

The technical conservation measures (3) were amended and supplemented (4) at the end of 1985, mainly to take into account the accession of Spain and Portugal to the Community.

In October 1986 the Council adopted a new regulation on technical conservation measures (5) that included all the amendments previously made to the 1983 regulation (3).

The most important points in this regulation concern net mesh size for fishing in the North Sea and the conditions for fishing in coastal areas. The Council also adopted an initial amendment to this new regulation (6).

The Council also adopted a basic regulation on technical conservation measures applying to the Baltic Sea (7), a regulation amending the technical conservation measures concerning fish stocks in the Antarctic (8), and a regulation defining the characteristics of fishing vessels (9).

c) Regulations made necessary by the accession to the Community of Portugal and Spain

The accession of Portugal and Spain to the Community in 1986 made it necessary, as stated above, to adjust technical conservation measures and to add new species and new quotas in the regulations on TACs and quotas. In addition, it required the adoption of regulations concerning mutual arrangements among the various Member States of the Community for the operation of their ships in waters not coming under their sovereignty or jurisdiction.

II. BILATERAL AND MULTILATERAL RELATIONS

On the basis of long-term bilateral agreements, the Community negotiated fishing arrangements for 1986 with several third countries.

The arrangements with Norway and Sweden specify catch possibilities in each area of the other party's zone and fix the total allowable catch (TAC) for certain shared stocks.

The arrangement with the Faroe Islands specifies catch possibilities in the other party's zone.

The trilateral arrangement between the EEC, Norway and Sweden specifies catch possibilities for each party in the Skagerrak and Kattegat.

Further to the fisheries agreement between the Community and Greenland, and on the basis of the protocol covering the activities of Community Member States' fishing vessels in Greenland waters up to 31st December 1989, the two parties negotiated an arrangement specifying the conditions to be applied in 1986.

The fisheries agreement between the Community and Madagascar, which was initialled in December 1984, was signed on 28th January 1986, came into force on 21st May 1986, and was amended in November 1986 in order to take into account the considerable increase in the Community fleet following the enlargement of the Community.

Owing to the expiry of the protocols to the fisheries agreements with certain ACP countries and to the provisions of the Act of Accession, fisheries agreements for the enlarged Community amending the earlier agreements with Guinea, Guinea Bissau, Equatorial Guinea and Senegal were initialled in 1986.

Since the negotiations with Sao Tomé and Principe on a new protocol applicable from 1st November 1986 had not resulted in an agreement by that date, fishing operations by ships flying the flag of a Community Member State in Sao Tomé waters were suspended.

New fisheries agreements with Gambia, the Seychelles and Mozambique were initialled. In addition, the Community continued negotiations with Dominica and also opened negotiations with Cap Verde.

The Community took part as a contracting party in the annual meetings of international organisations concerned with the conservation and management of marine resources. These included that of the North-West Atlantic Fisheries Organisation (NAFO), as well as those concerning the convention on future multilateral co-operation on North-Eastern Atlantic Fisheries (NEAFC), the International Baltic Sea Fishery Convention (IBSFC), the Convention for the Conservation of Salmon in the North Atlantic (NASCO) and the Convention for the Conservation of Antarctic Marine Living Resources (CCAMLR).

The Community attended, as an observer, the meeting of the International Commission for the Conservation of Atlantic Tunas (ICCAT), and the 7th special session of the International Commission for South-East Atlantic Fisheries (ICSEAF). It also took part in meetings of the regional organisations under the auspices of the FAO, such as the General Fisheries Council for the Mediterranean (GFCM) and the Committee for the Management of Indian Ocean Tuna.

III. COMMUNITY MARKET ORGANISATION

The main factors concerning the organisation of Community fishery product markets can be summed up as follows:

i) The information at present available shows that the Community of the Twelve's catches in 1986 were down from the previous year (apart from herring). Imports were up in terms of both volume and value, while exports rose less in volume and declined slightly in value. This implies a proportional decline in the trade balance in terms of volume and value.

ii) The white fish market, which was relatively stable in previous years, encountered some supply difficulties at both Community and world levels. White fish production as well as withdrawals from the market were down, with a resulting increase in average prices. The market situation with regard to pelagic species varied widely, partly because of the fluctuation in production from one year to another. The structure of the sardine and anchovy market changed considerably with the enlargement of the Community. Average prices for Atlantic sardine and especially herring declined greatly, while they were up for Mediterranean sardine, mackerel and anchovy. Overall withdrawals were down, apart from herring.

iii) Guide prices for the 1986 season were increased by 2.5 per cent on average, with a decrease of 3 per cent for herring, no change for dogfish and Atlantic sardine, and a 6 per cent increase for hake (10). With few exceptions, withdrawal and reference prices for fresh fish changed in line with guide prices (11) (12), while the reference prices for frozen products were raised by about 2 per cent on average (12).

iv) With the enlargement of the Community, certain temporary provisions were introduced to facilitate matching of prices. For instance, a guaranteed minimum price was set for Atlantic sardine and a compensatory allowance for Mediterranean sardine. Other special measures were aimed to ease the Spanish and Portuguese components smoothly into the Community market. In addition, some permanent adjustments were made, such as the setting of prices for certain new species (chub mackerel, monkfish, megrim, pomfret) (13), as well as the introduction of a new private storage premium system for Norway lobsters and edible crabs.

v) Certain technical measures were introduced in connection with Community market management. These included the application of common standards for certain fresh or chilled fish (14), carp and tuna (15) (16) (17). A regulation was also adopted to keep imports of common squid from Poland and the USRR to the reference price during a certain period (18).

IV. STRUCTURAL POLICY

As part of the common measure approved on 4th October 1983 (Council Regulation (EEC) 2908/83 (19) for restructuring, modernising and developing the fishing industry and for developing aquaculture, the Commission adopted several multi-annual guidance programmes submitted by Belgium, France, Germany, Portugal and Spain.

Pursuant to the same Regulation (EEC) No. 2908/83 (19), the Commission granted as the second 1985 instalment and the single 1986 instalment, a total of Ecus 117.56 million for 436 projects concerning new vessels, 926 projects for the modernisation of fishing vessels, 201 aquaculture investment projects and 6 artificial reef projects.

The breakdown of this aid by Member State is given in Table I in the Statistical Annex.

Further to Regulation (EEC) No. 2909/83 (19), the Commission gave its approval in 1986 to a French exploratory fishing voyage off the Faroe Islands.

Under Directive 515/83/EEC (19), the Commission decided that the Community would contribute Ecus 8.6 million towards the expenditures incurred by Belgium, Danemark, Germany, Greece, the Netherlands and the United Kingdom in carrying out certain measures to adjust capacity in the fisheries sector.

Under the same Directive, it approved the national measures applying the system for adjusting capacity in the fishing sector which were submitted by Spain and Portugal for the laying up or scrapping of certain fishing vessels.

Under Regulation (EEC) 355/77 (20) on common measures to improve the conditions under which agricultural and fishery products are processed and marketed, the Commission contributed Ecus 30.2 million to 129 investment projects in the fisheries sector in 1986.

Table II in the Statistical Annex gives a breakdown of this aid by Member State.

Under the system of structural aid for the conversion of sardine canning plants set up by Regulation (EEC) No. 3722/85 (21), the Commission granted Ecus 3.9 million to two projects for promoting sardine canning in France and Italy.

On 17th December 1986 the Council adopted a regulation on Community measures to improve and adapt structures in the fisheries and aquaculture sector (22).

This regulation defines structural policy in the fishing sector for a ten-year period. Implementing this policy involves a total estimated expenditure under the Community budget of Ecus 800 million in the first five-year period 1987-1991.

V. ENSURING COMPLIANCE WITH REGULATIONS CONCERNING RESOURCES

The Commission submitted a report to the Council on the implementation of the common policy on the conservation of resources approved on 25th January 1983. This report deals mainly with the inspection of fishing activities in each Member State.

The Community control system has been tightened (23). The amended regulation reiterates the Commission's right to put a stop to fishing once the TAC or the share available to the Community in a TAC is exhausted (without the quotas of all Member States being exhausted). It also provides for a procedure whereby alternative fishing possibilities may be granted by way of compensation to fishermen who are unable to use up their quotas because of overfishing by others. The tightening of the control system gives the Commission's fishery inspectors additional powers to check on-the-spot compliance by Member States with Community regulations on the conservation of resources.

NOTES AND REFERENCES

1. Council Regulation (EEC) No. 3721/85 of 20th December 1985: OJ No. L361 of 31.12.85, p. 5; Regulation (EEC) No. 3777/85 of 31.12.85: OJ No. L363 of 31.12.85, p. 1.

2. OJ No. L17 of 23.1.86, p. 4; L176 of 1.7.86, p. 3; L206 of 30.7.86, p. 4 and L300 of 24.10.86, p. 2.

3. Regulation (EEC) No. 171/83 of 25.1.83; OJ No. L24 of 27.1.83, p. 14.

4. OJ No. L363 of 31.12.85, p. 28.

5. Regulation (EEC) No. 3094/86 of 7.10.86; OJ No. L288 of 11.10.86, p. 1.

6. OJ No. L376 of 31.12.86, p. 1.

7. OJ No. L162 of 18.6.86, p. 1.

8. OJ No. L201 of 24.7.86, p. 2.

9. OJ No. L274 of 25.9.86, p. 1.

10. OJ No. L 344 of 21.12.85

11. OJ No. L 351 of 28.12.85

12. OJ No. L 351 of 28.12.85

13. OJ No. L 54 of 25.2.86

14. OJ No. L 351 of 28.12.85

15. OJ No. L 202 of 25.7.86

16. OJ No. 35 of 11.2.86 and OJ No. L 105 of 22.4.86

17. OJ No. L 211 of 1.8.86

18. OJ No. L 117 of 6.5.1986

19. OJ L 290 of 22.10.83

20. OJ L 51 of 23.2.77

21. OJ L 361 of 31.12.85, p. 38

22. Regulation (EEC) 4028/86, OJ L 376 of 31.12.86.

23. Regulation (EEC) No. 4027 of 31.12.86 -- OJ L376 of 31.12.86.

AUSTRALIA/AUSTRALIE

Table I/Tableau I

LANDINGS AND VALUES/DEBARQUEMENTS ET VALEURS
1985 & 1986

Quant. : '000 tons/tonnes (live weight/poids vif)
Val. : A$/$A million a)

	1985[b]		1986[c]	
	Quant.	Val.	Quant.	Val.
1. National landings in domestic ports/Débarquements nationaux dans les ports nationaux	155.2	521.5	151.9	592.3
For direct human consumption/Pour la consommation humaine directe	150.6	519.0	147.1	590.7
Finfish/Poisson	81.2	99.6	84.7	116.9
Tuna/Thon	13.0	14.1	13.5	16.5
Other/Autres	68.2	85.5	71.2	100.4
Shellfish/Coquillages[d]	69.4	419.4	62.4	473.3
Scallops/Coquilles St. Jacques	19.2	19.5	16.2	24.4
Prawns/Crevettes	20.5	164.0	19.1	195.0
Rock lobster/Langouste	14.1	172.0	13.0	165.6
Abalone/Ormeaux	7.7	35.4	7.1	59.7
For other purposes/Pour d'autres buts	4.6	2.5	4.8	1.6
2. National landings in foreign ports/Débarquements nationaux dans les ports étrangers[e]	3.6	7.0	2.7	5.2
Southern bluefin tuna/Thon rouge	3.6	7.0	2.7	5.2
3. Foreign landings in domestic ports/Débarquements étrangers dans les ports nationaux	-	-	-	-
Squid/Calmar	-	-	-	-
TOTAL 1 + 3	155.2	521.5	151.9	592.3

a) Wholesale prices at principal markets/Prix de gros aux principaux marchés.

b) 1985 equals financial year 1984/85/L'année 1985 est égale à l'année fiscale 1984/85.

c) 1986 equals financial year 1985/86/L'année 1986 est égale à l'année fiscale 1985/86.

d) Crustaceans, molluscs, squid, etc./Crustacés, mollusques, calmars, etc.

e) Represents domestic catch sold to foreign vessels/Représente les prises nationales vendues aux navires étrangers.

AUSTRALIA/AUSTRALIE

Table II/Tableau II

EXTERNAL TRADE IN FISH AND FISH PRODUCTS/ECHANGES INTERNATIONAUX DE POISSON ET PRODUITS DE LA PECHE
1985a) & 1986b)

Quant. : '000 tons/tonnes (product weight/poids du produit)
Val. : A/A million

	IMPORTS/IMPORTATIONS				EXPORTS/EXPORTATIONS			
	1985		1986		1985		1986	
	Quant.	Val. (cif)	Quant.	Val. (cif)	Quant.	Val. (fob)	Quant.	Val. (fob)
Total fish and fish products/Total poisson et produits de la pêche Fresh, chilled/Frais, sur glace	na 1.4	317.7 4.8	na 1.7	352.8 6.7	na 0.3	419.8 1.1	na 0.6	512.5 3.0
Frozen/Congelés Frozen fillets/Filets congelés	48.0 26.7	106.4 69.8	44.2 24.9	116.0 74.7	6.3 0.3	10.2 1.0	7.3 0.4	15.5 1.5
Salted, dried, smoked/Salés, séchés, fumés	4.7	15.1	4.6	17.5	d)	d)	d)	0.1
Canned/En conserve	15.5	64.4	15.6	74.6	0.1	0.8	0.1	1.1
Shellfish (live, fresh, chilled, salted dried, canned)/Coquillages (vivants, frais, sur glace, salés, séchés, en conserve)	16.7	107.7	17.6	116.6	25.0	387.3	24.8	462.5
Fish meal/Farine de poisson	12.7	5.8	10.0	4.2	d)	d)	d)	d)
Fish oil/Huile de poissonc)	0.6	0.6	0.7	0.8	0.1	0.3	0.1	0.1
Other/Autres	na	12.9	na	16.4	na	20.1	na	30.2

a) 1985 equals financial year 1984/85/L'année 1985 est égale à l'année fiscale 1984/85.
b) 1986 equals financial year 1985/86/L'année 1986 est égale à l'année fiscale 1985/86.
c) '000 litres.
d) Quantity less than 500 tons or value less than A$ 50 000/Quantité inférieure à 500 tonnes et valeur inférieure à $A 50 000.

AUSTRALIA/AUSTRALIE

Table III(a)/Tableau III(a)

IMPORTS BY MAJOR PRODUCTS AND BY COUNTRY/IMPORTATIONS PAR PRODUITS PRINCIPAUX ET PAR PAYS
1985[a] & 1986[b]

Quant. : '000 tons (product weight)/tonnes (poids du produit)
Val. : A/A million

	1985		1986	
	Quant.	Val.	Quant.	Val.
Total fish and fish products/Total poisson et produits de la pêche	na	317.7	na	352.8
Fresh, chilled/Frais, sur glace	1.4	4.8	1.7	6.7
Whole/Entiers	1.2	4.2	1.5	6.0
Fillets/Filets	0.2	0.6	0.2	0.6
New Zealand/Nouvelle-Zélande	1.3	4.6	1.6	6.0
United Kingdom/Royaume-Uni	d)	d)	0.1	0.2
Frozen/Congelés (excl. fillets/filets)	21.4	36.6	19.3	41.3
Hake/Merlu	1.1	2.1	0.4	2.4
Other whole fish/Autres, entiers	9.9	12.4	8.6	11.6
Fingers, sticks/Bâtonnets	3.4	5.8	2.6	5.5
New Zealand/Nouvelle-Zélande	4.9	8.4	5.0	11.0
Japan/Japon	2.3	2.7	3.1	11.1
South Africa/Afrique du sud	2.9	3.9	2.6	4.4
Frozen fillets/Filets congelés	26.7	69.8	24.9	74.7
Hake/Merlu	13.3	28.4	12.3	27.9
Other/Autres	13.3	41.4	12.7	46.9
New Zealand/Nouvelle-Zélande	7.8	25.4	6.9	25.7
South Africa/Afrique du sud	7.5	15.9	8.2	18.3
Japan/Japon	3.8	10.1	2.6	8.9
Salted, dried, smoked/Salés, séchés, fumés	4.7	15.1	4.6	17.5
South Africa/Afrique du sud	2.9	5.6	2.2	4.1
Norway/Norvège	0.2	0.8	0.2	1.1
Canada	d)	2.1	d)	3.6
Shellfish (live, fresh, chilled, salted, dried and canned Coquillages (vivants, frais, sur glace, salés, séchés, en conserve)	16.7	107.7	17.6	116.6
Prawn or shrimp/Crevettes	9.2	72.4	9.4	76.4
Lobster/Homard	d)	5.2	0.5	6.0
Crab/Crabe	0.7	3.8	0.6	3.2
Malaysia/Malaysie	3.6	24.4	3.7	26.1
Thailand/Thaïlande	3.8	22.9	3.1	17.2
New Zealand/Nouvelle-Zélande	2.3	11.3	2.8	13.5
Meal/Farine	12.7	5.8	10.0	4.2
Denmark/Danemark	5.1	2.7	2.0	1.1
Chile/Chili	1.8	0.7	3.8	1.6
Oil/Huile[c]	0.6	0.6	0.7	0.8
Japan/Japon	0.2	0.1	0.1	0.1
Norway/Norvège	0.1	0.1	0.1	0.1
Other/Autres	na	77.3	na	91.0
Canned/En conserve	15.5	64.4	15.6	74.6
Other/Autres	na	12.9	na	16.4
United States/Etats-Unis	na	25.7	na	25.7
Canada	na	14.1	na	23.0
Japan/Japon	na	9.6	na	5.7

a) 1985 equals financial year 1984/85/L'année 1985 est égale à l'année fiscale 1984/85.

b) 1986 equals financial year 1985/86/L'année 1986 est égale à l'année fiscale 1985/86.

c) '000 litres.

d) Quantity less than 500 tons or value less than A$ 50 000/Quantité inférieure à 500 tonnes et valeur inférieure à $A 50 000.

AUSTRALIA/AUSTRALIE

Table III(b)/Tableau III(b)

EXPORTS BY MAJOR PRODUCTS AND BY COUNTRY/EXPORTATIONS PAR PRODUITS PRINCIPAUX ET PAR PAYS
1985[a]) & 1986[b])

Quant. : '000 tons/tonnes (product weight/poids du produit)
Val. : A$/$A million

	1985		1986	
	Quant.	Val.	Quant.	Val.
Total fish and fish products/Total poisson et produits de la pêche	na	419.8	na	512.5
Fresh, chilled/Frais, sur glace	0.3	1.1	0.6	3.0
Whole/Entiers	0.3	1.0	0.6	2.9
Fillets/Filets	d)	d)	d)	0.1
Japan/Japon	d)	0.3	0.4	2.1
United States/Etats-Unis	d)	0.1	0.1	0.5
Frozen/Congelés (excl. fillets/filets)	5.9	9.2	6.9	14.0
Tuna/Thon	3.5	4.7	2.3	3.2
Whiting/Merlan	1.0	1.4	1.3	2.6
Italy/Italie	3.1	3.5	1.8	2.1
Japan/Japon	1.3	2.8	2.8	6.9
Frozen fillets/Filets congelés	0.3	0.8	0.4	1.5
Japan/Japon	0.3	0.8	0.3	1.1
United States/Etats-Unis	d)	d)	d)	0.2
Salted, dried, smoked/Salés, séchés, fumés	d)	d)	d)	0.1
Shellfish (live, fresh, chilled, salted, dried and canned)/Coquillages (vivants, frais, sur glace, salés, séchés, en conserve)	25.0	387.3	24.8	462.5
Prawn and shrimp/Crevettes	12.3	149.8	13.1	207.1
Lobster/Homard	5.9	157.3	5.5	145.6
Abalone/Ormeaux	4.3	55.0	4.2	86.9
Japan/Japon	14.2	178.8	15.5	262.3
United States/Etats-Unis	5.8	150.4	4.1	119.4
Meal/Farine	d)	d)	d)	d)
Oil/Huile[c])	0.1	0.3	0.1	0.1
Other/Autres	na	20.9	na	31.3
Japan/Japon	17.1	17.7	0.3	23.2

a) 1985 equals financial year 1984/85/L'année 1985 est égale à l'année fiscale 1984/85.

b) 1986 equals financial year 1985/86/L'année 1986 est égale à l'année fiscale 1985/86.

c) '000 litres.

d) Quantity less than 500 tons or value less than A$ 50 000/Quantité inférieure à 500 tonnes et valeur inférieure à $A 50 000.

BELGIUM/BELGIQUE

Table I(a)/Tableau I(a)

BELGIAN LANDINGS/DEBARQUEMENTS BELGES
1985/1986

Tons (live weight)/Tonnes (poids vif)

Species/Espèces	In Domestic Ports/ Dans les ports belges		In Foreign Ports/ Dans les ports étrangers		Total	
	1985	1986	1985	1986	1985	1986
A. WHITE FISH/POISSON MAIGRE						
Haddock/Eglefin	1 014	505	116	102	1 130	607
Cod/Morue	5 774	7 771	466	360	6 240	8 131
Saithe/Lieu noir	202	254	23	6	225	260
Whiting/Merlan	2 488	2 495	42	27	2 530	2 522
Plaice/Plie	8 979	7 467	3 024	2 558	12 003	10 025
Sole	4 323	4 747	105	30	4 428	4 777
Skate/Raie	2 071	1 735	126	125	2 197	1 860
Redfish/Sébaste	416	423	-	-	416	423
Others/Autres	8 033	6 680	977	715	9 010	7 395
TOTAL	33 300	32 077	4 879	3 923	38 179	36 000
B. PELAGIC FISH/POISSONS PELAGIQUES						
Herring/Hareng	3 482	415	0	-	3 482	415
Others/Autres	60	72	2	5	62	77
TOTAL	3 542	487	2	5	3 544	492
C. CRUSTACEANS AND MOLLUSCS/CRUSTACES ET MOLLUSQUES						
Shrimps/Crevettes	735	613	148	87	883	700
Norway lobster/Langouste	680	344	-	-	680	344
Others/Autres	1 327	1 421	7	10	1 334	1 431
TOTAL	2 742	2 378	155	97	2 897	2 475
GRAND TOTAL/TOTAL GENERAL	39 584	34 942	5 036	4 025	44 620	38 967

BELGIUM/BELGIQUE

Table I(b)/Tableau I(b)

VALUE OF BELGIAN LANDINGS/VALEUR DES DEBARQUEMENTS BELGES
1985/1986

'000 BF/FB

	In Domestic Ports/ Dans les ports belges		In Foreign Ports/ Dans les ports étrangers		Total	
	1985	1986	1985	1986	1985	1986
A. WHITE FISH/POISSON MAIGRE						
Haddock/Eglefin	35 962	21 411	3 849	3 700	39 811	25 111
Cod/Morue	312 085	360 500	24 014	23 000	336 099	383 500
Saithe/Lieu noir	7 122	9 418	1 007	230	8 129	9 648
Whiting/Merlan	74 614	75 979	1 101	800	75 715	76 779
Plaice/Plie	376 321	307 056	150 627	135 000	526 948	442 056
Sole	1 094 237	1 450 173	22 876	11 900	1 117 113	1 462 073
Skate/Raie	89 154	85 497	3 802	3 800	92 956	89 297
Redfish/Sébaste	24 603	24 646	-	-	24 603	24 646
Others/Autres	594 724	549 086	59 726	47 570	654 450	596 656
TOTAL	2 608 822	2 883 766	267 002	226 000	2 875 824	3 109 766
B. PELAGIC FISH/POISSONS PELAGIQUES						
Herring/Hareng	37 065	4 418	-	-	37 065	4 418
Others/Autres	947	1 237	7	28	954	1 265
TOTAL	38 012	5 655	7	28	38 019	5 683
C. CRUSTACEANS AND MOLLUSCS/CRUSTACES ET MOLLUSQUES						
Shrimps/Crevettes	73 021	68 281	9 288	8 850	82 309	77 131
Norway lobster/Langouste	98 359	74 659	1	-	98 360	74 659
Others/Autres	44 993	64 860	349	360	45 342	65 220
TOTAL	216 373	207 800	9 638	9 210	226 011	217 010
GRAND TOTAL/TOTAL GENERAL	2 863 207	3 097 221	276 647	235 238	3 139 854	3 332 459

BELGIUM/BELGIQUE

Table I(c)/Tableau I(c)

FOREIGN LANDINGS IN BELGIAN PORTS/DEBARQUEMENTS ETRANGERS DANS LES PORTS BELGES[a)]
1985/1986

Quant. : tons (product weight)/tonnes (poids du produit)
Val. : BF/FB

Species/Espèces	1985		1986	
	Quant.	Val.	Quant.	Val.
a) By vessels/Par des navires[b)]				
A. White fish/Poisson maigre				
Haddock/Eglefin	4	97 795	5	149 030
Cod/Morue	213	13 571 731	269	22 676 445
Plaice/Plie	25	999 491	57	2 802 643
Sole	6	1 633 743	14	5 200 376
Redfish/Sébaste	-	-	-	-
Others/Autres	91	4 605 934	54	3 977 818
Sub-total/Sous-total	339	20 908 694	399	34 806 312
B. Pelagic fish/Poisson pélagique				
Herring/Hareng	-	-	-	-
Others/Autres	-	-	-	-
Sub-total/Sous-total	-	-	-	-
C. Crustaceans and molluscs/Crustacés et mollusques				
Squid/Calmar	-	-	-	-
Shrimps/Crevettes	-	-	-	-
Others/Autres	49	5 032 294	49	6 427 363
Sub-total/Sous-total	49	5 032 294	49	6 427 363
TOTAL	388	25 940 928	448	41 233 675
b) Containers[c)]				
Haddock/Eglefin	28	1 854 907	2	89 946
Cod/Morue	187	11 131 168	40	1 902 893
Plaice/Plie	8	440 817	1	21 503
Redfish/Sébaste	49	2 510 376	39	1 428 332
Others/Autres	103	5 560 766	37	1 834 468
TOTAL	375	21 498 034	119	5 277 142
GRAND TOTAL/TOTAL GENERAL	763	47 439 022	567	46 510 817

a) Included in imports/Compris dans les importations.

b) Danish, Dutch and British vessels/Navires danois, hollandais et britanniques.

c) Icelandic containers/Conteneurs islandais.

BELGIUM/BELGIQUE

Table II/Tableau II

FINANCIAL AID TO BELGIAN FISHERIES/LES AIDES FINANCIERES AUX PECHES BELGES
1985 & 1986

A. Primary sector (classification according to management objective, i.e. socio-economic objectives)
   Secteur primaire (classification en fonction des objectifs de la gestion, c.à.d. des objectifs socio-économiques)

(BF/FB)

Type of aid/Nature de l'aide	1985	1986
a) Social aid/Aide à caractère social		
1. Unplanned programme (unforeseen conditions)/Programme non planifié (conditions imprévisibles)	-	-
2. Regional development/Développement régional	-	-
3. Social insurance/Assurance sociale :	-	-
- seasonal fishermen/pêcheurs saisonniers	-	-
- resource guarantees/garanties de resources	-	-
- fixed salaries for apprentices/rémunération des mousses :		
a) State aid/Aide de l'Etat	8 595 797	7 499 991
b) Aid from Province of West Flanders/Aide de la Province de la Flandre Occidentale	4 034 850	3 273 020
4. Tax incentives/Incitations fiscales	-	-
Sub-total/Sous-Total	12 630 647	10 773 011
b) Economic aid/Aide économique		
1. Operational subsidies/Subventions d'exploitation	-	-
2. Restructuring - modernisation or diversification/ Restructuration - modernisation ou diversification		
a) interest reductions for the building of new vessels/ bonifications d'intérêt pour la construction de nouveaux navires	8 806 831	8 003 768
loans/prêts : 1982 = 236 425 200		
1983 = 414 421 900		
1984 = 658 673 500		
1985 = 232 423 000		
1986 = 111 712 000		
b) subsidies for the scrapping of vessels/Subventions pour la démolition de navires	-	-
c) EEC decommissioning grants/primes d'arrêt définitif CEE (national share/contribution nationale)	12 000 000	21 000 000
d) upkeep of the fleet/assainissement de la flotte (liquidation VOZOR)	7 283 450	1 568 950
e) premium of the Province of West Flanders to encourage restructuring the fleet/prime de la Province de la Flandre occidentale pour encourager la restructuration de la flotte	15 316 790	27 000 000
3. Development/Développement :		
a) improving efficiency/amélioration de l'efficacité :		
- management/gestion : book-keeping premiums/primes de comptabilité	246 000	252 000
- fishing gear/engins de pêche	1 495 667	3 837 127
b) to reduce costs/encouragement à la réduction des coûts :		
- fuel economy/économie de combustibles		
1. premiums/primes	-	-
2. investments/investissements	5 098 000	2 828 000
c) experimental fisheries/pêches expérimentales	951 799	602 721
4. Withdrawals and institutional support/Retraits et aide institutionnelle	-	-
Sub-total/Sous-total	51 198 537	65 092 566
TOTAL - Primary sector/Secteur primaire	63 829 184	75 865 577

BELGIUM/BELGIQUE

Table II (cont'd)/Tableau II (suite)

FINANCIAL AID TO BELGIAN FISHERIES/LES AIDES FINANCIERES AUX PECHES BELGES
1985 & 1986

B. Processing, marketing, distribution (classification according to activities)/
Transformation, commercialisation, distribution (classification en fonction des activities)

(BF/FB)

Type of aid/Nature de l'aide	1985	1986
a) Investments/Investissements (renewal, modernisation/installations, modernisation)		
1. Modernisation, distribution/Modernisation, distribution	5 657 062	
2. Distribution (small and medium enterprises/petites et moyennes entreprises	9 950 000	
Sub-total/Sous-total	15 607 102	
b) Structural adjustments/Ajustements structurels	-	
c) Marketing support/Aide à la commercialisation - promotion and programmes/propagande en faveur de la consommation de poisson	3 000 000	2 879 000
TOTAL B	18 607 102	
Total of financial aid/Total des aides financières		
A. Primary sector/Secteur primaire	63 829 184	75 865 577
B. Processing, marketing, distribution/Transformation, commercialisation, distribution	18 607 102	
GRAND TOTAL/TOTAL GENERAL	82 436 286	

BELGIUM/BELGIQUE

Table III(a)/Tableau III(a)

EXTERNAL TRADE IN FISH AND FISH PRODUCTS/ECHANGES INTERNATIONAUX DE POISSON ET PRODUITS DE LA PECHE
1985 & 1986

Quant. : tons/tonnes
Val. : BF/FB million

	IMPORTS/IMPORTATIONS				EXPORTS/EXPORTATIONS			
	1985		1986		1985		1986	
	Quant.	Val.	Quant.	Val.	Quant.	Val.	Quant.	Val.
Fresh or chilled/Frais ou sur glace								
Sea fish/Poisson de mer	22 771	2 745.2	23 395	3 033.8	14 036	1 242.3	14 138	1 635.2
Freshwater fish/Poisson d'eau douce	11 702	2 303.5	10 058	1 513.2	2 028	382.7	1 552	238.2
Frozen fish/Poisson congelé	9 868	465.9	11 365	961.1	1 784	241.9	2 805	494.9
Frozen fillets/Filets congelés	10 769	1 108.5	11 870	1 323.6	3 279	443.4	2 935	427.0
Sub-total/Sous-total	20 637	1 574.4	23 235	2 284.7	5 063	685.3	5 740	921.9
Salted, dried, smoked/Salés, séchés, fumés	2 798	617.7	2 822	619.9	555	221.4	537	286.9
Shellfish/Coquillages	45 288	4 922.5	45 170	6 357.3	3 888	1 039.0	5 489	1 523.8
Canned/En conserve								
Fish/Poisson	24 293	3 845.8	32 474	5 629.9	5 022	950.4	7 892	1 802.1
Shellfish/Coquillages	20 697	2 937.4	23 397	3 163.7	3 975	563.5	3 588	716.8
	3 596	908.4	9 077	2 466.2	1 047	386.9	4 304	1 085.3
Total fish and shellfish/Total poisson et coquillages	127 489	16 009.1	137 154	19 438.8	30 592	4 521.1	35 348	6 448.1
Meal/Farine	42 520	902.9	42 833	782.6	21 503	430.9	5 105	91.1
Oil/Huile	44 236	908.6	32 000	436.7	173	7.3	201	11.4
Grand total/Total général	214 245	17 820.6	211 987	20 658.1	52 268	4 959.3	40 654	6 550.6
Total seafood (freshwater fish)/Total produits de la mer (poisson d'eau douce)	199 007	14 617.1	196 167	17 949.1	49 140	4 209.5	37 764	5 897.6
Total seafood (meal/oil)/Total produits de la mer (farine/huile)	112 251	12 805.6	121 334	16 729.8	27 464	3 771.3	32 458	5 795.1

Note: Imports - including foreign landings in Belgian ports/Importations - y compris les débarquements étrangers dans les ports belges.
Exports - not including Belgian landings in foreign ports/Exportations - non compris les débarquements belges dans les ports étrangers.
Salmon - although biologically a marine species, including under "freshwater fish" (for statistical standard reasons)/Saumon - bien qu'étant biologiquement une espèce marine, est inclus sous la rubrique "poisson d'eau douce" (pour des raisons statistiques).

BELGIUM/BELGIQUE

Table III(b)/Tableau III(b)

SEA FOOD: IMPORTS BY MAJOR PRODUCTS AND BY COUNTRY/
PRODUITS DE LA MER : IMPORTATIONS PAR PRINCIPAUX PRODUITS ET PAR PAYS
1985 & 1986

Quant. : tons/tonnes (product weight/poids du produit)
Val. : BF/FB million

		1985		1986	
		Quant.	Val.	Quant.	Val.
Total sea food/Total produits de la mer		199 007	14 617.1	196 167	17 949.1
a) Fresh, chilled/ Frais, sur glace		22 771	2 745.2	23 395	3 033.8
	Cod/Morue	6 261	707.8	8 527	1 172.3
	Herring/Hareng	1 523	63.6	1 602	55.2
	Sole	1 355	419.9	1 270	448.5
b) Frozen (excl. fillets)/Congelés (non compris les filets		9 565	380.1	8 908	459.8
	Herring/Hareng	4 333	117.4	4 417	116.2
	Mackerel/Maquereau	844	20.2	399	8.8
	Sprat	1 386	35.6	472	12.2
c) Frozen fillets/Filets congelés		10 769	1 108.5	11 870	1 323.6
	Cod/Morue	4 482	563.3	4 411	651.5
	Saithe/Lieu noir	2 799	167.6	2 107	149.0
	Hake/Merlu	813	49.0	1 502	74.1
a) + b) + c)					
	Netherlands/Pays-Bas	12 025	1 172.8	12 453	1 250.0
	Denmark/Danemark	8 125	977.6	8 278	1 153.2
	France	5 905	543.3	6 527	731.8
	Germany/Allemagne	3 711	411.9	2 937	433.9
d) Salted, dried, smoked/Salés, séchés, fumés		2 366	327.2	2 513	394.5
	Herring/Hareng	1 333	98.1	1 368	109.7
	Cod/Morue	364	84.5	259	50.2
	Netherlands/Pays-Bas	1 189	97.5	1 714	172.5
e) Canned fish/Poisson en conserve		17 896	2 413.7	20 401	2 694.6
f) Shellfish/Crustacés et coquillages		45 288	4 922.5	45 170	6 357.3
	Mussels/Moules	30 594	899.0	25 544	1 137.6
	Shrimps/Crevettes	9 360	2 653.7	12 004	3 446.8
	Lobster/Homard	1 167	586.0	805	433.5
	Netherlands/Pays-Bas	34 359	2 324.9	30 054	2 830.3
	France	1 796	316.3	2 604	528.1
g) Canned shellfish/Crustacés en conserve		3 596	908.4	9 077	2 466.2
h) Meal/Farine		42 520	902.9	42 833	782.6
	Netherlands/Pays-Bas	16 615	372.5	23 256	450.1
	Peru/Pérou	9 775	210.7	-	-
	Germany/Allemagne	6 000	100.4	6 987	100.1
	Denmark/Danemark	5 761	129.0	6 231	122.9
i) Oil/Huile		44 236	908.6	196 167	17 949.1

Note : Including foreign landings in Belgian ports/Y compris les débarquements étrangers dans les ports belges.

BELGIUM/BELGIQUE

Table III(c)/Tableau III(c)

SEA FOOD: EXPORTS BY MAJOR PRODUCTS AND BY COUNTRY/
PRODUITS DE LA MER : EXPORTATIONS PAR PRINCIPAUX PRODUITS ET PAR PAYS
1985 & 1986

Quant. : tons/tonnes (product weight/poids du produit)
Val. : BF/FB million

		1985 Quant.	1985 Val.	1986 Quant.	1986 Val.
Total sea food/Total produits de la mer		49 140	4 209.5	37 764	5 857.6
a) Fresh, chilled/Frais, sur glace		14 036	1 242.3	14 138	1 635.2
	Herring/Hareng	2 291	27.8	-	-
	Plaice/Plie	4 771	225.7	4 837	228.9
	Cod/Morue	1 358	75.4	3 283	161.8
	Sole	2 169	599.4	2 925	956.7
b) Frozen (excl. fillets)/Congelés (non compris les filets)		1 481	156.1	1 937	333.2
	Sole	546	147.4	726	231.9
c) Frozen fillets/Filets congelés		3 279	443.4	2 935	427.0
	Cod/Morue	1 147	169.3	965	161.0
	Saithe/Lieu noir	1 204	114.5	1 114	119.2
a) + b) + c)	Netherlands/Pays-Bas	11 271	925.2	9 833	1 251.6
	France	4 816	517.4	4 399	566.2
	Germany/Allemagne	1 122	97.7	716	97.8
d) Salted, dried, smoked/Salés, séchés, fumés		371	58.0	269	79.4
	France	149	17.9	192	26.5
	Germany/Allemagne	66	9.4	11	1.5
	Netherlands/Pays-Bas	28	4.5	44	6.7
e) Canned fish/Poisson en conserve		3 362	445.6	3 386	671.2
f) Shellfish/Crustacés et coquillages		3 888	1 039.0	5 489	1 523.8
	Shrimps/Crevettes	3 055	932	4 717	1 341.2
	France	1 610	322.0	1 615	424.6
	Germany/Allemagne	724	238.1	860	327.6
g) Canned shellfish/Crustacés en conserve		1 047	386.9	4 304	1 085.3
h) Meal/Farine		21 503	430.9	5 105	91.1
	Algeria/Algérie	19 125	379.2	1 660	30.8
	Libya/Libye	50	1.2	960	13.2
	Netherlands/Pays-Bas	815	15.9	980	13.8
	France	2 062	23.2	958	18.5
h) Oil/Huile		173	7.3	201	11.4

Note : Excluding Belgian landings in foreign ports/Non compris les débarquements belges dans les ports étrangers

CANADA

Table I/Tableau I

LANDINGS & VALUES/DEBARQUEMENTS ET VALEURS
1985 & 1986

Quant.: tons/tonnes (live weight/poids vif)
Val.: C$'000

	1985		1986	
	Quant.	Val.	Quant.	Val.
TOTAL LANDINGS/DEBARQUEMENTS TOTAUX	1 426 754	1 106 528	1 478 160	1 296 260
For direct human consumption/Pour la consommation humaine directe	1 338 507	1 044 526	1 399 352	1 239 978
Fish species/Différentes espèces de poisson				
Cod/Morue	480 197	188 171	471 329	217 275
Redfish/Sébaste	89 958	24 741	102 950	34 782
Flounder and sole/Flet et sole	89 845	30 569	88 083	38 249
Herring/Hareng	135 297	26 431	121 544	15 349
Mackerel/Maquereau	18 750	3 498	16 111	3 262
Salmon/Saumon	108 351	250 227	105 640	259 553
Other/Autres	254 827	180 327	317 527	243 898
Shellfish species/Différentes espèces de crustacés et coquillages				
Scallop/Coquille St. Jacques	46 453	61 139	55 800	72 600
Lobster/Homard	32 500	183 700	35 300	223 400
Squid/Calmar	268	141	50	27
Other/Autres	82 061	95 582	85 018	131 586
For other purposes (e.g. bait, roe and reduction to meal and oil)/Pour d'autres buts (appâts, rogue et réduction en farine et huile)	88 247	62 002	78 808	56 282
Species/Espèces				
Herring/Hareng	79 270	59 875	70 200	53 496
Mackerel/Maquereau	6 850	1 302	6 589	1 338
Squid/Calmar	122	49	312	168
Other/Autres	2 005	776	1 707	1 280

CANADA

Table II/Tableau II

NOMINAL CATCHES AND LANDED VALUES BY MAIN SPECIES/CAPTURES NOMINALES ET VALEURS AU DEBARQUEMENT PAR PRINCIPALES ESPECES
1985a)

Quant. : tons/tonnes (live weight/Poids vif)
Val. : C$'000

SPECIES/ESPECES	Atlantic Coast/Côte Atlantique		Pacific Coast/Côte Pacifique		CANADA	
	Quant.	Val.	Quant.	Val.	Quant.	Val.
TOTAL GROUNDFISH/POISSON DE FOND	765 000	295 010	61 051	45 134	826 051	340 144
Cod/Morue	478 000	187 000	2 435	1 229	480 345	188 229
Haddock/Eglefin	36 900	27 600	-	-	36 900	27 600
Redfish/Sébaste	72 200	15 500	17 814	9 253	90 014	24 753
Halibut/Flétan	3 800	12 000	6 256	13 778	10 056	25 778
Flatfish/Poisson plat	87 700	29 100	2 845	1 714	90 545	30 814
Turbot	19 000	5 700	765	164	19 765	5 864
Pollock/Goberge	49 000	11 200	1 689	358	46 689	11 558
Hake/Merlu	14 000	3 900	19 308	2 993	33 308	6 893
Cusk/Brosme	2 400	1 200	-	-	2 400	1 200
Catfish/Poisson-chat	3 800	900	-	-	3 800	900
Other/Autres	2 200	910	10 029	15 645	12 229	16 555
TOTAL PELAGIC AND OTHER FINFISH PELAGIQUES ET AUTRES	258 680	49 972	136 527	305 157	395 207	355 129
Herring/Hareng	188 800	28 900	25 767	57 406	214 567	86 306
Mackerel/Maquereau	25 600	4 800	-	-	25 600	4 800
Tuna/Thon	120	490	57	149	177	639
Alewife/Gaspareau	3 500	800	-	-	3 500	800
Eel/Anguille	550	1 160	-	-	550	1 160
Salmon/Saumon	860	3 700	107 565	246 671	108 425	250 371
Skate/Raie	50	2	370	56	420	58
Smelt/Eperlan	500	420	1	2	501	422
Capelin/Capelan	37 700	6 300	-	-	37 700	6 300
Other/Autres	1 000	3 400	2 767	873	3 767	4 273
TOTAL SHELLFISH/CRUSTACES ET COQUILLAGES	145 240	319 070	16 256	21 655	161 496	340 725
Clam	5 000	4 900	8 302	8 071	13 302	12 971
Oysters/Huîtres	1 650	1 450	3 420	2 613	5 070	4 063
Scallop/Coquille St. Jacques	46 400	61 000	53	139	46 453	61 139
Squid/Calmar	390	120	-	-	390	120
Lobster/Homard	32 500	183 700	-	-	32 500	183 700
Shrimps/Crevettes	13 500	19 300	1 192	4 559	14 692	23 859
Crab/Crabe	45 300	48 200	1 165	4 719	46 465	52 919
Other/Autres	500	400	2 124	1 554	2 624	1 954
MISCELLANEOUS/DIVERS	-	3 100	-	5 850	-	8 950
TOTAL SEA FISHERIES/PECHES MARITIMES	1 168 520	667 152	213 834	377 776	1 382 754	1 044 528
INLAND FISHERIES/PECHES INTERIEURESb)					44 000	61 600
GRAND TOTAL/TOTAL GENERAL					1 426 754	1 106 528

a) Prov.

b) Est.

CANADA

Table II (cont'd)/Tableau II (suite)

NOMINAL CATCHES AND LANDED VALUES BY MAIN SPECIES/CAPTURES NOMINALES ET VALEURS AU DEBARQUEMENT PAR PRINCIPALES ESPECES
1986[a]

Quant. : tons/tonnes (live weight/Poids vif)
Val. : C$'000

SPECIES/ESPECES	Atlantic Coast/Côte Atlantique		Pacific Coast/Côte Pacifique		CANADA	
	Quant.	Val.	Quant.	Val.	Quant.	Val.
TOTAL GROUNDFISH/POISSON DE FOND	772 800	362 100	82 700	58 720	855 500	420 820
Cod/Morue	468 000	215 300	3 600	2 100	471 600	217 400
Haddock/Eglefin	42 700	35 600	-	-	42 700	35 600
Redfish/Sébaste	79 000	20 500	24 300	14 400	103 300	34 900
Halibut/Flétan	3 600	14 600	5 600	18 600	9 200	33 200
Flatfish/Poisson plat	85 400	36 500	2 800	1 800	88 200	38 300
Turbot	19 100	10 600	900	200	20 000	10 800
Pollock/Goberge	50 000	17 900	600	200	50 600	18 100
Hake/Merlu	16 200	6 900	36 500	6 220	52 700	13 120
Cusk/Brosme	2 100	1 200	-	-	2 100	1 200
Catfish/Poisson-chat	3 400	900	-	-	3 400	900
Other/Autres	3 300	2 100	8 400	15 200	11 700	17 300
TOTAL PELAGIC AND OTHER FINFISH/ PELAGIQUES ET AUTRES	276 100	77 400	124 900	296 200	401 000	373 600
Herring/Hareng	177 500	34 500	16 600	40 000	194 100	74 500
Mackerel/Maquereau	22 700	4 600	-	-	22 700	4 600
Tuna/Thon	200	2 600	-	-	200	2 600
Alewife/Gaspareau	2 100	600	-	-	2 100	600
Eel/Anguille	200	400	-	-	200	400
Salmon/Saumon	1 200	4 700	104 500	255 000	105 700	259 700
Skate/Raie	100	0	500	300	600	300
Smelt/Eperlan	1 200	500	0	0	1 200	500
Capelin/Capelan	69 200	22 100	-	-	69 200	22 100
Other/Autres	1 700	7 400	3 300	900	5 000	8 300
TOTAL SHELLFISH/CRUSTACES ET COQUILLAGES	162 460	406 940	14 200	21 100	176 660	428 040
Clam	7 300	8 500	7 700	7 800	15 000	16 300
Oysters/Huîtres	1 900	2 600	3 600	2 800	5 500	5 400
Scallop/Coquille St. Jacques	55 800	72 400	-	200	55 800	72 600
Squid/Calmar	360	190	-	-	360	190
Lobster/Homard	35 300	223 400	-	-	35 300	223 400
Shrimps/Crevettes	18 000	30 300	1 100	3 500	19 100	33 800
Crab/Crabe	42 400	68 300	1 200	5 100	43 600	73 400
Other/Autres	1 400	1 250	600	1 700	2 000	2 950
MISCELLANEOUS/DIVERS	-	2 600	-	3 800	-	6 400
TOTAL SEA FISHERIES/PECHES MARITIMES	1 211 360	849 040	221 800	379 820	1 433 160	1 228 860
INLAND FISHERIES/PECHES INTERIEURES[b]					45 000	67 400
GRAND TOTAL/TOTAL GENERAL					1 478 160	1 296 260

a) Prov.

b) Est.

CANADA

Table III/Tableau III

FISHERY PRODUCTS AND VALUES BY MAIN PRODUCT GROUPS AND SPECIES/
PRODUITS DE LA PECHE ET VALEURS PAR PRINCIPAUX GROUPES DE PRODUITS ET ESPECES
1985 & 1986a)

Quant. : tons/tonnes (product weight/Poids du produit)
Val. : C$'000

PRODUCTS AND SPECIES/ PRODUITS ET ESPECES	1985						1986					
	Atlantic Coast/ Côte Atlantique		Pacific Coast/ Côte Pacifique		CANADA		Atlantic Coast/ Côte Atlantique		Pacific Coast/ Côte Pacifique		CANADA	
	Quant.	Val.	Quant.	Val.	Quant.	Val.	Quant.	Val.	Quant.	Val.	Quant.	Val.
SEAFISH/POISSON DE MER												
Fresh/Frozen, whole or dressed/Frais, congelé entier ou paré	127 100	136 800	73 641	265 364	200 741	402 164	124 130	155 970	90 900	267 150	215 030	423 120
Cod/Morue	25 000	31 500	200	232	25 200	31 732	22 980	36 860	270	350	23 250	37 210
Halibut/Flétan	2 800	17 400	4 694	18 035	7 494	35 435	2 580	20 760	4 780	29 480	7 360	50 240
Herring/Hareng	45 700	20 600	140	95	45 840	20 695	24 410	6 100	670	630	25 080	6 730
Mackerel/Maquereau	11 800	6 600	-	-	11 800	6 600	10 600	7 550	-	-	10 600	7 550
Salmon/Saumon	560	5 250	38 521	220 784	39 081	226 034	1 040	8 310	36 630	211 500	37 670	219 810
Capelin/Capelan	16 800	21 000	-	-	16 800	21 000	36 760	40 050	-	-	36 760	40 050
Fresh and frozen fillets/ Filets frais et congelés	122 000	437 800	6 611	25 200	128 611	463 000	131 020	562 320	7 210	28 260	138 230	590 580
Cod/Morue	61 500	218 300	560	2 102	62 060	220 402	65 300	274 830	820	3 760	66 120	278 590
Haddock/Eglefin	6 500	33 200	-	-	6 500	33 200	7 670	46 040	-	-	7 670	46 040
Redfish/Sébaste	17 100	53 600	-	-	17 100	53 600	20 510	79 680	-	-	20 510	79 680
Pollock/Goberge	5 950	13 100	381	1 220	6 331	14 320	5 890	16 120	160	530	6 050	16 650
Flounder/Flet & Sole	18 500	92 600	740	4 254	19 240	96 854	20 900	118 850	560	4 650	21 460	123 500
Herring/Hareng	5 450	3 400	-	-	5 450	3 400	4 130	2 810	-	-	4 130	2 810
Frozen blocks/Blocs congelés	54 700	148 700	57	133	54 757	148 833	67 160	239 820	160	420	67 320	240 240
Cod/Morue	46 000	123 300	8	21	46 008	123 321	57 000	206 340	40	130	57 040	206 470
Haddock/Eglefin	1 650	7 250	-	-	1 650	7 250	2 180	11 010	-	-	2 180	11 010
Redfish/Sébaste	150	250	-	-	150	250	450	910	-	-	450	910
Pollock/Goberge	3 084	5 264	10	16	3 094	5 280	4 000	10 360	10	20	4 010	10 380
Flounder/Flet & Sole	3 406	5 264	16	43	3 422	5 307	2 070	8 270	10	40	2 780	8 310
Herring/Hareng	51	20	-	-	51	20	190	240	-	-	190	240
Smoked/Fumé	3 900	11 200	819	12 080	4 719	23 280	6 730	21 580	790	10 650	7 520	32 230
Bloater/Hareng bouffi	2 500	3 750	-	-	2 500	3 750	3 300	4 900	-	-	3 300	4 900
Salmon/Saumon	50	1 300	747	11 547	797	12 847	70	1 440	700	9 800	770	11 240
Salted/Salé	42 100	104 200	-	-	42 100	104 200	36 160	112 260	-	-	36 160	112 260
Cod/Morue	33 800	87 400	-	-	33 800	87 400	25 600	90 700	-	-	25 600	90 700
Cured & pickled/ Mariné et en saumure	20 300	21 700	937	7 514	21 237	29 214	10 540	11 350	650	5 600	11 190	16 950
Herring/Hareng	16 000	17 400	147	889	16 147	18 289	7 100	7 050	-	-	7 100	7 050
Mackerel/Maquereau	650	950	-	-	650	950	1 420	2 340	-	-	1 420	2 340
Canned/En conserve	26 500	117 000	41 431	258 905	67 951	375 905	17 920	95 350	42 060	273 410	59 980	366 760
Herring/Hareng & sardine	18 800	83 700	-	-	18 800	83 700	16 000	87 480	-	-	16 000	87 480
Salmon/Saumon	-	-	41 378	258 646	41 378	258 646	-	-	42 000	273 000	42 000	273 000
Meal/Farine	53 100	16 100	7 267	3 299	60 367	19 399	51 500	17 650	5 940	2 960	57 440	20 650
Groundfish/Poisson de fond	43 200	12 500	-	-	43 200	12 500	43 000	14 190	-	-	43 000	14 190
Herring/Hareng	9 100	3 350	4 552	2 121	13 652	5 471	8 500	3 500	2 900	1 500	11 400	5 000
Oil/Huileb)	6 900	2 700	1 277	1 001	8 177	3 701	9 060	3 670	1 350	840	10 410	4 510
Groundfish/Poisson de fond	2 900	1 100	-	-	2 900	1 100	3 330	1 430	-	-	3 330	1 430
Herring/Hareng	4 000	1 600	452	242	4 452	1 842	5 730	2 180	300	300	6 030	2 480
Roe/Rogue	5 800	40 200	5 850	109 514	11 660	149 714	6 220	45 130	4 580	88 840	10 800	133 970
Herring/Hareng	5 200	39 100	3 787	97 491	8 987	136 591	4 600	39 000	2 550	73 700	7 150	112 700
Other seafish products/ Autres produits de la mer	34 700	94 300	13 379	14 612	48 079	108 912	46 310	155 970	13 160	10 260	59 470	166 230

243

CANADA

Table III (cont'd)/Tableau III (suite)

FISHERY PRODUCTS AND VALUES BY MAIN PRODUCT GROUPS AND SPECIES/
PRODUITS DE LA PECHE ET VALEURS PAR PRINCIPAUX GROUPES DE PRODUITS ET ESPECES
1985 & 1986[a]

Quant. : tons/tonnes (product weight/Poids du produit)
Val. : C$'000

PRODUCTS AND SPECIES/ PRODUITS ET ESPECES	1985						1986					
	Atlantic Coast/ Côte Atlantique		Pacific Coast/ Côte Pacifique		CANADA		Atlantic Coast/ Côte Atlantique		Pacific Coast/ Côte Pacifique		CANADA	
	Quant.	Val.	Quant.	Val.	Quant.	Val.	Quant.	Val.	Quant.	Val.	Quant.	Val.
SHELLFISH/CRUSTACES ET COQUILLAGES												
Fresh & frozen in shell/Frais & congelés en carapace	39 900	206 500	3 972	12 503	43 872	219 403	42 730	272 780	4 470	16 740	47 200	289 520
Squid/Calmar	40	20	-	-	40	20	50	50	-	-	50	50
Lobster/Homard	17 500	134 100	-	-	17 500	134 100	19 510	164 350	-	-	19 510	164 350
Crab/Crabe	15 200	57 000	783	4 427	15 983	61 427	14 300	89 870	970	7 560	15 270	97 430
Shrimp/Crevette	3 800	11 000	358	2 561	4 158	13 561	4 280	12 280	290	1 870	4 570	14 150
Fresh & frozen, shucked/Frais & congelés, écaillés	17 100	254 500	1 456	10 133	18 566	264 633	19 350	310 170	1 400	10 250	20 750	320 420
Scallop/Coquille St. Jacques	5 700	89 300	-	-	5 700	89 300	6 800	101 800	-	-	6 800	101 800
Squid/Calmar	-	-	-	-	-	-	-	-	-	-	-	-
Lobster/Homard	2 500	67 800	-	-	2 500	67 800	3 200	90 300	-	-	3 200	90 300
Crab/Crabe	6 000	74 800	48	1 183	6 048	75 983	3 500	38 750	60	1 880	3 500	38 750
Shrimp/Crevette	2 300	25 500	209	2 669	2 509	28 169	3 300	36 650	170	2 100	3 500	38 750
Canned/En conserve	1 550	21 000	2	15	1 552	21 015	980	16 820	-	-	980	16 820
Clam	800	6 550	0	1	800	6 551	530	5 980	-	-	530	5 980
Lobster/Homard	250	5 650	-	-	250	5 650	260	7 330	-	-	260	7 330
Crab/Crabe	500	8 800	1	12	501	8 812	140	3 370	-	-	140	3 370
Other shellfish products/Autres produits de crustacés et coquillages	12 600	21 000	650	4 828	13 250	25 828	10 910	12 370	370	3 720	11 280	12 670
MISC./DIVERS[c]	22 400	3 200	-	-	22 400	3 200	23 200	3 440	-	-	23 200	3 440
TOTAL SEA FISHERIES/ PECHE MARITIME	590 750	1 646 600	157 389	725 501	748 139	2 372 101	603 920	2 034 630	173 040	719 110	776 960	2 753 730
INLAND FISHERIES/ PECHES INTERIEURES[d]					34 300	135 000					36 600	128 000
GRAND TOTAL/TOTAL GENERAL					782 439	2 507 101					813 560	2 881 730

a)   Prov.

b)   Including seal oil/Y compris huile de phoque.

c)   Excluding number of seals/Non compris le nombre des phoques.

d)   Est.

CANADA

Table IV(a)/Tableau IV(a)

CANADIAN EXPORTS BY MAIN PRODUCT GROUPS AND COUNTRIES/
EXPORTATIONS CANADIENNES PAR PRINCIPAUX GROUPES DE PRODUITS ET PAR PAYS
1985 & 1986

Quant. : tons/tonnes (product weight/poids du produit)
Val. : C$'000

PRODUCTS AND COUNTRIES/PRODUITS ET PAYS	1985				1986			
	Quant.	Val.	Quant. (%)	Val. (%)	Quant.	Val.	Quant. (%)	Val. (%)
SEAFISH/POISSON DE MER								
Whole or dressed, fresh/Entier ou paré, frais	89 281	107 048	100	100	77 453	125 063	100	100
United States/Etats-Unis	79 552	100 248	85	94	70 332	82 953	91	96
USRR/URSS	10 076	1 813	11	2	3 628	604	5	0
Whole or dressed, frozen/Entier ou paré, congelé	65 256	231 125	100	100	84 246	332 383	100	100
United States/Etats-Unis	12 305	42 331	19	18	11 587	43 624	14	13
EEC/CEE	16 865	70 880	26	31	15 409	73 529	28	22
Japan/Japon	25 991	98 098	40	42	47 768	192 295	57	58
Fresh fillets/Filets frais	26 565	101 755	100	100	32 546	153 367	100	100
United States/Etats-Unis	26 227	100 861	99	99	32 307	152 507	99	99
Frozen fillets/Filets congelés	72 614	264 362	100	100	75 397	327 106	100	100
United States/Etats-Unis	64 007	248 101	88	94	66 132	303 919	88	93
EEC/CEE	4 825	8 596	7	3	5 968	14 264	8	4
Frozen blocks/Blocs congelés	55 642	147 711	100	100	60 263	209 549	100	100
United States/Etats-Unis	46 933	129 670	84	88	55 260	194 040	92	93
EEC/CEE	8 062	16 257	14	11	4 025	12 140	7	6
Smoked/Fumé	7 171	14 426	100	100	8 234	18 996	100	100
United States/Etats-Unis	866	3 866	12	27	910	5 276	11	28
Dominican Republic/République Dominicaine	2 071	2 720	29	19	2 924	4 152	36	22
Haiti	2 583	3 164	36	22	2 066	2 470	25	13
Salted and dried/Salé et séché	39 769	105 780	100	100	43 505	145 686	100	100
United States/Etats-Unis	9 242	34 232	23	32	9 348	39 593	21	27
EEC/CEE	14 929	32 678	38	31	12 194	44 644	28	31
Puerto Rico/Porto-Rico	4 616	12 979	12	12	5 720	19 813	13	14
Cured and pickled/Salé, séché, fumé et saumuré	19 672	18 751	100	100	25 560	23 834	100	100
United States/Etats-Unis	8 136	10 286	41	55	8 357	11 704	33	49
EEC/CEE	1 024	720	5	4	516	539	2	2
Haiti	3 492	2 543	18	14	4 217	2 505	16	11
Canned/En conserve	27 632	118 158	100	100	37 922	191 501	100	100
United States/Etats-Unis	3 530	13 708	13	12	4 498	19 900	12	52
EEC/CEE	8 516	49 176	31	42	18 813	111 979	50	58
Australia/Australie	3 553	19 315	13	16	2 659	17 689	7	9
New Zealand/Nouvelle-Zélande	1 340	6 200	5	5	1 789	8 454	5	4
Meal/Farine	26 448	8 287	100	100	20 007	9 276	100	100
United States/Etats-Unis	11 835	6 302	45	76	10 061	6 559	50	71
EEC/CEE	3 493	920	13	11	1 373	513	7	6
Oil/Huile	6 236	3 544	100	100	5 243	2 875	100	100
United States/Etats-Unis	5 663	3 239	91	91	5 143	2 814	98	98
EEC/CEE	-	-	-	-	-	-	-	-
Roe/Rogue	12 650	152 225	100	100	12 243	164 587	100	100
Japan/Japon	11 424	147 022	90	97	10 504	153 578	86	93
Other seafish products/Autres produits de la mer	20 456	14 111	100	100	16 385	9 393	100	100
United States/Etats-Unis	19 252	11 568	94	82	15 793	7 513	96	80

CANADA

Table IV(a) (cont'd)/Tableau IV(a) (suite)

CANADIAN EXPORTS BY MAIN PRODUCT GROUPS AND COUNTRIES/
EXPORTATIONS CANADIENNES PAR PRINCIPAUX GROUPES DE PRODUITS ET PAR PAYS
1985 & 1986

Quant. : tons/tonnes (product weight/poids du produit)
Val. : C$'000

PRODUCTS AND COUNTRIES/PRODUITS ET PAYS	1985				1986			
	Quant.	Val.	Quant. (%)	Val. (%)	Quant.	Val.	Quant. (%)	Val. (%)
SHELLFISH/CRUSTACES ET COQUILLAGES								
In shell and peeled, fresh and frozen/ En carapace et décortiqués, frais et congelés	51 990	462 958	100	100	56 649	582 114	100	100
United States/Etats-Unis	35 546	351 505	69	77	38 528	440 193	68	76
EEC/CEE	5 116	49 041	10	11	7 261	66 745	13	11
Japan/Japon	9 758	56 121	19	12	8 188	53 121	14	9
Canned/En conserve	695	9 022	100	100	439	7 388	100	100
United States/Etats-Unis	246	3 864	35	43	129	2 309	29	31
EEC/CEE	308	3 965	44	44	142	3 150	32	43
Japan/Japon	34	474	5	5	93	1 208	21	16
MISCELLANEOUS/DIVERS[a]	6 744	10 769	100	100	9 323	15 675	100	100
United States/Etats-Unis	1 736	2 403	26	22	5 284	4 120	57	26
EEC/CEE	4 811	4 778	71	44	3 609	4 129	39	26
TOTAL SEA FISHERIES/TOTAL PRODUITS DE LA MER	528 821	1 770 091			565 415	2 318 793		
FRESHWATER FISH/POISSONS D'EAU DOUCE								
Whole or dressed, fresh and frozen/Entiers ou parés, frais et congelés	19 839	42 689	100	100	17 269	44 044	100	100
United States/Etats-Unis	13 287	32 204	67	75	10 916	33 188	63	75
EEC/CEE	793	2 116	4	5	995	2 897	6	7
Japan/Japon	4 591	6 410	23	15	3 815	4 762	22	11
Fillets and blocks, fresh and frozen/ Filets et blocs, frais et congelés	7 744	46 177	100	100	8 171	59 307	100	100
United States/Etats-Unis	6 218	38 284	80	83	6 652	49 035	81	83
EEC/CEE	1 224	5 499	16	12	1 262	8 651	15	15
TOTAL EXPORTS - ALL GROUPS/EXPORTATIONS TOTALES - TOUS GROUPES	556 404	1 858 898			590 855	2 422 143		
SUMMARY BY MAIN COUNTRIES/RESUME PAR PRINCIPAUX PAYS								
United States/Etats-Unis	340 377	1 133 770	61	61	350 499	1 431 749	59	59
EEC/CEE	72 977	248 424	13	14	74 035	353 513	123	14
Other European countries/Autres pays européens	35 407	40 228	6	2	35 740	61 947	6	3
Central and South America/Amérique centrale et du Sud	28 065	61 700	5	3	32 981	81 196	6	3
Japan/Japon	54 253	319 892	10	17	77 109	432 800	13	18
All other countries/Tous les autres pays	25 325	54 879	5	3	20 492	60 939	3	3
TOTAL ALL COUNTRIES/TOTAL TOUS PAYS	556 404	1 858 898	100	100	590 855	2 422 143	100	100

a)   Excluding seal skins which are reported in numbers/Non compris les peaux de phoques qui sont rapportées en nombre.

CANADA

Table IV(b)/Tableau IV(b)

EXPORTS OF FROZEN FILLETS BY SPECIES AND COUNTRIES/
EXPORTATIONS DE FILETS CONGELES PAR ESPECES ET PAR PAYS
1985 & 1986a)

Quant. : tons/tonnes (product weight/poids du produit)
Val. : C$'000

	1985		1986	
	Quant.	Val.	Quant.	Val.
COD FILLETS/FILETS DE MORUE	36 980	136 750	38 107	170 412
United States-Etats-Unis	34 624	130 797	34 669	158 264
EEC/CEE	2 197	5 402	3 146	10 792
OCEAN PERCH FILLETS/FILETS DE SEBASTE	9 643	31 219	11 998	49 074
United States/Etats-Unis	9 249	30 018	11 487	47 414
EEC/CEE	-	-	-	-
SOLE AND FLOUNDER FILLETS/FILETS DE SOLE ET DE FLETAN	10 488	54 213	10 642	62 058
United States/Etats-Unis	10 293	53 190	10 300	59 625
EEC/CEE	174	910	219	867
HERRING FILLETS/FILETS DE HARENG	4 715	3 406	3 481	2 666
United States/Etats-Unis	255	341	186	207
EEC/CEE	2 313	1 752	2 497	1 979
OTHER FILLETS/AUTRES FILETS	10 788	38 774	11 161	42 869
United States/Etats-Unis	9 586	33 765	9 490	38 409
EEC/CEE	143	532	106	626
TOTAL FILLETS/TOTAL DES FILETS	72 614	264 362	75 397	327 106
United States/Etats-Unis	64 007	248 101	66 132	303 919
EEC/CEE	4 825	8 596	5 968	14 264

a) Est.

CANADA

Table IV(c)/Tableau IV(c)

CANADIAN IMPORTS BY MAIN PRODUCT GROUPS AND COUNTRIES/
IMPORTATIONS CANADIENNES PAR PRINCIPAUX GROUPES DE PRODUITS ET PAR PAYS
1985 & 1986

Quant. : tons/tonnes (product weight/poids du produit)
Val. : C$'000

PRODUCTS AND COUNTRIES/PRODUITS ET PAYS	1985 Quant.	1985 Val.	1985 Quant. (%)	1985 Val. (%)	1986 Quant.	1986 Val.	1986 Quant. (%)	1986 Val. (%)
SEAFISH/POISSON DE MER								
Fresh or frozen/Frais ou congelé	51 732	122 076	100	100	56 206	154 934	100	100
United States/Etats-Unis	32 712	82 249	63	67	38 812	104 082	69	67
EEC/CEE	7 503	21 311	15	14	10 352	29 445	18	19
Japan/Japon	3 331	5 427	6	4	1 443	4 801	3	3
Steaks, blocks, etc. fresh or frozen/ Steaks, blocs, etc. frais ou congelés	5 369	14 351	100	100	6 250	21 334	100	100
United States/Etats-Unis	5 029	13 459	94	94	5 770	19 856	92	93
Smoked/Fumé	382	1 855	100	100	450	2 546	100	100
United States/Etats-Unis	137	956	36	52	132	877	29	34
EEC/CEE	220	761	58	41	276	1 445	61	57
Salted or dried/Salé ou seché	1 518	5 614	100	100	1 278	5 212	100	100
United States/Etats-Unis	383	1 137	25	20	446	1 315	35	25
Norway/Norvège	440	1 539	29	27	203	910	16	17
Hong Kong	308	1 426	20	26	188	1 130	15	22
Cured or pickled/Salé, seché, fumé ou saumuré	410	681	100	100	410	620	100	100
United States/Etats-Unis	18	34	4	5	26	50	6	8
EEC/CEE	269	474	63	67	283	434	69	70
Canned/En conserve[a]	16 308	71 663	100	100	21 188	91 978	100	100
United States/Etats-Unis	4 219	20 414	26	28	2 321	10 811	11	12
Japan/Japon	4 602	20 804	28	29	5 491	28 376	26	31
Philippines	2 479	7 591	15	11	963	3 087	5	3
Thailand/Thaïlande	2 529	7 634	16	11	7 500	25 071	35	27
Fiji/Fidji	563	2 527	3	4	675	2 954	3	3
Meal/Farine	742	193	100	100	2 994	1 323	100	100
United States/Etats-Unis	742	193	100	100	1 640	709	55	54
Oil/Huile	359	620	100	100	468	880	100	100
United States/Etats-Unis	124	283	35	46	185	272	40	31
Norway/Norvège	80	173	22	28	98	224	21	25
Other seafish products/Autres produits de la mer	5 802	8 227	100	100	6 129	12 967	100	100
United States/Etats-Unis	2 996	3 416	52	42	5 427	7 655	89	59
Japan/Japon	407	1 271	7	15	371	1 979	6	15
SHELLFISH/CRUSTACES ET COQUILLAGES								
Fresh or frozen/Frais ou congelés	26 957	218 556	100	100	29 678	269 807	100	100
United States/Etats-Unis	16 114	125 710	59	58	18 496	146 166	62	54
Hong Kong	2 731	23 051	10	11	2 641	27 582	9	10
Cuba	1 117	18 998	4	9	663	14 144	2	5
Ecuador/Equateur	242	3 090	1	1	428	6 509	1	2
Canned/En conserve	8 868	42 113	100	100	9 167	42 640	100	100
United States/Etats-Unis	1 022	9 603	12	23	942	8 801	10	21
Japan/Japon	183	567	2	1	43	229	-	1
South Korea/Corée du sud	1 553	8 784	18	21	1 746	9 295	19	22
Thailand/Thaïlande	4 294	15 253	48	36	4 517	15 258	49	36
Other shellfish products/Autres produits de crustacés et coquillages	14 855	1 509	100	100	15 023	1 539	100	100
United States/Etats-Unis	14 847	1 508	100	100	15 023	1 539	100	100

CANADA

Table IV(c) (cont'd)/Tableau IV(c) (suite)

CANADIAN IMPORTS BY MAIN PRODUCT GROUPS AND COUNTRIES/
IMPORTATIONS CANADIENNES PAR PRINCIPAUX GROUPES DE PRODUITS ET PAR PAYS
1985 & 1986

Quant. : tons/tonnes (product weight/poids du produit)
Val. : C$'000

PRODUCTS AND COUNTRIES/PRODUITS ET PAYS	1985				1986			
	Quant.	Val.	Quant. (%)	Val. (%)	Quant.	Val.	Quant. (%)	Val. (%)
TOTAL SEA FISHERIES/TOTAL PECHES MARITIMES	133 302	487 458			149 241	605 780		
FRESHWATER FISH/POISSON D'EAU DOUCE								
Fresh or frozen/Frais ou congelé	2 487	8 374	100	100	3 130	10 695	100	100
United States/Etats-Unis	2 225	7 455	89	89	2 769	9 488	88	89
TOTAL IMPORTS ALL GROUPS/IMPORTATIONS TOTALES TOUS LES GROUPES	135 789	495 832			152 371	616 475		
SUMMARY BY MAIN COUNTRIES/RESUME PAR PRINCIPAUX PAYS								
United States/Etats-Unis	80 569	266 416	59	54	91 986	311 620	60	51
EEC/CEE	10 419	39 606	8	8	13 907	50 559	9	8
Other European countries/Autres pays européens	4 721	10 462	3	2	3 100	9 474	2	2
Central and South America/Amérique centrale et du Sud	9 730	47 927	7	10	4 507	43 350	3	7
Japan/Japon	9 452	36 401	7	7	8 276	45 156	5	7
All other countries/Tous les autres pays	20 898	95 020	15	19	30 595	156 316	20	25
TOTAL ALL COUNTRIES/TOTAL TOUS PAYS	135 789	495 832	100	100	152 371	616 475	100	100

a) Excluding quantity of sardines and anchovy reported in number of boxes/Non compris les quantités de sardines et d'anchois qui sont rapportées en nombre de boîtes.

DENMARK/DANEMARK

Table I/Tableau I

TOTAL DANISH LANDINGS AND FOREIGN LANDINGS IN DANISH PORTS/
DEBARQUEMENTS DANOIS TOTAUX ET DEBARQUEMENTS ETRANGERS DANS LES PORTS DANOIS

1985 & 1986[a]

Landings/Débarquements : metric tons/tonnes métriques
Val. : '000 DKr/KrD

	1985		1986[a]	
	Quant.	Val.	Quant.	Val.
1. National landings in domestic ports/Débarquements nationaux dans les ports danois				
Plaice/Plie	40 666	307 147	40 159	334 902
Common sole/Sole commune	943	47 420	953	58 610
Cod/Morue	165 502	1 120 800	146 517	1 200 930
Haddock/Eglefin	25 829	178 923	20 026	135 209
Saithe/Lieu noir	8 323	33 510	10 264	51 229
Whiting/Merlan	2 446	10 065	1 967	8 710
Hake/Merlu	1 761	24 358	1 805	27 329
Herring/Hareng	101 606	248 589	79 633	186 895
Mackerel/Maquereau	16 112	40 332	22 120	45 990
Salmon/Saumon	1 472	63 102	848	27 625
Norway lobster/Langoustine	2 988	140 354	2 914	152 747
Other/Autres	115/724	432 011	137 823	515 673
For other purposes/Pour d'autres buts[b]	1 244 003	786 448	1 364 135	716 341
TOTAL 1	1 727 375	3 433 059	1 829 164	3 462 190
2. National landings in foreign ports/Débarquements nationaux dans les ports étrangers				
Plaice/Plie	806	6 983	638	5 127
Cod/Morue	6 448	66 512	6 421	66 099
Salmon/Saumon	24	848	0	15
Other/Autres	3 007	57 169	1 160	21 034
For other purposes/Pour d'autres buts[b]	1 059	351	686	341
TOTAL 2	11 344	131 863	8 905	92 616
TOTAL 1 + 2	1 738 719	3 564 922	1 838 069	3 554 806
3. Foreign landings in domestic ports/Débarquements étrangers dans les ports nationaux				
Cod/Morue	19 568	110 166	22 543	152 788
Herring/Hareng	30 612	82 354	39 535	105 734
Mackerel/Maquereau	32 014	71 880	15 573	30 662
Others/Autres	29 166	178 183	25 053	183 757
For other purposes/Pour d'autres buts[b]	74 898	54 757	81 655	48 653
TOTAL 3	186 258	497 340	184 359	521 594
TOTAL 1 + 3	1 913 633	3 930 399	2 013 523	3 983 784

a) Provisional figures/Chiffres provisoires.

b) Reduction to meal and oil, pet food, bait, etc./Réduction en farine et huile, alimentation des animaux domestiques, appâts, etc.

DENMARK/DANEMARK

Table II(a)/Tableau II(a)

EXTERNAL TRADE OF MAJOR FISH PRODUCTS AND EXPORT VALUE BY SELECTED COUNTRIES/
ECHANGES INTERNATIONAUX DES PRINCIPAUX PRODUITS DE LA PECHE ET VALEUR DES EXPORTATIONS PAR PAYS SELECTIONNES
1985 & 1986

Quant. : tons/tonnes
Val. : '000 DKr/KrD

	1985		1986		EXPORTS BY VALUE/EXPORTATIONS EN VALEUR, 1986							
	Quant.	Val.	Quant.	Val.	GERMANY/ ALLEMAGNE	FRANCE	U.K./ R.U.	ITALY/ ITALIE	B.L.E.U./ U.E.B.L.	SPAIN/ ESPAGNE	SWEDEN/ SUEDE	U.S./ E.U.	OTHER/ AUTRES
Fresh, chilled/Frais sur glace (whole/entier)													
Herring/Hareng	75 964	1 056 417	72 506	1 132 945	243 773	221 810	81 548	144 497	295 059	42 181	65 778	1 368	36 931
Cod/Morue	14 697	71 591	13 365	65 677	23 422	6 654	1 869	3	33 068	4	340	-	318
Saithe/Lieu noir	28 698	433 059	27 309	472 720	48 356	124 302	54 316	44 909	175 710	15 543	112	1	9 472
Haddock/Eglefin	6 692	58 123	6 052	62 982	25 620	17 972	41	5 389	7 909	892	1 532	-	3 626
Mackerel/Maquereau	3 990	49 946	4 106	53 918	22 599	470	15 825	-	9 399	-	5 165	- 70	390
Plaice/Plie	2 634	15 058	2 589	15 255	7 697	211	43	1 969	411	120	4 320	-	485
Other flatfish/Autres poissons plats	7 545	95 835	7 566	97 170	59 024	33	5 201	772	15 576	77	15 700	8	778
European hake/Merlu d'Europe	4 670	171 525	4 709	202 953	22 272	35 544	1 782	51 868	32 006	14 466	30 877	1 285	12 851
	1 652	37 757	1 679	40 847	231	28 628	1	1 105	1 367	7 948	402	-	1 164
Fresh, chilled/Frais, sur glace (fillets/filets)													
Herring/Hareng	72 663	735 393	64 977	707 987	314 650	93 579	3 709	30 756	83 007	57 935	28 721	4 538	68 264
Cod/Morue	52 982	272 878	47 304	235 461	186 978	13 667	1 780	-	14 320	-	426	-	8 290
Other gadoids/Autres gadidés	12 380	305 600	10 333	290 566	44 725	69 594	1 290	30 581	54 495	51 255	904	46	8 290
Flatfish/Poissons plats	4 033	64 098	3 958	71 479	40 123	5 401	594	54	6 062	2 711	1 698	3 411	37 676
	1 750	62 221	1 723	68 716	3 200	3 844	23	40	1 264	3 746	24 461	1 020	11 425
Frozen/Congelés (whole/entier)													
Herring/Hareng	23 206	247 848	25 877	356 799	95 561	28 066	68 361	2 669	26 530	18 390	8 194	23 425	134 048
Halibut/Flétan	2 934	17 955	2 358	13 343	1 714	93	-	-	11 409	-	-	-	127
Greenland halibut/Flétan noir	399	13 484	363	13 631	4 576	196	3 165	-	666	10	1 442	-	3 575
Cod/Morue	1 952	28 524	2 667	44 562	18 179	1 483	435	257	3 557	415	1 161	1 529	17 546
Mackerel/Maquereau	3 593	48 140	7 720	122 655	23 525	10 058	59 096	112	7 978	22	54	16 137	5 646
	8 858	46 933	5 101	25 963	2 622	2 334	150	511	161	93	273	-	19 818
Frozen fillets/Filets congelés													
Herring/Hareng	93 332	1 956 587	100 007	2 319 343	447 712	189 799	426 124	266 078	34 674	7 695	151 945	634 266	202 513
Cod/Morue	6 280	31 935	6 939	35 714	14 577	304	120	-	13 069	-	227	0	17 416
Saithe/Lieu noir	47 424	1 136 768	55 459	1 407 512	134 710	109 011	403 717	164 247	12 062	660	86 108	461 206	35 790
Haddock/Eglefin	19 504	225 197	17 884	281 212	236 746	13 847	1 826	0	2 018	185	5 003	15 796	5 790
Flatfish/Poissons plats	4 607	132 989	3 902	106 221	916	1 960	8 187	-	1 244	67	1 682	88 093	4 071
	10 418	337 143	11 817	408 055	21 580	44 413	7 625	99 759	15 580	6 756	53 607	66 150	92 587
Salted, dried, smoked/Salés, séchés, fumés													
Herring/Hareng, whole, headed, in brine/Hareng, entier, etêté, en saumure	26 305	865 496	25 783	972 670	269 878	80 693	6 233	225 651	64 125	86 099	10 551	2 009	227 431
	4 473	31 308	3 440	28 337	5 754	1 243	9	461	16 956	123	719	5	3 068

251

DENMARK/DANEMARK

Table II(a) cont'd/Tableau II(a)suite

EXPORTS BY VALUE/EXPORTATIONS EN VALEUR, 1986

	1985		1986		GERMANY/ ALLEMAGNE	FRANCE	U.K./ R.U.	ITALY/ ITALIE	B.L.E.U./ U.E.B.L.	SPAIN/ ESPAGNE	SWEDEN/ SUEDE	U.S./ E.U.	OTHER/ AUTRES
	Quant.	Val.	Quant.	Val.									
Cod, dried and/or salted/ Morue, séchée et/ou salée	9 669	217 194	10 993	289 207	3 810	26 598	1 096	87 985	79	49 439	190	-	120 009
Cod fillets, salted/Filets de morue salés	1 591	50 883	913	31 114	-	3 433	0	21 915	-	4 416	775	-	576
Smoked salmon/Saumon fumé	2 849	365 769	3 483	417 414	203 618	42 069	1 074	73 093	23 329	1 968	2 448	1 579	68 237
Smoked trout/Truite fumée	548	32 030	467	29 083	7 597	1 182	3 109	1 328	8 065	304	1 473	246	5 780
Salted fish roe/Rogue de poisson salé	619	19 171	857	27 780	17 578	9	288	-	0	2 437	987	-	6 481
Shellfish (live, fresh, chilled, etc.)/Coquillages (vivants, frais, sur glace, etc.)	73 549	1 324 113	74 963	1 582 986	32 269	335 602	87 647	245 172	73 446	41 326	173 207	9 057	585 261
Pink shrimp/Crevette rose	33 734	967 496	38 311	1 175 798	9 290	275 535	77 341	42 054	15 853	23 663	167 743	1	564 318
Common and other shrimp/ Crevettes communes et autres	1 942	56 663	2 634	69 210	7 876	7 213	3 110	2 757	34 559	304	3 853	248	9 291
Norway lobster/Langoustine	3 926	242 726	4 016	281 406	5 115	42 460	5 641	189 365	4 344	16 557	465	8 582	8 878
Canned or prepared/En conserve ou préparés	68 208	1 912 288	73 477	2 282 707	344 991	307 048	626 528	269 953	107 914	21 445	204 843	31 718	368 267
Caviar substitutes/Succédanés	3	4 421	0	0	469	103	-	247	-	-	6	1	112
de caviar	962	65 188	987	75 353	1 283	20 031	4 826	12 638	2 681	1 275	10 410	4 146	18 526
Salmon preparations/Prépara- tions de saumon	176	21 720	227	27 169	17 550	257	391	2 286	556	2	296	13	5 818
Herring preparations/ Préparations de hareng	6 223	56 896	8 960	74 700	24 949	2 869	4 215	61	18 471	283	11 723	772	11 358
Mackerel - various/ Maquereau divers	10 407	184 776	10 546	191 822	48 721	1 257	45 846	8 310	32 438	283	44 151	72	10 742
Fish fillets, breaded portions/ Filets de poisson, portions panées	11 445	291 691	12 615	353 292	25 862	13 809	123 675	57 748	6 913	482	45 367	279	79 157
Fish roe/Rogue de poisson	2 391	31 103	1 813	27 501	0	795	21 298	4	13	2 304	515	55	2 518
Shrimps, not canned/Crevettes non en conserve	7 787	516 325	9 423	754 921	125 748	67 471	240 817	164 480	7 120	9 950	43 134	4 830	91 481
Mussels, canned/Moules en conserve	6 280	113 046	6 384	117 254	11 314	52 735	9 203	1 686	1 841	7	14 167	1 408	24 891
Herring, canned/Hareng en conserve	5 590	116 933	4 774	102 673	58 397	7 673	6 878	442	1 968	266	5 178	2 328	19 542
Crab, canned/Crabe en conserve	30	1 506	3	634	3	543	-	22	0	0	51	-	16
Shrimp, canned/Crevettes, en conserve	1 133	87 517	1 478	116 598	7 065	24 139	53 160	7 553	7 685	50	713	72	16 161
Tuna, bonito/Thon bonite	36	871	23	520	112	0	-	2	21	-	61	-	325
Other, canned/Autres, en conserve	2 407	64 303	2 764	78 433	1 157	36 781	11 811	1 708	1 701	3 916	646	1 438	19 275
Other, frozen/Autres congelés	10 587	296 305	8 834	273 112	16 025	44 773	97 974	1 997	16 317	2 698	19 553	12 591	61 185
Fish oil/Huile de poisson	61 059	193 107	62 437	101 221	44 018	4 935	12 401	5 304	10 198	-	621	-	23 747
Fish meal/Farine de poisson	217 794	895 400	224 661	815 115	17 291	17 767	67 356	70 327	62 991	0	66 667	493	512 223

252

DENMARK/DANEMARK

Table II(b)/Tableau II(b)

EXTERNAL TRADE OF MAJOR FISH PRODUCTS AND EXPORT QUANTITY BY SELECTED COUNTRIES/
ECHANGES INTERNATIONAUX DES PRINCIPAUX PRODUITS DE LA PECHE ET QUANTITE DES EXPORTATIONS PAR PAYS SELECTIONNES
1985 & 1986

Quant. : tons/tonnes
Val. : '000 DKr/KrD

	1985		1986		EXPORTS BY QUANTITY/EXPORTATIONS EN QUANTITE							
	Quant.	Val.	Quant.	Val.	GERMANY/ ALLEMAGNE	FRANCE	U.K./ R.U.	ITALY/ ITALIE	B.L.E.U./ U.E.B.L.	SPAIN/ ESPAGNE	SWEDEN/ SUEDE	U.S./ E.U.	OTHER/ AUTRES
Fresh, chilled/Frais sur glace (whole/entier)													
Herring/Hareng	75 964	1 056 417	72 506	1 132 945	19 948	11 447	8 653	4 620	21 479	1 544	3 551	22	1 241
Cod/Morue	14 597	71 591	13 365	65 677	4 744	988	1 058	0	6 450	0	85	-	39
Saithe/Lieu noir	28 698	433 059	27 309	472 720	2 784	6 328	5 540	1 664	9 877	769	8	0	339
Haddock/Eglefin	6 692	58 123	6 052	62 982	3 257	1 310	12	214	877	47	210	-	126
Mackerel/Maquereau	3 990	49 946	4 106	53 918	1 323	35	1 320	-	713	-	677	7	32
Plaice/Plie	2 634	15 058	2 589	15 255	1 669	39	7	181	79	2	545	-	68
Other flatfish/Autres poissons	7 545	95 835	7 566	97 170	4 323	2	512	59	1 693	6	928	0	43
plats	4 670	171 525	4 709	202 953	395	1 106	69	1 073	841	234	688	15	288
European hake/Merlu d'Europe	1 652	37 757	1 679	40 847	11	1 170	0	41	76	339	18	-	24
Fresh, chilled/Frais, sur glace (fillets/filets)													
Herring/Hareng	72 663	735 393	64 977	707 987	45 779	5 533	397	948	5 074	2 029	996	118	4 999
Cod/Morue	52 982	272 878	47 304	235 461	40 243	2 618	299	-	2 577	-	75	-	1 491
Other gadoids/Autres gadidés	12 380	305 600	10 333	290 566	1 769	2 520	59	943	1 871	1 798	35	2	1 491
Flatfish/Poissons plats	4 033	64 098	3 958	71 479	2 448	243	37	4	318	114	78	98	1 337
	1 750	62 221	1 723	68 716	105	114	1	1	37	109	744	17	619
Frozen/Congelés (whole/entier)													
Herring/Hareng	23 206	247 848	25 877	356 799	6 463	1 983	3 756	193	2 774	701	427	2 257	8 589
Halibut/Flétan	2 934	17 955	2 358	13 343	322	22	-	-	1 989	-	-	-	26
Greenland halibut/Flétan noir	399	13 484	363	13 631	121	9	81	-	27	0	31	-	93
Cod/Morue	1 952	28 524	2 667	44 562	1 158	60	25	13	188	38	66	37	1 082
Mackerel/Maquereau	3 593	48 140	7 720	122 655	1 226	747	354	3	368	1	3	1 819	198
	8 858	46 933	5 101	25 963	542	560	25	122	53	22	43	-	3 733
Frozen fillets/Filets congelés													
Herring/Hareng	93 332	1 956 587	100 007	2 319 343	26 983	7 710	16 908	8 092	1 601	201	5 773	24 886	13 825
Cod/Morue	6 280	31 935	6 939	35 714	3 050	56	18	-	410	-	28	-	3 377
Saithe/Lieu noir	47 424	1 136 768	55 459	1 407 512	5 665	4 630	15 960	5 065	454	21	3 518	19 102	1 044
Haddock/Eglefin	19 504	225 197	17 884	281 212	15 405	819	121	0	135	6	244	794	359
Flatfish/Poissons plats	4 607	132 989	3 902	106 221	53	100	328	-	44	3	60	3 154	160
	10 418	337 143	11 817	408 055	656	1 256	239	2 933	494	169	1 629	1 722	2 619
Salted, dried, smoked/Salés, séchés, fumés													
Herring, whole, headed, in brine/Hareng, entier, etêté,	26 305	865 496	25 783	972 670	4 541	2 279	158	5 610	2 993	2 993	418	21	6 769
en saumure	4 473	31 308	3 440	28 337	726	110	0	14	2 352	5	72	0	161

253

DENMARK/DANEMARK

Table II(b) cont'd/Tableau II(b) suite

Quant. : tons/tonnes
Val. : '000 DKr/KrD

EXPORTS BY QUANTITY/EXPORTATIONS EN QUANTITE

	1985		1986		GERMANY/ ALLEMAGNE	FRANCE	U.K./ R.U.	ITALY/ ITALIE	B.L.E.U./ U.E.B.L.	SPAIN/ ESPAGNE	SWEDEN/ SUEDE	U.S./ E.U.	OTHER/ AUTRES
	Quant.	Val.	Quant.	Val.									
Cod, dried and/or salted/ Morue, séchée et/ou salée	9 669	217 194	10 993	289 207	173	1 416	30	2 889	5	1 677	8	-	4 794
Cod fillets, salted/Filets de morue salés	1 591	50 883	913	31 114	-	121	0	620	-	122	34	-	16
Smoked salmon/Saumon fumé	2 849	365 769	3 483	417 414	1 623	345	29	612	200	21	27	12	615
Smoked trout/Truite fumée	548	32 030	467	29 083	101	20	65	15	121	20	26	4	94
Salted fish roe/Rogue de poisson salé	619	19 171	857	27 780	387	0	24	-	0	58	72	-	316
Shellfish (live, fresh, chilled, etc.)/Coquillages (vivants, frais, sur glace, etc.)	73 549	1 324 113	74 963	1 582 986	14 744	12 586	3 309	4 649	16 845	1 254	6 237	79	15 260
Pink shrimp/Crevette rose	33 734	967 496	38 311	1 175 798	396	10 699	3 140	1 754	679	943	6 052	0	14 649
Common and other shrimp/ Crevettes communes et autres	1 942	56 663	2 634	69 210	111	143	65	38	1 934	20	92	4	228
Norway lobster/Langoustine	3 926	242 726	4 016	281 406	71	683	79	2 671	67	258	10	70	107
Canned or prepared/En conserve ou préparés	68 208	1 912 288	73 477	2 282 707	12 932	10 577	17 240	5 261	6 823	565	9 065	682	10 332
Caviar	3	4 421	0	469	0	-	0	-	-	-	0	0	0
Caviar substitutes/Succédanés de caviar	962	65 188	987	75 815	19	292	83	172	25	15	123	44	214
Salmon preparations/Préparations de saumon	176	21 720	227	27 169	139	10	3	14	18	0	3	0	40
Herring preparations/ Préparations de hareng	6 223	56 896	8 960	74 700	3 269	280	200	2	2 821	15	1 648	23	702
Mackerel - various/ Maquereau divers	10 407	184 776	10 546	191 822	2 552	55	2 458	349	1 880	14	2 708	2	528
Fish fillets, breaded portions/ Filets de poisson, portions panées	11 445	291 691	12 615	353 292	946	533	4 422	1 909	321	17	1 836	7	2 623
	2 391	31 103	1 813	27 501	0	36	1 509	0	0	114	37	2	115
Fish roe/Rogue de poisson	7 787	516 325	9 423	754 921	1 543	945	3 009	1 947	114	132	591	45	1 098
Shrimps, not canned/Crevettes non en conserve	6 280	113 046	6 384	117 254	642	2 921	674	81	114	0	730	55	1 166
Mussels, canned/Moules en conserve	5 590	116 933	4 774	102 673	2 872	308	307	13	125	10	287	75	778
Herring, canned/Hareng en conserve	30	1 506	3	634	0	2	-	0	0	0	0	-	0
Crab, canned/Crabe en conserve	1 133	87 517	1 478	116 598	85	413	608	84	107	0	14	0	167
Shrimp, canned/Crevettes, en conserve	36	871	23	520	4	0	-	0	1	-	3	-	15
Tuna, bonito/Thon bonite	2 407	64 303	2 764	78 433	61	1 158	486	35	58	149	41	69	705
Other, canned/Autres, en conserve	10 587	296 305	8 834	273 112	469	1 573	3 140	61	575	98	754	280	1 885
Other, frozen/Autres congelés													
Fish oil/Huile de poisson	61 059	193 107	62 437	101 221	25 108	3 631	6 961	2 066	7 207	-	232	-	17 232
Fish meal/Farine de poisson	217 794	895 400	224 661	815 115	4 735	5 154	20 554	19 565	18 423	0	18 530	160	137 539

DENMARK/DANEMARK

Table III(a)/Tableau III(a)

EXTERNAL TRADE OF MAJOR FISH PRODUCTS AND IMPORT VALUE BY SELECTED COUNTRIES/
ECHANGES INTERNATIONAUX DES PRINCIPAUX PRODUITS DE LA PECHE ET VALEUR DES IMPORTATIONS PAR PAYS SELECTIONNES
1985 & 1986

Quant. : tons/tonnes
Val. : '000 DKr/KrD

	1985		1986		IMPORTS BY VALUE/IMPORTATIONS EN VALEUR							
	Quant.	Val.	Quant.	Val.	B.L.E.U./ U.E.B.L.	GERMANY/ ALLEMAGNE	FAROES/ FEROE	SWEDEN/ SUEDE	NORWAY/ NORVEGE	ICELAND/ ISLANDE	GREENL'D/ GROENL'D	POLAND/ POLOGNE	OTHER/ AUTRES
Fresh, chilled/Frais sur glace (whole/entier)													
Herring/Hareng	168 662	748 367	178 860	952 659	187 745	111 176	77 722	242 207	201 853	1 134	184	60 193	70 443
Cod/Morue	60 098	160 636	68 752	168 770	67	1 050	7 109	77 478	67 514	-	-	2	15 550
Saithe/Lieu noir	42 273	290 881	50 838	418 244	59 772	66 203	60 848	146 129	13 481	865	-	56 407	14 539
Haddock/Eglefin	16 245	68 679	15 615	86 606	529	18 642	4 705	5 108	53 697	-	-	59	3 867
Mackerel/Maquereau	3 914	27 346	7 077	58 425	1 059	6 105	1 064	4 899	43 788	24	-	46	1 390
Plaice/Plie	31 870	73 304	17 071	34 486	1 153	104	1 374	408	1 352	-	-	0	30 094
Other flatfish/Autres poissons	9 940	74 682	13 654	118 782	105 977	9 488	32	110	758	4	-	18	2 395
plats	741	16 027	1 197	23 206	13 790	4 102	1 549	658	1 346	188	119	270	1 184
European hake/Merlu d'Europe	145	1 921	139	1 878	292	323	35	243	843	-	-	0	142
Fresh, chilled/Frais, sur glace (fillets/filets)													
Herring/Hareng	3 134	51 477	4 555	70 662	51 737	555	1 425	5 756	10 294	-	603	-	583
Cod/Morue	1 172	7 335	2 065	10 642	47	430	187	356	9 332	-	-	-	290
Other gadoids/Autres gadidés	398	8 843	328	6 612	138	3	137	5 397	436	-	501	-	290
Flatfish/Poissons plats	84	1 146	85	1 620	1 410	118	2	-	91	-	-	-	1
	1 414	33 170	1 912	48 914	48 914	-	-	-	-	-	-	-	-
Frozen/Congelés (whole/entier)	12 889	135 222	31 022	321 941	9 182	6 699	43 290	4 790	32 368	5 737	43 576	586	173 841
Herring/Hareng	29	200	126	403	-	36	-	87	280	-	-	-	-
Halibut/Flétan	382	10 201	418	14 278	-	18	11 413	-	30	259	1 664	-	895
Greenland halibut/Flétan noir	3 391	46 404	2 977	46 149	-	1 644	12 729	2	2 875	238	28 655	-	5
Cod/Morue	3 212	28 485	19 306	190 262	6 577	514	1 289	2 513	11 141	933	2 886	497	163 911
Mackerel/Maquereau	2 274	9 163	3 390	12 642	215	3 013	24	-	6 934	-	-	-	2 456
Frozen fillets/Filets congelés	20 321	262 932	24 471	405 746	7 016	5 733	205 215	5 676	57 575	4 285	93 033	391	200 561
Herring/Hareng	466	2 594	952	5 003	-	2 771	-	-	869	-	-	-	1 363
Cod/Morue	1 796	39 413	5 191	112 601	370	1 180	19 363	584	37 432	3 101	31 208	-	14 363
Saithe/Lieu noir	12 443	123 253	12 063	161 805	-	297	151 562	23	8 992	-	0	391	540
Haddock/Eglefin	63	1 613	675	13 116	148	117	2 490	-	9 564	45	-	-	753
Flatfish/Poissons plats	2 095	44 367	2 521	67 156	6 109	330	212	69	-	126	60 208	-	102
Salted, dried, smoked/Salés, séchés, fumés													
Herring/Hareng, whole, headed, in brine/Hareng, entier, etêté,	14 458	262 482	12 420	246 908	417	564	168 800	199	49 169	796	22 143	-	4 821
en saumure	439	3 681	208	1 921	-	-	-	-	1 917	0	3	-	-

255

DENMARK/DANEMARK

Table III(a) cont'd/Tableau III(a) suite

Quant. : tons/tonnes
Val. : '000 DKr/KrD

IMPORTS BY VALUE/IMPORTATIONS EN VALEUR

	1985		1986		B.L.E.U./ U.E.B.L.	GERMANY/ ALLEMAGNE	FAROES/ FEROE	SWEDEN/ SUEDE	NORWAY/ NORVEGE	ICELAND/ ISLANDE	GREENL'D/ GROENL'D	POLAND/ POLOGNE	OTHER/ AUTRES
	Quant.	Val.	Quant.	Val.									
Cod, dried and/or salted/ Morue, séchée et/ou salée	8 513	157 773	8 710	177 098	-	-	107 806	7	46 232	17	19 484	-	3 552
Cod fillets, salted/Filets de morue salés	591	18 538	91	2 855	-	-	2 839	-	16	-	-	-	-
Smoked salmon/Saumon fumé	66	6 772	88	8 223	-	183	7 794	15	167	-	9	-	54
Smoked trout/Truite fumée	22	2 124	28	2 147	-	-	1 974	-	172	-	0	-	-
Salted fish roe/Rogue de poisson salé	186	4 115	157	2 783	384	345	-	-	-	779	688	-	587
Shellfish (live, fresh, chilled etc.)/Coquillages (vivants, frais, sur glace, etc.)	41 165	1 027 411	47 121	1 201 736	2 464	13 904	128 653	50 992	3 466	8 044	937 942	1	130 834
Pink shrimp/Crevette rose	38 054	938 177	41 680	1 087 408	0	594	121 577	34	1 371	7 814	931 328	-	1 785
Common and other shrimp/ Crevettes communes et autres	629	14 665	1 036	21 154	1 366	9 720	1 974	-	210	-	0	-	7 884
Norway lobster/Langoustine	965	45 297	1 077	56 284	11	928	3 722	50 905	272	220	-	0	224
Canned or prepared/En conserve ou préparés	17 797	706 793	19 199	991 478	4 112	10 646	10 787	5 211	122 702	159 025	548 060	1	130 834
Caviar	3	2 709	2	1 785	-	-	-	-	-	-	-	-	1 785
Caviar substitutes/Succédanés de caviar	1 222	37 237	1 409	48 440	-	970	-	57	2 468	22 007	495	-	22 443
Salmon preparations/Préparations de saumon	42	1 547	63	1 952	82	19	19	75	3	155	12	-	1 585
Herring preparations/ Préparations de hareng	485	3 777	543	5 182	3	316	-	187	1 532	2 706	3	-	436
Mackerel - various/ Maquereau divers	20	435	10	231	-	228	-	1	1	-	-	-	1
Fish fillets, breaded portions Filets de poisson, portions panées	107	2 751	438	11 028	779	7	525	71	1 226	-	8 404	-	16
Fish roe/Rogue de poisson	507	6 314	35	841	-	-	271	15	-	527	-	-	27
Shrimps, not canned/Crevettes non en conserve	9 547	511 070	11 240	793 910	130	1 787	9 345	425	115 790	130 878	516 728	-	18 829
Mussels, canned/Moules en conserve	5	255	124	2 015	1 870	7	-	-	-	-	-	-	138
Herring, canned/Hareng en conserve	311	6 146	330	7 141	14	5 702	-	1 419	3	-	-	-	4
Crab, canned/Crabe en conserve	107	5 164	157	5 372	12	-	-	3	134	-	-	-	5 224
Shrimp, canned/Crevettes, en conserve	531	27 642	343	24 789	0	334	-	387	1 467	2 753	19 219	-	629
Tuna, bonito/Thon bonite	2 255	53 986	3 365	60 619	-	858	-	2	-	-	-	-	59 759
Other, canned/Autres, en conserve	83	3 491	69	2 798	12	308	39	711	58	-	17	-	1 653
Other, frozen/Autres congelés	823	18 608	161	4 252	134	10	193	42	-	-	2 710	-	1 163
Fish oil/Huile de poisson	2 103	7 390	1 704	4 378	-	950	5	2 068	1 342	12	-	-	-
Fish meal/Farine de poisson	16 805	59 796	20 639	65 628	98	6	5 182	8 608	2 121	48 781	595	-	236

DENMARK/DANEMARK

Table III(b)/Tableau III(b)

EXTERNAL TRADE OF MAJOR FISH PRODUCTS AND IMPORT QUANTITY BY SELECTED COUNTRIES/
ECHANGES INTERNATIONAUX DES PRINCIPAUX PRODUITS DE LA PECHE ET QUANTITE DES IMPORTATIONS PAR PAYS SELECTIONNES
1985 & 1986

Quant. : tons/tonnes
Val. : '000 DKr/KrD

	1985		1986		IMPORTS BY QUANTITY/IMPORTATIONS EN QUANTITE							
	Quant.	Val.	Quant.	Val.	B.L.E.U./ U.E.B.L.	GERMANY/ ALLEMAGNE	FAROES/ FEROE	SWEDEN/ SUEDE	NORWAY/ NORVEGE	ICELAND/ ISLANDE	GREENL'D/ GROENL'D	POLAND/ POLOGNE	OTHER/ AUTRES
Fresh, chilled/Frais sur glace (whole/entier)													
Herring/Hareng	168 662	748 367	178 860	952 659	20 323	15 299	12 166	49 667	46 507	104	7	8 805	25 982
Cod/Morue	60 098	160 636	68 752	168 770	24	440	2 695	30 052	27 878	-	-	2	7 672
Saithe/Lieu noir	42 273	290 881	50 838	418 244	6 522	8 642	7 768	17 162	1 208	83	-	7 724	1 728
Haddock/Eglefin	16 245	18 679	15 615	86 606	69	3 176	678	807	10 232	-	-	13	640
Mackerel/Maquereau	3 914	27 346	7 077	58 425	157	941	130	598	5 047	2	-	9	193
Plaice/Plie	31 870	73 304	17 071	34 486	410	22	713	151	484	-	-	0	15 291
Other flatfish/Autres poissons plats	9 940	74 682	13 654	118 782	12 093	1 169	6	13	79	1	-	4	289
	741	16 027	1 197	23 206	449	148	67	55	106	6	5	321	42
European hake/Merlu d'Europe	145	1 921	139	1 878	25	27	6	19	54	-	-	0	8
Fresh, chilled/Frais, sur glace (fillets/filets)													
Herring/Hareng	3 134	51 477	4 555	70 662	2 061	91	152	425	1 670	-	24	-	264
Cod/Morue	1 172	7 335	2 065	10 642	18	80	89	155	1 590	-	-	-	132
Other gadoids/Autres gadidés	398	8 843	328	6 612	9	0	7	270	23	-	19	-	132
Flatfish/Poissons plats	84	1 146	85	1 620	70	10	0	-	5	-	-	-	0
	1 414	33 170	1 912	48 914	1 912	-	-	-	-	-	-	-	-
Frozen/Congelés (whole/entier)													
Herring/Hareng	12 889	135 222	31 022	321 941	880	1 035	2 452	515	4 490	433	2 799	67	17 959
Halibut/Flétan	29	200	126	403	-	5	-	30	91	-	-	-	-
	382	10 201	418	14 278	-	0	332	2	-	9	47	-	29
Greenland halibut/Flétan noir	3 391	46 404	2 977	46 149	-	137	927	0	188	21	1 704	-	0
Cod/Morue	3 212	28 485	19 306	190 262	635	91	103	269	978	121	262	56	16 790
Mackerel/Maquereau	2 274	9 163	3 390	12 642	72	592	6	-	2 063	-	-	-	656
Frozen fillets/Filets congelés													
Herring/Hareng	20 321	262 932	24 471	405 746	291	734	14 086	290	2 915	197	3 837	33	20 043
Cod/Morue	466	2 694	952	5 003	-	495	-	-	143	-	-	-	314
Saithe/Lieu noir	1 796	39 413	5 191	112 601	13	49	923	284	1 567	145	1 522	-	688
Haddock/Eglefin	12 443	123 253	12 063	161 805	-	21	11 328	1	637	-	0	33	43
Flatfish/Poissons plats	63	1 613	675	13 116	7	18	178	-	445	2	-	-	26
	2 095	44 367	2 521	67 156	257	21	8	5	-	6	2 220	-	4
Salted, dried, smoked/Salés, séchés, fumés													
Herring, whole, headed, in brine/Hareng, entier, etêté,	14 458	262 482	12 420	246 908	11	13	8 174	9	2 233	21	1 694	-	264
en saumure	439	3 681	208	1 921	-	-	-	-	208	0	1	-	-

257

DENMARK/DANEMARK

Table III(b) cont'd/Tableau III(b) suite

Quant.: tons/tonnes
Val.: '000 DKr/KrD

IMPORTS BY QUANTITY/IMPORTATIONS EN QUANTITE

	1985		1986		B.L.E.U./ U.E.B.L.	GERMANY/ ALLEMAGNE	FAROES/ FEROE	SWEDEN/ SUEDE	NORWAY/ NORVEGE	ICELAND/ ISLANDE	GREENL'D/ GROENL'D	POLAND/ POLOGNE	OTHER/ AUTRES
	Quant.	Val.	Quant.	Val.									
Cod, dried and/or salted/ Morue, séchée et/ou salée	8 513	157 773	8 710	177 098	-	-	5 067	0	1 981	1	1 527	-	134
Cod fillets, salted/Filets de morue salés	591	18 538	91	2 855	-	-	89	-	2	-	-	-	-
Smoked salmon/Saumon fumé	66	6 772	88	8 223	-	1	80	0	3	-	0	-	3
Smoked trout/Truite fumée	22	2 124	28	2 147	-	-	27	1	2	-	0	-	-
Salted fish roe/Rogue de poisson salé	186	4 115	157	2 783	10	10	-	-	-	21	23	-	92
Shellfish (live, fresh, chilled etc.)/Coquillages (vivants, frais, sur glace, etc.)	41 165	1 027 411	47 121	1 201 736	181	3 136	6 533	993	217	243	33 727	1	2 090
Pink shrimp/Crevette rose	38 054	938 177	41 680	1 087 408	0	6	6 425	2	74	239	33 636	-	1 297
Common and other shrimp/ Crevettes communes et autres	629	14 665	1 036	21 154	114	794	27	-	3	-	0	-	98
Norway lobster/Langoustine	965	45 297	1 077	56 284	1	17	60	987	6	4	-	0	2
Canned or prepared/En conserve ou préparés	17 797	706 793	19 199	991 478	211	422	162	174	1 966	2 574	8 179	0	5 511
Caviar	3	2 709	2	1 785	-	-	-	-	-	-	-	-	2
Caviar substitutes/Succédanés de caviar	1 222	37 237	1 409	48 440	-	24	-	1	75	596	14	-	699
Salmon preparations/Préparations de saumon	42	1 547	63	1 952	2	0	0	1	0	3	0	-	56
Herring preparations/ Préparations de hareng	485	3 777	543	5 182	0	19	-	22	193	228	0	-	80
Mackerel - various/ Maquereau divers	20	435	10	231	-	10	-	0	0	-	-	-	0
Fish fillets, breaded portions Filets de poisson, portions panées	107	2 751	438	11 028	30	0	22	2	45	-	338	-	0
	507	6 314	35	841	-	-	8	0	-	26	-	-	1
Fish roe/Rogue de poisson													
Shrimps, not canned/Crevettes non en conserve	9 547	511 070	11 240	793 910	3	34	107	4	1 624	1 670	7 491	-	308
Mussels, canned/Moules en conserve	5	255	124	2 015	122	0	-	-	-	-	-	-	2
Herring, canned/Hareng en conserve	311	6 146	330	7 141	1	281	-	48	0	0	-	-	0
	107	5 164	157	5 372	0	-	-	0	3	3	-	-	154
Crab, canned/Crabe en conserve													
Shrimp, canned/Crevettes, en conserve	531	27 642	343	24 789	0	5	-	27	23	50	219	-	18
	2 555	53 986	3 365	60 619	-	38	-	0	-	-	-	-	3 327
Tuna, bonito/Thon bonite													
Other, canned/Autres, en conserve	83	3 491	69	2 798	0	7	1	20	2	-	2	-	36
	823	18 608	161	4 252	4	0	11	2	-	-	105	-	37
Other, frozen/Autres congelés													
Fish oil/Huile de poisson	2 103	7 390	1 704	4 378	-	332	0	1 315	54	3	-	-	-
Fish meal/Farine de poisson	16 805	59 796	20 639	65 628	34	0	1 643	2 843	509	15 420	163	-	26

258

GREENLAND/GROENLAND

Table I(a)/Tableau I(a)

LANDINGS AND VALUES/DEBARQUEMENTS ET VALEURS
1985 & 1986

Quant. : tons, whole fish/tonnes, poisson entier
Val. : '000 DKr/KrD

			1985		1986	
			Quant.	Val.	Quant.	Val.
1. **National landings in domestic ports/Débarquements nationaux dans les ports domestiques**			55 352			
	For direct human consumption/Pour la consommation humaine directe					
	Finfish/Poisson	Atlantic cod/Morue atlantique	10 049		5 211	
		Atlantic redfish/Sébaste atlantique	143		1 166	
		American plaice/Plie américaine	9		-	
		Greenland halibut/Flétan noir	9 135		6 857	
		Atlantic halibut/Flétan atlantique	89		42	
		Greenland cod/Morue du Groenland	6 571		-	
		Roundnose grenadier	34		31	
		Wolffish	1 767		1 097	
		Lumpfish/Lompe	419		-	
		Herring/Hareng	1		-	
		Atlantic salmon/Saumon atlantique	863		-	
		Capelin/Capelan	991		-	
		Dogfish/Aiguillat	11		-	
		Skate/Raie	102		-	
		Char/Omble	131		-	
		Other/Autres	44		-	
	Shellfish/Coquillages	Northern prawn/Crevette nordique	24 015		29 525	
		Clams	973		-	
2. **National landings in foreign ports/Débarquements nationaux dans les ports étrangers**			30 652		33 708	
	For direct human consumption/Pour la consommation humaine directe					
		Atlantic cod/Morue atlantique	2 292		0	
		Northern prawn/Crevette nordique	28 338		33 708	
		Other/Autres	22		0	
3. **Foreign landings in domestic ports/Débarquements étrangers dans les port nationaux**			63 421		66 213	
	For direct human consumption/Pour la consommation humaine directe					
		Atlantic cod/Morue atlantique	113		308	
		Atlantic redfish/Sébaste atlantique	7 170		11 100	
		Other/Autres	188		435	
	For other purposes/Pour d'autres buts					
		Capelin/Capelan	55 950		54 370	

Note: Catches taken by vessels below 80 grt (private) and vessels owned by the the Greenland Homerule/Prises des navires de moins de 80 tjb (privés) et des navires appartenant au "Greenland Homerule".

Catches taken by vessels above 80 grt (private)/Prises des navires de plus de 80 tbj (privés).

Catches taken by vessels chartered by the Greenland Homerule/Prises des navires du "Greenland Homerule".

GREENLAND/GROENLAND

Table I(b)/Tableau I(b)

NATIONAL LANDINGS IN DOMESTIC PORTS/DEBARQUEMENTS NATIONAUX DANS LES PORTS NATIONAUX
1985

Quant.: tons/tonnes
Val.: '000DKr/KrD

Species/Espèces	"Greenland Homerule" enterprises/entreprises		Private entreprises/ Entreprises privées	
	Quant.	Val.	Quant.	Val.
Atlantic cod/Morue atlantique	6 040	20 794		
Greenland cod/Morue du Groënland	4 826	8 813		
Blue whiting/Merlan bleu	0	0		
Roundnose grenadier	27	65		
Atlantic halibut/Flétan atlantique	67	823		
Greenland halibut/Flétan noir	5 681	21 322		
American plaice/Plie américaine	1	2		
Atlantic redfish/Sébaste atlantique	76	138		
Lumpfish/Lompe	226	410		
Wolffish	1 409	3 118		
Atlantic salmon/Saumon atlantique	278	3 233		
Char/Omble	59	442		
Capelin/Capelan	672	1 162		
Herring/Hareng	1	3		
Dogfish/Aiguillat	12	8		
Skate/Raie	61	44		
Groundfish/Poisson de fond	44	11		
Other/Autres	2	3		
Clams	893	3 150		
Crab/Crabe	0	0		
Northern prawn/Crevette nordique	24 459	182 044		
TOTAL	44 834	245 586		

GREENLAND/GROENLAND

Table II/Tableau II

EXTERNAL TRADE IN FISH AND FISH PRODUCTS/ECHANGES INTERNATIONAUX DE POISSON ET PRODUITS DE LA PECHE
1985 & 1986a)

Quant. : tons/tonnes
Val. : '000 DKr/KrD

	IMPORTS/IMPORTATIONS				EXPORTS/EXPORTATIONS			
	1985		1986		1985		1986	
	Quant.	Val.	Quant.	Val.	Quant.	Val.	Quant.	Val.
Total fish and fish products/Total poisson et produits de la pêche	342	7 486	200	5 238	47 256	1 413 106	35 977	1 138 678
Fresh, chilled/Frais, sur glace	0	67	0	39	0	0)		
Frozen/Congelés	81	761	55	1 115	4 293	96 935)		
Frozen fillets/Filets congelés	25	867	14	505	5 160	108 456)	6 698	144 003
Salted, dried, smoked/Salés, séchés, fumés	36	1 447	20	833	3 771	58 408)		
Canned/En conserve	169	3 604	96	2 332	0	0)		
Shellfish (live, fresh, chilled salted, dried and canned/Crustacés et coquillages (vivants, frais, sur glace, salés séchés et en conserve)b)	11	545	15	413	33 762	1 148 853	29 279	994 675
Meal/Farine	0	2	0	0	270	454	0	0
Oil/Huile	0	0	0	0	0	0	0	0
Other/Autres	20	193	0	1	0	0	0	0

a) Jan.-Sept.

b) If not included above/S'ils ne sont pas compris ci-dessus.

GREENLAND/GROENLAND

Table III/Tableau III

EXPORTS BY MAJOR PRODUCTS AND BY COUNTRY/
EXPORTATIONS PAR PRINCIPAUX PRODUITS ET PAR PAYS
1985 & 1986a)

Quant. : tons/tonnes
Val. : '000 DKr/KrD

	1985		1986	
	Quant.	Val.	Quant.	Val.
Total fish and fish products/Total poisson et produits de la pêche	47 256	1 413 106	35 977	1 138 678
Fresh, chilled/Frais, sur glace	0	0)	-	-
Frozen/Congelés				
Denmark/Danemark	4 293	96 935	-	-
United States/Etats-Unis	4 173	95 675	-	-
United Kingdom/Royaume-Uni	105	883	-	-
Canada	10	277	-	-
	5	94	-	-
Frozen fillets/Filets congelés	5 160	108 456	6 698	144 003
Denmark/Danemark	2 765	56 705	-	-
United Kingdom/Royaume-Uni	1 173	23 533	-	-
United States/Etats-Unis	955	22 697	-	-
Germany/Allemagne	140	2 535	-	-
Canada	125	3 976	-	-
Salted, dried, smoked/Salé, séché, fumé	3 771	58 408	-	-
Denmark/Danemark	3 419	54 712	-	-
Portugal	352	3 696	-	-
Shellfish (live, fresh, chilled, salted, dried and canned)/Coquillages (vivants, frais, sur glace, salés, séchés et en conserve)	33 762	1 148 853	29 279	994 675
Denmark/Danemark	33 596	1 140 221	-	-
United States/Etats-Unis	157	7 844	-	-
Canada	9	788	-	-
Meal/Farine	270	454	-	-
Denmark/Danemark	270	454	-	-

a) Jan.-Sept.

FINLAND/FINLANDE

Table I/Tableau I

LANDINGS AND VALUES/DÉBARQUEMENTS ET VALEURS
1985 & 1986a)

Quant.: tons/tonnes (live weight/poids vif)
Val.: Mk/mkF '000

	1985		1986	
	Quant.	Val.	Quant.	Val.
National landings in domestic ports/Débarquements nationaux dans les ports nationaux				
For direct human consumption/Pour la consommation humaine directe	59 838	207 713	53 615	195 600
Finfish/Poisson Baltic herring/Hareng de la Baltique	40 010	83 314	35 000	75 000
Salmon/Saumon	976	28 633	900	25 000
Vendace/Corégone blanc	3 709	24 410	3 700	25 000
White fish/Poisson maigre	1 709	22 580	1 700	23 000
Cod/Morue	6 244	13 836	5 100	12 000
Other/Autres	7 175	31 440	7 200	32 000
Shellfish/Crustacés et coquillages				
Crayfish/Ecrevisses (Astacus astacus L.)	15	3 500	15	3 600
For other purposes/Pour d'autres buts (b) :				
Baltic herring/hareng de la Baltique	58 180	37 817	63 800	41 500
TOTAL	118 018	245 530	117 415	237 100

a) Preliminary estimates/Chiffres provisoires.

b) Reduction to meal and oil, pet food, bait, etc./Réduction en farine et huile, alimentation des animaux domestiques, appâts, etc.

FINLAND/FINLANDE

Table II /Tableau II

EXTERNAL TRADE IN FISH AND FISH PRODUCTS/ECHANGES INTERNATIONAUX DE POISSON ET PRODUITS DE LA PECHE
1985 & 1986

Quant. : tons/tonnes
Val. : Mk/mkF '000

	IMPORTS/IMPORTATIONS				EXPORTS/EXPORTATIONS			
	1985		1986		1985		1986	
	Quant.	Val.	Quant.	Val.	Quant.	Val.	Quant.	Val.
Total fish and fish products/Total poisson et produits de la pêche	350 138	708 634	30 696	684 888	6 717	91 867	2 247	27 585
Fresh, chilled/Frais, sur glace	2 476	8 055	2 347	7 756	4 661	57 630	1 395	8 130
Frozen/Congelés Excluding fillets/Non compris les filets	1 695	13 429	2 782	19 740	1 205	16 187	403	3 875
Frozen fillets/Filets congelés	7 207	84 171	7 782	99 022	354	3 779	51	1 357
Salted, dried, smoked/Salés, séchés, fumés	2 911	21 897	3 252	24 528	59	321	5	174
Canned/En conserve	10 447	119 269	12 141	142 703	123	4 399	143	6 115
Shellfish (live, fresh, chilled, salted, dried, canned)/Coquillages (vivants, frais, sur glace, salés, séchés, en conserve) (a)	1 693	43 034	1 839	69 147	42	1 033	29	998
Sub-total: Human consumption/ Sous-total: Consommation humaine	26 429	289 855	30 143	362 896	6 444	83 349	2 026	20 649
Meal/Farine	89 222	219 307	67 863	159 785	-	-	1	4
Oil/Huile	4 154	12 489	5 716	12 729	-	-	28	63
Other/Autres	230 335	186 983	196 975	149 478	272	8 518	191	6 869

a) If not included above/S'ils ne sont pas compris ci-dessus.

FINLAND/FINLANDE

Table III(a)/Tableau III(a)

IMPORTS BY MAJOR PRODUCTS AND BY COUNTRY/
IMPORTATIONS PAR PRINCIPAUX PRODUITS ET PAR PAYS
1985 & 1986

Quant. : tons/tonnes
Val. : Mk/mkF '000

	PRODUCTS/COUNTRIES/PRODUITS/PAYS	1985 Quant.	1985 Val.	1986 Quant.	1986 Val.
	Total fish and fish products/Total poisson et produits de la pêche	350 138	708 635	300 696	684 888
1.	Fresh, chilled/Frais, sur glace	2 476	8 055	2 347	7 756
	Clupeidae, other than Baltic herring or sprat/autres que hareng de la Baltique ou sprat	1 682	3 844	2 054	5 261
	Flatfish/Poisson plat	16	323	26	610
	Clupeidae, other than Baltic herring (strömming)/ autres que le hareng de la Baltique	179	835	89	462
	Sweden/Suède	2 221	6 188	2 251	6 193
	Denmark/Danemark	190	1 151	70	1 177
	Norway/Norvège	48	608	22	312
2.	Frozen, excl. fillets/Congelés, non compris les filets	1 695	13 429	2 782	19 740
	White fish/Poisson maigre	796	8 200	1 248	11 552
	Saltwater fish, other than clupeidae, flatfish or gadidae/ Poisson de mer salé, autre que clupeidae, poisson plat ou gadidae	297	1 852	383	2 476
	Gadidae	206	1 384	268	1 917
	Canada	806	8 334	1 301	12 099
	Norway/Norvège	449	2 709	560	3 705
	Sweden/Suède	302	777	745	1 887
3.	Frozen fillets/Filets congelés	7 207	84 171	7 782	99 022
	Gadidae	5 739	63 689	5 380	63 467
	Flatfish/Poisson plat	554	7 373	1 064	17 387
	Saltwater fish, other than flatfish, gadidae or clupeidae/ Poisson de mer salé, autre que poisson plat, gadidae ou clupeidae	750	6 142	1 150	9 962
	Norway/Norvège	6 078	71 944	4 566	60 987
	Germany/Allemagne	313	2 267	1 215	11 382
	Iceland/Islande	256	2 965	860	10 851
6.	Shellfish (live, fresh, chilled, salted, dried and canned)/ Coquillages (vivants, frais, sur glace, salés, séchés et en conserve)	1 693	43 034	1 839	69 147
	Shrimps, not in airtight containers/Crevettes, non conservées en récipients étanches	943	27 296	1 163	49 684
	Shrimps/Crevettes	227	4 830	178	5 759
	Shrimps, in airtight containers/Crevettes, en récipients étanches	70	2 658	67	3 452
	Norway/Norvège	1 103	29 518	1 186	49 410
	Sweden/Suède	81	3 027	125	4 891
	Denmark/Danemark	214	2 935	234	4 155
7.	Meal/Farine	89 220	219 307	67 863	159 785
	Denmark/Danemark	21 368	56 216	32 067	84 900
	Iceland/Islande	40 772	91 059	30 670	61 982
	Norway/Norvège	25 874	69 410	5 098	12 747
8.	Oil/Huile	4 154	12 489	5 716	12 729
	Iceland/Islande	148	661	3 016	7 896
	Norway/Norvège	4 006	11 828	1 205	3 165
	Denmark/Danemark	-	-	1 476	1 578
4.5.9.	Other/Autres	243 693	328 149	212 368	316 708
	Fish waste/Déchets	191 131	157 739	171 985	130 453
	Frozen fish other than herring, not in airtight containers/ Poisson congelé, autre que le hareng, non conservé en récipients étanches	3 032	34 087	3 000	38 720
	Tuna in airtight containers/Thon en récipients étanches	1 894	26 466	2 578	29 836
	Norway/Norvège	77 745	97 218	71 011	85 847
	Sweden/Suède	21 689	59 497	22 806	63 153
	Denmark/Danemark	33 472	36 356	23 069	30 122

FINLAND/FINLANDE

Table III(b) /Tableau III(b)

EXPORTS BY MAJOR PRODUCTS AND BY COUNTRY/
EXPORTATIONS PAR PRINCIPAUX PRODUITS ET PAR PAYS
1985 & 1986

Quant. : tons/tonnes
Val. : Mk/mkF '000

	PRODUCTS/COUNTRIES/PRODUITS/PAYS	1985 Quant.	1985 Val.	1986 Quant.	1986 Val.
	Total fish and fish products/Total poisson et produits de la pêche	6 717	91 867	2 247	27 585
1.	Fresh, chilled/Frais, sur glace	4 661	57 630	1 395	8 130
	Gadidae	2 844	10 692	1 262	5 246
	Salmonidae other than salmon or white fish/autres que le saumon ou le poisson maigre	1 683	44 484	83	2 379
	Denmark/Danemark	3 698	36 856	1 077	4 589
	United States/Etats-Unis	63	2 123	47	1 468
2.	Frozen, excl. fillets/Congelés, non compris les filets	1 205	16 187	403	3 875
	Salmonidae other than salmon or white fish/autres que le saumon ou le poisson maigre	467	11 713	90	2 316
	Gadidae	607	2 126	268	1 156
	Denmark/Danemark	272	3 802	268	2 299
	Sweden/Suède	272	6 196	109	1 399
3.	Frozen fillets/Filets congelés	354	3 779	51	1 357
	Salmonidae other than salmon or white fish/autres que le saumon ou le poisson maigre	30	1 203	15	606
	Freshwater fish other than salmon, white fish or other salmonidae/Poisson d'eau douce, autres que le saumon, le poisson maigre ou autres salmonidae	4	114	8	348
	Gadidae	132	830	24	289
	Sweden/Suède	54	1 464	16	601
	Switzerland/Suisse	5	150	9	365
	Denmark/Danemark	123	977	24	289
6.	Shellfish (live, fresh, chilled, salted, dried and canned)/ Coquillages (vivants, frais, sur glace, salés, séchés et en conserve)	42	1 033	29	998
	Freshwater crayfish/Ecrevisse d'eau douce	24	458	23	773
	Shrimps/Crevettes	17	562	5	818
	Sweden/Suède	42	1 020	19	793
	Norway/Norvège	0	12	9	187
7.	Meal/Farine	-	-	1	4
	Sweden/Suède	-	-	1	4
8.	Oil/Huile	-	-	28	63
	Norway/Norvège	-	-	28	63
4.5.9.	Other/Autres	454	13 238	339	13 158
	Prepared fish roe other than genuine caviar/Rogue preparée, autre que le caviar	59	2 658	63	4 359
	Salted, in brine or dried roes other than sturgeon or gadidae, in barrels/Rogues, salées, en saumure ou sechées, autre que l'esturgeon ou gadidae en conteneurs	92	3 002	90	3 369
	Fresh, chilled or frozen roes of fish other than sturgeon or gadidae/Rogue, fraiche, sur glace ou congelée, autre que l'esturgeon ou gadidae	175	5 463	76	3 227
	Japan/Japon	241	6 678	183	7 371
	Sweden/Suède	46	3 125	45	3 215
	United States/Etats-Unis	41	1 241	51	1 276

FRANCE

Table I/Tableau I

MAIN IMPORTED PRODUCTS BY COUNTRY OF ORIGIN/
PRINCIPAUX PRODUITS IMPORTES PAR PAYS DE PROVENANCE
1986

Quant. : tons/tonnes
Val. : '000 FF

	1986			
	Quant.	%	Val.	%
Total imports/Total des importations	551 233.5	100.00	9 303 048.0	100.00
Total for the following species/ Total des espèces ci-dessous	387 599.7	70.31	6 132 549.5	65.92
Total salmon/saumon	41 680.4	7.56	1 345 150.4	14.46
United States/Etats-Unis	12 058.0	28.93	322 232.0	23.96
Norway/Norvège	10 972.8	26.33	462 263.0	34.37
Canada	6 908.4	16.57	190 146.0	14.14
Total cod/morue	40 022.5	7.26	833 386.5	8.96
Denmark/Danemark	15 241.4	38.08	289 331.0	34.72
Norway/Norvège	7 292.9	18.22	182 976.0	21.96
Netherlands/Pays-Bas	4 226.2	10.56	74 666.0	8.96
Germany/Allemagne	2 967.0	7.41	65 895.0	7.91
Total tuna/thon	57 718.0	10.47	789 961.0	8.49
Senegal	18 810.8	32.59	306 352.0	38.78
Ivory Coast/Côte d'Ivoire	17 441.9	30.22	278 249.0	35.22
Total other fish/autres poissons	42 281.8	7.67	903 793.0	9.72
Denmark/Danemark	5 212.4	9.37	123 828.0	7.66
Germany/Allemagne	5 110.4	11.97	87 635.0	7.27
United Kingdom/Royaume-Uni	5 061.7	12.33	65 728.0	13.70
Netherlands/Pays-Bas	4 845.9	12.09	106 299.0	9.70
Norway/Norvège	3 960.9	11.46	69 200.0	11.76
Total shrimp/crevettes	35 255.4	6.40	1 376 775.0	14.80
Greenland/Groënland	7 296.6	20.70	170 274.0	12.37
Senegal	4 977.9	14.12	272 765.0	19.81
Netherlands/Pays-Bas	3 629.5	10.29	74 539.0	5.41
Faroe Islands/Iles Féroé	2 875.7	8.16	56 829.0	4.13
Denmark/Danemark	1 740.6	4.94	24 497.0	1.78
Total sole	4 660.6	0.85	218 116.1	2.34
Netherlands/Pays-Bas	2 080.4	44.64	112 497.0	51.58
United Kingdom/Royaume-Uni	1 282.9	27.53	61 524.0	28.21
Senegal	370.6	7.95	6 823.0	3.13
Total mussels/moules	40 469.3	7.34	154 763.0	1.66
Netherlands/Pays-Bas	19 640.4	48.53	86 742.0	56.05
Spain/Espagne	9 519.0	23.52	35 117.0	22.69
Ireland/Irlande	5 096.4	12.59	10 691.0	6.91
Total mackerel/maquereau	26 936.7	4.89	102 359.5	1.10
United Kingdom/Royaume-Uni	12 786.4	47.47	38 746.0	37.85
Ireland/Irlande	8 289.4	30.77	27 729.0	27.09
Total sardines	31 510.3	5.72	227 942.0	2.45
Italy/Italie	17 398.6	55.22	88 170.0	38.68
Morocco/Maroc	11 347.6	36.01	113 858.0	49.95
Portugal	1 836.4	5.83	20 497.0	8.99
Total meal/farine	67 864.7	12.17	180 382.0	1.94
Chile/Chili	34 444.4	51.36	87 190.0	48.36
Iceland/Islande	16 573.3	24.71	46 357.0	25.71
Denmark/Danemark	6 641.6	9.90	20 433.0	11.33
Norway/Norvège	3 929.2	5.86	11 600.0	6.43

FRANCE

Table II/Tableau II

EXTERNAL TRADE/COMMERCE EXTERIEUR, 1986

Quant. : tons/tonnes
Val. : '000 FF

	IMPORTS/IMPORTATIONS		EXPORTS/EXPORTATIONS		% of exports over imports/ Taux de couverture	
	Quant.	Val.	Quant.	Val.	Quant.	Val.
TOTAL	551 234	9 303 048	204 858	3 044 000	37.16	32.72
Fresh, chilled/Frais, sur glace	96 705	1 641 805	52 747	1 208 537	54.54	73.61
Frozen/Congelés	151 171	2 078 005	97 241	612 464	64.32	29.47
Frozen fillets/Filets congelés	56 538	874 445	6 438	104 951	11.39	12.00
Cured/Salés, séchés, fumés	17 140	470 153	4 033	157 119	23.53	33.42
Canned/En conserve	73 642	2 061 316	8 295	221 033	11.26	10.72
Shellfish and crustaceans/ Crustacés et coquillages	108 228	2 709 179	22 977	736 276	21.23	27.18
Meal/Farine	67 065	180 302	8 206	25 508	12.24	14.15
Oil/Huile	13 257	29 886	8 458	26 046	63.80	87.15

FRANCE

Table III/Tableau III

PRODUCTION, 1985 & 1986

Quant. : tons/tonnes
Val. : '000 FF

	1985		1986		1986/1985	
	Quant.	Val.	Quant.	Val.	Quant.	Val.
I. FRESH/FRAIS						
A. FISH/POISSON	371 360	3 764 124	353 238	3 917 172	95	104
Diadromous fish/Poissons diadromes	2 425	81 407	1 954	82 210	81	101
Eel/Anguille	2 131	73 007	1 405	34 018	66	47
Sea fish/Poisson de mer	368 935	3 682 717	351 284	3 834 962	95	104
Cod/Morue	19 294	221 049	23 681	247 235	123	112
Haddock/Eglefin	10 932	77 364	9 683	75 240	89	97
Pollack/Lieu jaune	5 633	74 523	6 360	87 287	113	117
Saithe/Lieu noir	55 826	289 813	57 233	328 800	103	113
Ling/Lingue	12 090	100 068	10 363	96 907	86	97
Blue ling/Lingue bleue	11 159	73 292	12 389	102 513	111	140
Whiting/Merlan	28 247	186 779	24 238	192 197	86	103
Hake/Merlu	19 857	417 792	20 203	463 608	102	111
Norway pout/Tacaud norvégien	6 190	22 531	6 623	26 029	107	116
Anglerfish/Baudroie	19 487	368 025	17 252	391 068	89	106
Red gurnard/Grondin rouge	4 342	37 398	3 935	32 753	91	88
Horse mackerel/Chinchard	6 909	15 495	5 520	16 126	80	104
Dogfish/Aiguillat	9 769	53 141	8 706	59 094	89	111
Brown cat shark/Roussette	5 879	22 685	5 466	24 618	93	109
Black sea bream/Griset	1 866	34 282	2 161	38 731	116	113
Pink sea bream/Dorade rose	615	21 133	417	15 947	68	75
Redfish/Sébaste	4 020	27 551	4 800	39 447	119	143
Megrim/Cardine	6 501	87 975	5 370	85 084	83	97
Plaice/Plie	5 860	29 003	6 261	33 097	107	114
Sole	6 090	254 685	5 701	275 009	94	108
Thornback ray/Raie bouclée	2 191	23 299	2 985	35 120	136	151
Anchovy/Anchois	4 528	33 552	4 033	30 773	89	92
Herring/Hareng	14 859	34 023	8 023	19 863	54	58
Sprat	1 798	2 983	701	1 349	39	45
Sardine	26 693	75 127	27 045	78 641	101	105
Albacore/Germon	1 574	33 465	1 111	25 162	71	75
Mackerel/Maquereau	17 124	71 635	14 796	61 269	86	86
Tropical tunas/Thons tropicaux						
Bluefin tuna/Thon rouge	3 202	48 209	4 008	50 818	125	105
B. CRUSTACEANS/CRUSTACES	24 977	654 658	24 269	713 648	97	109
Spinous spider crab/Araignée de mer	3 208	40 190	2 609	39 202	81	98
Edible crab/Tourteau	8 923	112 427	7 828	114 319	88	102
Norway lobster/Langoustine	8 394	279 946	8 806	301 512	105	108
Common shrimp/Crevette grise	762	25 325	1 361	34 008	179	134
C. MOLLUSCS/MOLLUSQUES	39 208	445 683	38 042	478 414	97	107
Scallops/Coquilles St. Jacques	9 898	147 104	8 088	144 406	82	98
Cockle/Coque	5 194	20 153	6 160	25 526	119	127
Quahaug/Praire	2 086	31 712	1 769	36 052	85	114
Squid/Encornet-calmar	3 284	69 077	882	9 794	27	14
Cuttlefish/Seiche	9 138	90 245	6 830	73 705	75	82
D. FARMED MOLLUSCS/MOLLUSQUES D'ELEVAGE	128 941	1 189 404	147 881	1 656 146	115	139
Common oysters/Huîtres plates	1 602	45 147	1 569	51 720	98	115
Oysters/Huîtres creuses	77 360	867 199	106 159	1 363 545	137	157
Mussels/Moules	49 979	277 058	40 154	240 880	80	87
E. ECHINODERMS/ECHINODERMES	475	6 262	307	3 711	65	59
F. AQUATIC PLANTS/PLANTES AQUATIQUES	60 676	14 727	68 778	14 863	113	101
TOTAL A + B + C + E	436 020	4 870 727	415 855	5 112 945	95	105
TOTAL A + B + C + D + E	564 961	6 060 131	563 736	6 769 091	100	112
TOTAL I = A + B + C + D + E + F	625 637	6 074 858	632 512	6 783 954	101	112

FRANCE

Table III(cont'd)/Tableau III(suite)

PRODUCTION, 1985 & 1986

Quant. : tons/tonnes
Val. : '000 FF

			1985		1986		1986/1985	
			Quant.	Val.	Quant.	Val.	Quant.	Val.
II.	FROZEN/CONGELES							
	A. FISH/POISSON		93 129	731 600	119 665	705 192	128	96
	Cod/Morue		10 815	158 998	10 952	211 945	101	133
	Blue whiting/Merlan bleu		870	5 585				
	Yellowfin tuna/Albacore		42 632					
	Skipjack/Listao		44 325	515 000	106 528	469 581	240	91
	Bigeye tuna/Patudo		1 426					
	Kerguelen fish/Poisson des Kerguelen		302	2 117	1 492	9 571	494	452
	Horse mackerel/Chinchard		1 502	7 266				
	Albacore/Germon		560					
	Haddock/Eglefin		949	16 904	693	14 095	73	83
	B. CRUSTACEANS/CRUSTACES		1 432	78 984	1 722	92 970	120	118
	Shrimp/Crevette		1 068	41 011	1 371	57 643	128	141
	Crawfish/Langouste rose		364	37 973	351	35 327	96	93
	TOTAL II = A + B		94 561	810 584	121 387	798 162	128	99
III.	SALTED PRODUCTS/PRODUITS SALES							
	Cod/Morue		223	3 353	308	5 507	101	164
	TOTAL III		223	3 353	308	5 507	138	164
IV.	INDUSTRIAL PRODUCTS/PRODUITS INDUSTRIELS							
	Meal/Farine		1 690	4 600	1 757	4 793	104	104
	TOTAL IV		1 690	4 600	1 757	4 793	104	104
	FISH/POISSONS A	I	371 360	3 764 124	353 238	3 917 172	95	104
		II	93 129	731 600	119 665	705 192	128	96
		III	223	3 353	308	5 507	138	164
	CRUSTACEANS/CRUSTACES B	I	24 977	654 658	24 269	713 648	97	109
		II	1 432	78 984	1 722	92 970	120	118
	MOLLUSCS/MOLLUSQUES C	I	39 208	445 683	38 042	478 414	97	107
	MOLLUSCS/MOLLUSQUES D	I	128 941	1 189 404	147 881	1 656 146	115	139
	ECHINODERMS/ECHINODERMES E	I	475	6 262	307	3 711	65	59
	INDUSTRIAL PRODUCTS/PRODUITS INDUSTRIELS F	I	60 676	14 727	68 776	14 863	113	101
		IV	1 690	4 600	1 757	4 793	104	104
	GRAND TOTAL/TOTAL GENERAL		722 111	6 893 395	755 964	7 592 416	105	110

GERMANY/ALLEMAGNE

Table I/Tableau I

LANDINGS AND VALUES/DEBARQUEMENTS ET VALEURS
1985 & 1986

Quant. : '000 t (catch weight/poids des prises)
Val. : Million DM

	1985		1986	
	Quant.	Val.	Quant.	Val.
1. National landings in domestic ports/Débarquements nationaux dans les ports nationaux				
For direct human consumption/Pour la consommation humaine directe	139.8	215.6	106.3	184.4
Fresh/Frais	65.7	101.8	61.2	101.0
Cod/Morue	23.6	38.9	20.0	37.5
Haddock/Eglefin	0.8	1.3	1.0	1.6
Saithe/Lieu noir	16.8	17.9	17.5	22.3
Redfish/Sébaste	11.9	26.1	10.6	23.7
Plaice/Plie	0.8	1.7	0.7	1.4
Herring/Hareng	7.6	3.8	7.9	4.2
Other/Autres	4.2	12.1	3.5	10.3
Frozen (all species)/Congelées (toutes espèces)	74.1	113.8	45.1	83.4
For other purposes/Pour d'autres buts	8.5	9.4	7.8	4.9
Fodder shrimps/Crevettes pour l'alimentation des animaux	2.8	0.3	3.3	0.4
Fish meal/Farine de poisson	9.2a)	7.3	5.7a)	3.9
Fish oil/Huile de poisson	2.0a)	1.6	0.7a)	0.3
Other/Autres (confiscated, withdrawn/retirés)	4.2	0.2	3.5	0.3
Shellfish (crustaceans and molluscs)/Coquillages (crustacés et mollusques)	38.5	49.2	45.6	51.3
Shrimps/Crevettes	14.5	37.1	13.2	35.4
Mussels/Moules	23.1	7.8	31.7	11.0
Other/Autres	0.9	4.3	0.7	4.9
TOTAL 1	186.8	274.2	159.7	240.6
2. National landings in foreign ports/Débarquements nationaux dans les ports étrangers				
For direct human consumption/Pour la consommation humaine directe	14.5b)	23.2b)	16.2b)	24.7b)
For other purposes/Pour d'autres buts	-	-	-	-
TOTAL 2	14.5b)	23.2b)	16.2b)	24.7b)
3. Foreign landings in domestic ports/Débarquements étrangers dans les ports nationaux				
For direct human consumption/Pour la consommation humaine directe	42.2	76.7	45.1	87.7
For other purposes/Pour d'autres buts	-	-	-	-
TOTAL 3	42.2	76.7	45.1	87.7
TOTAL 1 + 3	229.0	350.9	204.8	328.3

a) Mainly included in frozen, except 1 540 t (1985) and 1 000 t (1986)/Compris principalement dans les produits congelés, sauf 1 540 tonnes (1985) et 1 000 t (1986).

b) Provisional/Provisoires.

GERMANY/ALLEMAGNE

Table II/Tableau II

EXTERNAL TRADE IN FISH AND FISH PRODUCTS/ECHANGES INTERNATIONAUX DE POISSON ET PRODUITS DE LA PECHE
1985 & 1986

Quant. : '000 t (product weight/poids du produit)
Val. : Million DM

	IMPORTS/IMPORTATIONS				EXPORTS/EXPORTATIONS			
	1985		1986		1985		1986	
	Quant.	Val.	Quant.	Val.	Quant.	Val.	Quant.	Val.
Total fish and fish products/Total poisson et produits de la pêche	1 042	2 330.1	1 061	2 354.1	322	766.9	307	721.4
Fresh, chilled/Frais, sur glace	172	561.0	176	591.1	11	49.7	14	55.9
Frozen/Congelés	138	447.7	163	552.2	44	190.8	37	169.5
Frozen fillets/Filets congelés	(71)	(252.4)	(94)	(354.7)	(27)	(137.3)	(23)	(121.8)
Salted, dried, smoked/Salés, séchés, fumés	25	137.1	24	152.4	2	13.6	1	16.0
Canned/En conserve	53	312.2	63	329.6	35	227.9	39	253.4
Shellfish (live, fresh, chilled, salted, dried and canned)/Coquillages (vivants, frais, sur glace, salés, séchés et en conserve)a)	39	293.5	50	325.6	17	76.9	16	64.7
Meal (until Nov.)/Farine (jusqu'à nov.)	365	294.0	415	275.5	211	197.5	198	153.0
Oil (until Nov.)/Huile (jusqu'à nov.)	249	257.9	170	102.4	2	1.9	2	1.4
Other/Autres	1	26.7	0	25.3	0	8.6	0	7.5

a) If not included above/S'ils ne sont pas compris ci-dessus.

GERMANY/ALLEMAGNE

Table III(a)/Tableau III(a)

IMPORTS BY MAJOR PRODUCTS AND BY COUNTRY/IMPORTATIONS PAR PRINCIPAUX PRODUITS ET PAR PAYS
1985 & 1986

Quant. : '000 t
Val. : Million DM

	1985		1986	
	Quant.	Val.	Quant.	Val.
Total fish and fish products/Total poisson et produits de la pêche	1 042	2 330.1	1 061	2 354.1
Fresh, chilled/Frais, sur glace	172	561.0	176	591.1
Herring/Hareng	67	92.4	66	83.1
Redfish/Sébaste	19	45.6	22	51.5
Saithe/Lieu noir	18	29.2	21	39.6
Denmark/Danemark	52	74.1	46	60.2
Iceland/Islande	15	35.0	16	38.1
France	8	13.0	6	9.3
Frozen, excl. fillets/Congelés, non compris les filets	67	195.3	69	197.5
Herring/Hareng	25	28.5	29	31.0
Mackerel/Maquereau	14	16.5	12	11.9
Halibut/Flétan	5	15.0	5	18.4
Netherlands/Pays-Bas	11	8.1	10	7.4
Netherlands/Pays-Bas	5	5.2	4	4.2
Norway/Norvège	2	5.2	2	8.0
Frozen fillets/Filets congelés	71	252.4	94	354.7
Saithe/Lieu noir	39	124.3	35	136.2
Cod/Morue	7	40.6	11	72.4
Redfish/Sébaste	4	21.2	4	23.3
Norway/Norvège	18	53.8	13	46.5
Denmark/Danemark	3	15.2	6	40.3
Faroe Islands/Iles Féroé	2	12.5	2	11.2
Salted, dried, smoked/Salés, séchés, fumés	25	137.1	24	152.4
Salted herring/Hareng salé	13	32.6	12	29.1
Other fish fillets, dried, salted/ Autres filets séchés, salés	4	19.1	4	21.3
Mackerel, smoked/Maquereau, fumé	3	9.1	2	7.2
Netherlands/Pays-Bas	10	26.5	9	23.8
Iceland/Islande	1	6.1	3	12.0
Netherlands/Pays-Bas	2	8.3	2	6.6
Fish canned/Poisson en conserve	53	312.2	63	329.6
Bonito/Bonite	14	78.7	21	91.5
Herring/Hareng	16	61.2	17	61.0
Sardines	7	44.4	9	51.3
Thailand/Thaïlande	10	53.7	15	62.1
Netherlands/Pays-Bas	6	26.2	7	29.9
Portugal	3	19.9	5	28.7

GERMANY/ALLEMAGNE

Table III(a) (cont'd)/Tableau III(a) (suite)

IMPORTS BY MAJOR PRODUCTS AND BY COUNTRY/IMPORTATIONS PAR PRINCIPAUX PRODUITS ET PAR PAYS
1985 & 1986

Quant. : '000 t
Val. : Million DM

		1985		1986	
		Quant.	Val.	Quant.	Val.
Shellfish (live, fresh, chilled, salted, dried and canned)/ Coquillages (vivants, frais, sur glace, salés, séchés et en conserve)		39	293.5	50	325.6
	Common mussels/Moules communes	14	6.1	22	10.3
	Molluscs, canned/Mollusques en conserve	7	45.5	7	41.8
	Other crustaceans, canned/Autres crustacés en conserve	7	95.6	6	98.0
	Denmark/Danemark	12	2.2	20	5.4
	Spain/Espagne	2	12.3	2	13.5
	Iceland/Islande	2	22.8	2	27.3
Meal/Farine		365	294.0	415	275.5
	Chile/Chili	168	136.2	203	135.8
	Peru/Pérou	127	96.2	184	120.0
Oil/Huile		249	257.9	170	102.4
	Japan/Japon	82	70.5	44	27.4
	Peru/Pérou	27	24.2	35	18.4
Other/Autres		1	26.7	0	25.3
	Fish liver, fish roe, etc./Foie, rogue de poisson, etc. Toyfish	1	26.7	0	25.3
	Denmark/Danemark	0	1.2	0	1.1
	Singapore/Singapour	0	9.7	0	8.8

GERMANY/ALLEMAGNE

Table III(b)/Tableau III(b)

EXPORTS BY MAJOR PRODUCTS AND BY COUNTRY/EXPORTATIONS PAR PRINCIPAUX PRODUITS ET PAR PAYS
1985 & 1986

Quant. : '000 t
Val. : Million DM

	1985		1986	
	Quant.	Val.	Quant.	Val.
Total fish and fish products/Total poisson et produits de la pêche	322	766.9	307	721.4
Fresh, chilled/Frais, sur glace	11	49.7	14	55.9
Cod/Morue	4	9.2	5	11.5
Cod fillets/Filets de morue	2	15.1	2	14.2
Saithe/Lieu noir	1	1.7	2	2.9
Denmark/Danemark	2	3.4	3	6.8
France	2	10.9	1	10.2
Netherlands/Pays-Bas	0	0.6	2	2.7
Frozen, excl. fillets/Congelés, non compris les filets	17	53.5	14	47.7
Mackerel/Maquereau	8	16.5	8	11.3
Shark/Requin	1	13.0	2	20.0
Redfish/Sébaste	2	7.5	0	1.7
Czechoslavakia/Tchécoslovaquie	7	11.7	4	5.9
Italy/Italie	1	11.7	2	18.6
Japan/Japon	2	5.4	0	1.1
Frozen fillets/Filets congelés	27	137.3	23	121.8
Cod/Morue	12	73.0	8	57.6
Saithe/Lieu noir	4	16.5	4	17.7
Hake/Merlu	1	5.6	2	6.9
United Kingdom/Royaume-Uni	4	24.8	3	19.8
France	2	6.5	2	7.9
Austria/Autriche	1	4.5	1	3.6
Salted, dried, smoked/Salés, séchés, fumés	2	13.6	1	16.0
Other fish, salted, dried/Autres poissons salés, séchés	1	4.4	1	3.6
Fish livers and roes, dried, in brine or smoked/Foie et rogue de poisson, séchés, en saumûre ou fumés	0	1.1	0	2.5
Other fish, smoked, prepared/Autres poissons, fumés, préparés	0	1.1	0	1.1
Angola	0	0.9	0	0.8
Spain/Espagne	0	0.7	0	1.3
Austria/Autriche	0	0.8	0	0.6
Fish canned/Poisson en conserve	35	227.9	39	253.4
Herring/Hareng	7	35.6	7	37.1
Other fish fillets/Autres filets de poisson	9	47.4	7	36.5
Mackerel/Maquereau	1	4.8	1	6.0
Austria/Autriche	2	9.9	2	10.1
France	2	14.0	2	13.0
Austria/Autriche	0	1.3	0	1.5

GERMANY/ALLEMAGNE

Table III(b) (cont'd)/Tableau III(b) (suite)

EXPORTS BY MAJOR PRODUCTS AND BY COUNTRY/EXPORTATIONS PAR PRINCIPAUX PRODUITS ET PAR PAYS
1985 & 1986

Quant. : '000 t
Val. : Million DM

		1985		1986	
		Quant.	Val.	Quant.	Val.
Shellfish (live, fresh, chilled, salted, dried and canned)/ Coquillages (vivants, frais, sur glace, salés, séchés et en conserve)		17	76.9	16	64.7
	Common mussel/Moule commune	6	1.6	7	2.2
	Shrimp of genus Crangon/ Crevettes du type Crangon	4	13.7	4	10.9
	Other molluscs, prepared/Autres mollusques préparés	3	7.8	3	8.2
	Netherlands/Pays-Bas	6	1.4	6	2.0
	Netherlands/Pays-Bas	4	13.7	3	10.8
	Netherlands/Pays-Bas	2	6.1	2	6.9
Meal, of which fish meal (until Nov.)/Farine, dont farine de poisson (jusqu'à nov.)		211	197.5	198	153.0
	Hungary/Hongrie	38	37.8	47	36.2
	Czechoslovakia/Tchécoslovaquie	44	39.6	38	28.2
Oil/Huile		2	1.9	2	1.4
	Netherlands/Pays-Bas	1	0.8	0	0.3
	Switzerland/Suisse	0	0.6	0	0.6
Other/Autres		0	8.6	0	7.5
	Fish roe etc./Rogue de poisson etc.	0	2.2	0	6.9
	Toy fish	0	6.4	0	0.5
	Denmark/Danemark	0	0.4	0	0.4
	Austria/Autriche	0	3.2	0	0.4

GREECE/GRECE

Table I/Tableau I

LANDINGS AND VALUES/DÉBARQUEMENTS ET VALEURS
1985 & 1986

Quant. : '000tons/tonnes
Val. : Dr million

	1985		1986	
	Quant.	Val.	Quant.	Val.
1. National landings in domestic ports/Débarquements nationaux dans les ports nationaux	130.5	41 804	138.1	50 345
For direct human consumption/Pour la consommation directe Seafish/Poisson de mer	105.0	34 125	112.0	40 544
Finfish/Poisson Frozen fish and crustaceans/ Poisson et crustacés congelés	14.0	4 340	14.5	5 510
Freshwater fish/Poisson d'eau douce	6.5	1 975	6.6	2 553
Shellfish/Coquillages Oysters, mussels/Huîtres, moules	2.5	325	2.2	374
For other purposes/Pour d'autres buts Sponges/Eponges	0.1	264	0.1	300
TOTAL 1	128.0	41 029	135.3	49 281
2. National landings in foreign ports/Débarquements nationaux dans les ports étrangers For direct human consumption/Pour la consommation directe Frozen fish and crustaceans/ Poisson et crustacés congelés	2.5	775	2.8	1 064
TOTAL 2	2.5	775	2.8	1 064

GREECE/GRECE

Table II/Tableau II

EXTERNAL TRADE IN FISH AND FISH PRODUCTS/ECHANGES INTERNATIONAUX DE POISSON ET PRODUITS DE LA PECHE
1985 & 1986

Quant. : tons/tonnes
Val. : Dr '000

	IMPORTS/IMPORTATIONS				EXPORTS/EXPORTATIONS			
	1985		1986		1985		1986	
	Quant.	Val.	Quant.	Val.	Quant.	Val.	Quant.	Val.
Total fish and fish products/Total poisson et produits de la pêche	45 310.0	10 156 144.0	60 276.0	19 177 571.0	11 802.4	4 556 088.0	13 760.0	6 232 145.0
Fresh, chilled/Frais, sur glace	4 569.3	773 518.0	3 684.0	878 109.0	1 151.9	531 313.0	1 353.0	789 325.0
Frozen/Congelés	4 945.9	749 133.0	16 429.0	2 852 976.0	486.2	199 785.0	695.0	310 685.0
Frozen fillets/Filets congelés	1 061.3	259 868.0	1 643.0	460 540.0	1.3	379.0	3.0	1 260.0
Salted, dried, smoked/Salés, séchés, fumés	9 475.1	2 723 994.0	7 925.0	2 997 022.0	2 933.0	593 394.0	4 424.0	1 116 205.0
Canned/En conserve	5 299.4	1 543 908.0	4 887.0	2 064 690.0	856.5	435 949.0	1 013.0	648 461.0
Shellfish (live, fresh, chilled salted, dried, canned)/Coquillages (vivants, frais, sur glace, salés, séchés, en conserve)	13 469.2	2 939 289.7	23 783.0	5 156 414.0	5 805.6	2 769 869.0	6 245.0	3 890 936.0
Meal/Farine	-	-	-	-	-	-	-	-
Oil/Huile	-	-	-	-	-	-	-	-
Other/Autres	6 489.8	1 171 433.3	1 925.0	767 870.0	174.9	25 399.0	27.0	475 272.0

GREECE/GRECE

Table III/Tableau III

IMPORTS BY MAJOR PRODUCTS AND BY COUNTRY/IMPORTATIONS PAR PRINCIPAUX PRODUITS ET PAR PAYS
1986

Quant. : tons/tonnes
Val. : Dr '000

	1986	
	Quant.	Val.
TOTAL FISH AND FISH PRODUCTS/TOTAL POISSON ET PRODUITS DE LA PECHE	60 275	15 177 571
I. Fresh, chilled/Frais, sur glace	3 684	878 109
Freshwater fish/Poisson d'eau douce	567	106 140
Seawater fish/Poisson de mer	654	118 399
Other/Autres	2 463	653 570
Turkey, Yugoslavia/Turquie, Yougoslavie		
France		
Argentina/Argentine		
II. Frozen, excluding fillets/Congelés, non compris les filets	16 429	2 852 976
Hake, mackerel, sea bream, other/Merlu, maquereau, dorade, autres		
Argentina, Netherlands, France, South Africa/Argentine, Pays-Bas, France, Afrique du Sud		
III. Frozen fillets/Filets congelés	1 643	460 540
Hake, flatfish, other/Merlu, poisson plat, autres		
Norway, Argentina, Italy, Netherlands/Norvège, Argentine, Italie, Pays-Bas		
IV. Shellfish (live, fresh, chilled, salted, dried, canned)/Coquillages	23 783	5 156 414
(vivants, frais, sur glace, salés, séchés, en conserve)		
Molluscs, crustaceans, other/Mollusques, crustacés, autres		
Netherlands, Italy, USSR, Faroe Islands/Pays-Bas, Italie, URRS, Iles Féroé		
V. Meal/Farine		
VI. Oil/Huile		
VII. Other/Autres	14 737	5 829 532

ICELAND/ISLANDE

Table I/Tableau I

LANDINGS, VALUES AND UTILISATION BY SPECIES/DEBARQUEMENTS, VALEURS ET UTILISATION PAR ESPECES

1986

Quant. : tons/tonnes (landed weight/poids débarqué)
Val. : IKr/krI million

	Frozen/ Congelés	Salted/ Salés	Dried/ Séchés	Direct consump./ Consom. directe	Meal & Oil Farine & Huile	On ice/ sur glace	1986 Quant. Total	1985 Quant. Total	Quant./Total Index/Indice 1985 = 100
Cod/Morue	172 671	135 402	4 424	51 794	70	1 369	365 730	319 819	114
Haddock/Eglefin	28 590	38	47	14 131	5	4 505	47 316	47 111	100
Saithe/Lieu noir	42 080	12 566	6	9 178	14	16	63 860	54 774	117
Redfish/Sébaste	66 339	-	-	19 268	251	131	85 989	91 035	94
Ling/Lingue									
Blue Ling/Lingue bleue	2 406	860	2	1 392	1	50	4 711	4 316	109
Tusk/Brosme	1 284	575	148	525	10	8	2 550	3 025	84
Catfish/Loup Atlantique	9 872	3	88	1 913	21	222	12 119	9 498	128
Halibut/Flétan	764	-	2	625	-	226	1 618	1 642	99
Greenland halibut/Flétan noir									
Plaice/Plie	25 943	32	-	4 973	87	3	31 038	28 818	108
Herring/Hareng	5 001	-	-	7 617	7	73	12 698	13 094	97
Capelin/Capelan	17 356	35 726	-	42	12 633	-	65 757	48 829	135
Lobster/Homard	6 051	-	271	62 283	826 316	3 093	898 014	997 761	90
Shrimp/Crevettes	2 562	-	-	-	2	-	2 564	2 391	107
Scallop/Coquilles	33 422	-	-	-	8	2 381	35 811	24 933	144
St-Jacques	16 374	-	-	-	55	-	16 429	17 232	95
Miscellaneous/Divers	2 895	74	14	1 317	214	542	5 056	4 247	119
TOTAL 1986	433 611	185 276	5 002	175 058	839 694	12 619	1 651 260		
TOTAL 1985	409 816	171 086	5 143	141 157	931 130	10 193		1 668 525	
Index/Indice 1985 = 100	106	108	97	124	90	124	99		99

ICELAND/ISLANDE

Table II/Tableau II

EXPORTS OF FISHERY PRODUCTS SELECTED BY MAJOR COUNTRIES/ EXPORTATIONS DE PRODUITS DE LA PECHE SELECTIONNES PAR PRINCIPAUX PAYS
1983-1986

'000 tons/tonnes (product weight/poids du produit)

	1983	1984	1985	1986
TOTAL	335.4	485.8	694.2	715.0
Wet salted groundfish/Poisson de fond salé	43.4	36.6	43.7	45.4
Portugal	24.5	18.1	25.2	26.6
Spain/Espagne	8.2	9.1	9.9	10.1
Greece/Grèce	3.8	3.1	3.2	3.1
Italy/Italie	3.3	2.8	2.8	3.1
Stockfish	6.5	0.4	0.9	7.2
Nigeria	6.0	-	0.1	5.7
Italy/Italie	0.3	0.3	0.4	0.8
Fresh and chilled fish/Poisson frais et sur glace	44.0	67.7	165.1	153.0
United Kingdom/Royaume-Uni	17.9	23.8	52.5	63.5
Germany/Allemagne	21.0	23.1	21.0	27.4
Faroe Islands/Iles Féroé	3.0	19.1	59.0	19.1
United States/Etats-Unis	1.8	1.1	1.3	1.0
Herring frozen/Hareng congelé	9.1	8.9	5.9	8.4
United Kingdom/Royaume-Uni	4.8	3.1	2.9	2.9
Czechoslovakia/Tchécoslovaquie	1.1	1.3	0.5	0.9
Fish frozen whole/Poisson entier congelé	11.7	14.4	16.7	20.4
Japan/Japon	-	3.2	5.7	9.4
USSR/URSS	5.9	6.5	5.2	5.3
Germany/Allemagne	2.8	2.0	2.0	1.8
Frozen fillets/Filets congelés	113.6	107.4	115.4	115.4
United States/Etats-Unis	67.1	64.7	66.5	58.6
United Kingdom/Royaume-Uni	16.2	14.4	20.7	25.5
USSR/URSS	19.6	14.3	15.6	14.0
France	5.2	6.3	6.6	9.3
Germany/Allemagne	4.4	4.8	3.9	5.2
Shrimp frozen/Crevettes congelées	2.9	5.4	8.5	12.4
United Kingdom/Royaume-Uni	1.6	1.9	2.9	4.0
Meal/Farine	48.5	134.9	159.9	149.5
Oil/Huile	8.1	59.9	126.6	97.9

IRELAND/IRLANDE

Table I/Tableau I

LANDINGS AND VALUES/DEBARQUEMENTS ET VALEURS
1985 & 1986

Quant. : tons/tonnes (live weight/poids vif)
Val. : Ir£/£Ir

	1985		1986	
	Quant.	Val.	Quant.	Val.
1. National landings in domestic ports/Débarquements nationaux dans les ports nationaux				
For direct human consumption/Pour la consommation humaine directe				
Finfish/Poisson				
Demersal/Démersaux	44 252	23 232 162	35 482	23 096 433
Pelagic/Pélagiquew	93 063	12 224 557	105 738	13 998 490
Sub-total/Sous-total	137 315	35 456 719	141 220	37 094 923
Shellfish/Crustacés et coquillages				
Crustaceans/Crustacés	9 288	10 422 186	10 056	11 969 431
Molluscs/Mollusques	13 711	3 871 564	14 394	4 995 588
Sub-total/Sous-total	22 999	14 293 750	24 450	16 965 019
For other purposes/Pour d'autres buts				
Fishmeal/Farine de poisson	25 452	1 004 165	23 009	750 963
Withdrawals/Retraits	7 078	1 010 063	8 227	931 348
Sub-total/Sous-total	32 530	2 014 228	31 236	1 682 311
TOTAL 1	192 844	51 764 697	196 906	55 742 253
2. National landings in foreign ports/Débarquements nationaux dans les ports étrangers				
For direct human consumption/Pour la consommation humaine directe				
Demersal/Démersaux	2 855	4 616 131	4 092	5 983 598
Pelagic/Pélagiques	5 825	574 385	9 373	481 289
Shellfish/Coquillages	72	111 218	260	415 995
For other purposes/Pour d'autres buts				
TOTAL 2	8 752	5 301 734	13 725	6 880 877
3. Foreign landings in domestic ports/Débarquements étrangers dans les port nationaux				
For direct human consumption/Pour la consommation humaine directe				
Demersal/Démersaux	143	65 229	518	408 427
Pelagic/Pélagiques	4 077	525 933	2 252	270 288
Shellfish/Coquillages	-	-	26	51 141
For other purposes/Pour d'autres buts				
TOTAL 3	4 220	591 162	2 796	729 856
TOTAL 1 + 3				

IRELAND/IRLANDE

Table II/Tableau II

EXTERNAL TRADE IN FISH AND FISH PRODUCTS/ECHANGES INTERNATIONAUX DE POISSON ET PRODUITS DE LA PECHE[a]

1985 & 1986

Quant. : tons (product weight)/tonnes (poids du produit)
Val. : '000 Irf/£Ir

	IMPORTS/IMPORTATIONS				EXPORTS/EXPORTATIONS			
	1985		1986		1985		1986	
	Quant.	Val.	Quant.	Val.	Quant.	Val.	Quant.	Val.
Total fish and fish products/Total poisson et produits de la pêche	37 619	37 498	35 827	37 689	154 784	94 743	147 526	95 557
Fresh, chilled/Frais, sur glace	12 336	3 608	12 045	3 849	36 851	22 039	33 794	24 546
Frozen/Congelés	1 128	1 773	1 137	2 129	74 722	32 673	66 349	24 138
Frozen fillets/Filets congelés	759	1 380	540	905	11 223	6 212	12 767	6 806
Salted, dried, smoked/Salés, séchés, fumés	1 261	1 380	1 238	2 097	12 310	8 932	10 796	10 257
Canned/En conserve	525	1 958	391	661	3	6	2	4
Shellfish (live, fresh, chilled, salted, dried, canned)/Crustacés et coquillages (vivants, frais, sur glace, salés, séchés et en conserve)	1 134	3 931	1 309	5 838	15 029	22 151	16 714	27 145
Meal/Farine	11 946	4 349	10 546	3 564	1 204	394	1 868	467
Oil/Huile	1 236	644	1 262	1 018	2 174	528	4 131	648
Other/Autres	7 294	18 972	7 359	17 628	1 268	1 808	1 105	1 546

a) Does not include foreign landings/Non compris les débarquements étrangers.

IRELAND/IRLANDE

Table III(a)/Tableau III(a)

IMPORTS BY MAJOR PRODUCTS AND BY COUNTRY/IMPORTATIONS PAR PRINCIPAUX PRODUITS ET PAR PAYS

1985 & 1986

Quant. : '000 tons/tonnes
Val. : Irf/£Ir '000

	1985 Quant.	1985 Val.	1986 Quant.	1986 Val.
Total fish and fish products/Total poisson et produits de la pêche	37.6	37.4	35.8	37.6
Fresh/chilled/Frais, sur glace				
Mackerel/Maquereau	6.3	0.9	5.6	1.0
Herring/Hareng	5.1	1.1	5.0	0.7
Salmon/Saumon	0.1	0.6	0.1	0.5
Whiting/Merlan			0.4	0.2
Northern Ireland/Irlande du Nord	8.4	2.2	6.5	1.7
Great Britain/Grande-Bretagne	3.7	1.0	5.1	1.2
Netherlands/Pays-Bas	-	-	-	-
Frozen (excl. fillets)/Congelés (non compris les filets)				
Salmon/Saumon	0.3	1.0	0.6	1.5
Sprat	0.3	0.1	0.1	0.1
Mackerel/Maquereau	0.2	0.1	0.1	0.1
United States/Etats-Unis	0.1	0.2	0.2	0.7
Northern Ireland/Irlande du Nord	0.3	0.2	0.1	0.1
Great Britain/Grande-Bretagne	3.7	1.0	0.2	0.4
Frozen fillets/Filets congelés				
Plaice/Plie	0.2	0.5	0.2	0.4
Cod/Morue	0.1	0.3	0.2	0.2
Mackerel/Maquereau	-	-	0.1	0.1
Netherlands/Pays-Bas	0.2	0.4	0.1	0.3
Great Britain/Grande-Bretagne	0.3	0.6	0.1	0.3
Shellfish/Crustacés				
Prawns/Crevettes	0.5	0.2	0.5	2.1
Crab/Crabe	0.2	0.2	0.1	0.2
Shrimps/Crevettes	0.1	0.6	0.1	0.7
Northern Ireland/Irlande du Nord	0.4	1.0	0.5	1.5
Great Britain/Grande-Bretagne	0.1	0.1	0.3	1.1
France	-	-	0.1	1.0
Meal/Farine				
Norway/Norvège	4.9	1.8	11.0	4.0
Faroe Islands/Iles Féroé	3.4	1.3	1.4	0.4
Great Britain/Grande-Bretagne	3.0	1.1	5.4	1.7
Denmark/Danemark	0.1	0.1	1.6	0.7
Norway/Norvège	1.6	1.0	-	-
			1.4	0.4
Oil/Huile				
	1.2	0.6	1.3	1.0
Netherlands/Pays-Bas	0.7	0.6	0.1	0.1
Great Britain/Grande-Bretagne	0.1	0.1	1.2	1.0
Japan/Japon	-	-	-	-
United States/Etats-Unis	-	-	-	-
Other/Autres				
Prepared or preserved/préparé ou préservé	7.3	18.9	7.2	17.6
Great Britain/Grande Bretagne	4.5	11.4	4.3	10.4
Denmark/Danemark	0.7	1.3	0.6	1.1
Northern Ireland/Irlande du Nord	0.6	1.3	0.7	1.1
Canada	-	-	0.7	1.9

IRELAND/IRLANDE

Table III(b)/Tableau III(b)

EXPORTS BY MAJOR PRODUCTS AND BY COUNTRY/EXPORTATIONS PAR PRINCIPAUX PRODUITS ET PAR PAYS[a]

1985 & 1986

Quant. : '000 tons/tonnes
Val. : Ir£/£Ir '000

	1985		1986	
	Quant.	Val.	Quant.	Val.
Total fish and fish products/Total poisson et produits de la pêche	154.7	94.7	148.0	96.0
Fresh/chilled/Frais, sur glace				
Mackerel/Maquereau	13.9	2.6	12.0	2.0
Herring/Hareng	4.1	0.7	5.1	1.4
Dogfish/Aiguillat	5.6	5.6	5.0	2.0
Salmon/Saumon	-	-	1.0	5.0
Northern Ireland/Irlande du Nord	11.3	3.7	10.4	3.9
France	5.7	6.5	4.9	7.2
Great Britain/Grande Bretagne	12.5	7.8	9.9	4.4
Frozen (excluding fillets)/Congelés (non compris les filets)				
Mackerel/Maquereau	62.6	25.7	47.0	14.0
Herring/Hareng	36.0	1.0	4.8	1.3
Sprat	4.9	15.0	6.0	0.5
Nigeria	31.0	15.0	9.8	3.1
USSR/URSS	1.0	0.1	2.2	0.4
Egypt/Egypte	9.8	3.6	5.4	3.2
Netherlands/Pays-Bas	7.7	2.5	13.9	3.2
Japan/Japon			6.4	4.1
Frozen fillets/Filets congelés				
Mackerel/Maquereau	5.9	3.0	5.4	3.0
Herring/Hareng	4.2	1.9	7.0	3.0
Whiting/Merlan	0.3	0.3	0.3	0.4
Germany/Allemagne	5.4	2.6	5.9	2.8
France	3.5	2.1	2.4	1.3
Czechoslovakia/Tchécoslovaquie	0.6	0.3	1.3	0.7
Shellfish/Crustacés				
Mussels/Moules	6.1	1.8	7.6	2.6
Periwinkles/Bigorneaux	0.1	0.1	2.4	1.5
Shrimps/Crevettes	2.3	7.3	0.1	0.5
Prawns/Crevettes			2.5	8.6
France	8.2	10.0	10.2	14.6
Great Britain/Grande-Bretagne	1.6	2.6	1.9	3.3
Northern Ireland/Irlande du Nord	1.1	2.2	1.8	2.5
Netherlands/Pays-Bas	2.1	0.9	0.6	0.5
Meal/Farine	1.1	0.3	1.8	0.4
Northern Ireland/Irlande du Nord	1.0	0.3	1.8	0.4
Oil/Huile	2.2	0.5	4.1	0.6
Great Britain/Grande Bretagne	1.4	0.4	3.3	0.4
Northern Ireland/Irlande du Nord	0.6	0.1	0.7	0.2
Other/Autres				
Smoked, salted, in brine/Fumés, salés, en saumure	12.6	8.9	10.7	10.2
Germany/Allemagne	4.1	3.6	5.5	4.2
Netherlands/Pays-Bas	1.0	0.6	1.0	0.8
France	2.0	1.0	1.2	0.9

a)   Does not include landings into foreign ports/Non compris les débarquements dans les ports étrangers.

ITALY/ITALIE

Table I/Tableau I

LANDINGS AND VALUES/DEBARQUEMENTS ET VALEURS

1985a) & 1986a)

Quant. : tons/tonnes (landed/débarquées)
Val. : Million L

	1985		1986	
	Quant.	Val.	Quant.	Val.
NATIONAL LANDINGS IN DOMESTIC PORTS/DEBARQUEMENTS NATIONAUX DANS LES PORTS NATIONAUX				
For direct human consumption/Pour la consommation humaine directea)b)	270 108	1 102 998	276 984	1 234 730
Fish/Poisson : Anchovies, sardines, mackerel/Anchois, sardines, maquereau	96 243	163 218	88 830	159 587
Tuna/Thon	2 683	9 878	4 023	15 695
Other/Autresc)	171 183	923 902	184 123	1 059 448
Shellfish/Coquillages : Molluscs/Mollusquesd)	93 398	297 738	102 900	356 065
Squid, octopus, cuttlefish/Calmar, poulpe, sèche	31 534	170 680	38 458	217 265
Other/Autres	61 684	127 658	64 442	138 800
Crustaceans/Crustacése)	32 107	272 948	33 770	346 578
TOTAL	395 613	1 673 684	413 564	1 937 373

a) Provisional figures/Chiffres provisoires.

b) Includes quantities used in the canning industry to obtain salted, dried and smoked products, etc./Y compris les quantités utilisées dans la conserverie pour obtenir des produits salés, séchés, fumés, etc.

c) Including frozen products from the Atlantic/Y compris les produits congelés de la pêche océanique.
   26 350 tons/tonnes - 1985, 29 650 tons/tonnes - 1986

d) Including frozen products from the Atlantic/Y compris les produits congelés de la pêche océanique.
   9 970 tons/tonnes - 1985, 12 180 tons/tonnes - 1986

e) Including frozen products from the Atlantic/Y compris les produits congelés de la pêche océanique.
   2 640 tons/tonnes - 1985, 2 940 tons/tonnes - 1986
   TOTAL 38 960 tons/tonnes - 1985, 44 770 tons/tonnes - 1986

ITALY/ITALIE

Table II/Tableau II

EXTERNAL TRADE IN FISH AND FISH PRODUCTS/ECHANGES INTERNATIONAUX DE POISSON ET PRODUITS DE LA PECHE
1985 & 1986a)

Quant. : tons/tonnes
Val. : Million L

	IMPORTS/IMPORTATIONS				EXPORTS/EXPORTATIONS			
	1985		1986		1985		1986	
	Quant.	Val.	Quant.	Val.	Quant.	Val.	Quant.	Val.
Total fish and fish products/Total poisson et produits de la pêche	600 170	1 891 546	490 302	1 684 350	156 665	270 560	93 550	224 510
Fresh, chilled/Frais, sur glace[b]	44 264	300 010	37 636	271 355	32 056	70 445	20 629	58 716
Frozen/Congelés[c]	177 368	439 382	149 884	335 752	21 360	34 856	18 323	32 632
Frozen fillets/Filets congelés	26 345	109 663	25 691	122 401	337	2 491	416	1 875
Salted, dried, smoked/Salés, séchés, fumés[d]	44 601	289 400	31 273	234 436	1 180	5 261	1 101	5 037
Canned/En conserve	38 128	142 790	27 904	146 418	12 592	63 277	9 705	54 355
Shellfish (live, fresh, chilled, frozen, salted, dried and canned)/Coquillages (vivants, frais, sur glace, congelés, salés, séchés et en conserve)[e]	140 452	507 021	121 914	518 463	40 401	58 529	29 296	58 846
Meal/Farine[f]	109 248	74 152	85 931	47 358	45 014	33 989	22 008	13 052
Oil/Huile	16 408	12 419	9 231	6 310	26	28	44	19
Other/Autres[g]	3 356	16 709	838	1 857	3 699	1 684	8	378

a) Jan./Nov.

b) Including fillets, liver and eggs, etc./Y compris les filets, le foie et les oeufs, etc.

c) Including tuna, liver and eggs, etc./Y compris le thon, le foie et les oeufs, etc.

d) Including fillets, meal, liver and eggs, etc./Y compris les filets, la farine, le foie et les oeufs, etc.

e) Including all molluscs and crustaceans (also canned)/Y compris tous les mollusques et crustacés (également en conserve).

f) Including food industry waste and food for animals/Y compris les résidus de l'industrie alimentaire et les aliments préparés pour les animaux.

g) Coral and sponges/Coraux et éponges.

ITALY/ITALIE

Table III(a)/Tableau III(a)

IMPORTS BY MAJOR PRODUCTS AND BY COUNTRY/IMPORTATIONS PAR PRINCIPAUX PRODUITS ET PAR PAYS
1985 & 1986a)

Quant. : tons/tonnes
Val. : Million L

	1985		1986	
	Quant.	Val.	Quant.	Val.
Total fish and fish products/Total poisson et produits de la pêche	600 170	1 891 546	490 302	1 684 350
Fresh, chilled/Frais, sur glaceb)	44 264	300 010	37 636	271 355
Shark/Requin	1 953	12 529	1 932	12 451
France	1 930	12 383	1 784	11 836
Mackerel/Maquereau	3 660	6 672	2 451	4 503
France	3 130	6 123	1 910	3 771
Spain/Espagne	227	250	481	538
Anchovies/Anchois	170	331	81	175
France	88	189	19	30
Spain/Espagne	-	-	45	97
Other/Autres	32 090	245 556	26 239	213 658
Denmark/Danemark	7 564	46 862	6 513	44 182
France	10 055	82 993	7 248	56 530
Netherlands/Pays-Bas	6 331	54 727	4 467	47 204
Fillets/Filets	305	1 341	282	1 034
Liver, eggs/Foie, oeufs, etc.	1	148	21	184
Frozen (excl. fillets)/Congelés, non compris les filetsc)	177 368	439 382	149 884	335 752
Tuna/Thon	104 021	240 035	102 933	180 416
Spain/Espagne	29 404	66 206	28 019	50 756
Panama	13 400	29 137	21 568	34 471
France	10 396	23 686	9 923	18 070
Seychelles	11 022	24 579	16 132	29 447
Shark/Requin	5 224	25 154	4 107	18 045
Germany/Allemagne	1 330	9 621	1 441	10 958
Argentina/Argentine	2 099	5 015	1 123	2 238
Cod/Morue	6 146	7 265	25	136
Hake/Merlu	21 180	27 646	11 587	14 027
Spain/Espagne	707	1 149	852	1 386
South Africa/Afrique du Sud	8 634	12 103	7 175	7 934
USSR/URSS	8 633	8 260	1 003	856
Sole	5 761	40 290	5 220	53 722
Netherlands/Pays-Bas	5 528	39 146	4 986	52 780
Other/Autres	21 221	52 513	13 777	40 991
Mauritania/Mauritanie	5 243	13 529	2 953	7 832
Morocco/Maroc	2 071	5 301	3 480	8 391
Panama	4 436	8 286	1 456	2 968
Liver, eggs/Foie, oeufs, etc.	68	841	80	1 180
Frozen fillets/Filets congelés	26 345	109 663	25 691	122 401
Cod/Morue	4 736	22 555	4 720	26 862
Denmark/Danemark	2 909	15 812	3 464	21 074
Plaice/Plie	13 470	60 456	14 867	75 837
Netherlands/Pays-Bas	11 638	50 384	11 948	57 626
Salted, dried, smoked/Salés, séchés, fumésd)	44 601	289 400	31 273	234 436
Cod and saithe (incl. fillets)/ Morue et lieu noir (y compris filets)	30 284	200 814	20 937	168 955
Iceland/Islande	3 840	18 721	3 371	22 751
Norway/Norvège	13 837	116 469	8 405	94 226
Faroe Islands/Iles Féroé	2 513	10 549	1 770	8 480
Denmark/Danemark	4 292	21 465	3 066	17 652

ITALY/ITALIE

Table III(a) (cont'd)/Tableau III(a) (suite)

Quant. : tons/tonnes
Val. : Million L

	1985		1986	
	Quant.	Val.	Quant.	Val.
Liver, eggs/Foie, oeufs, etc.	23	440	20	449
Meal/Farine	32	87	2	15
Canned/En conserve	38 128	142 790	27 904	146 418
Tuna/Thon	4 255	30 861	5 440	34 286
Portugal	2 457	16 154	1 194	8 289
Germany/Allemagne	-	-	2 108	15 704
France	1 299	7 224	1 929	9 259
Mackerel/Maquereau	6 868	28 679	5 501	23 972
Portugal	2 217	10 805	1 436	7 718
Morocco/Maroc	1 878	8 994	2 124	9 868
Shellfish (live, fresh, chilled, frozen, salted, dried, canned)/Coquillages (vivants, frais, sur glace, congelé, salés, séchés et en conserve)e)	140 452	507 021	121 915	518 463
Mussels/Moules	17 297	15 516	9 057	7 724
Spain/Espagne	15 056	12 716	6 567	4 758
Shrimps/Crevettes	16 054	136 654	10 930	113 663
Cuba	1 672	14 562	1 816	19 152
Argentina/Argentine	5 054	51 472	3 211	34 546
Denmark/Danemark	2 099	7 951	418	2 698
Frozen squid/Calmar congelé	52 230	129 859	42 955	96 207
Poland/Pologne	13 449	25 943	13 242	16 548
Spain/Espagne	6 657	14 431	1 878	3 875
Thailand/Thaïlande	9 895	34 951	9 798	32 497
South Africa/Afrique du sud	2 095	9 683	2 357	9 996
USSR/URSS	4 461	6 931	4 565	5 566
Mauritania/Mauritanie	-	-	2 643	4 088
Frozen cuttlefish/Seiche congelée	10 912	30 784	9 706	31 265
Mauritania/Mauritanie	1 519	5 114	1 461	5 380
France	5 205	13 634	4 127	11 777
Frozen octopus/Poulpe congelée	22 040	57 650	23 111	82 301
Panama	3 627	8 909	2 051	6 972
Morocco/Maroc	3 144	8 655	4 306	16 533
Spain/Espagne	2 620	6 467	2 890	11 124
Mauritania/Mauritanie	4 309	13 680	4 356	17 580
South Africa/Afrique du sud	-	-	2 928	10 864
Meal/Farinef)	109 248	74 152	85 931	47 358
Chile/Chili	77 987	51 497	57 731	29 466
Denmark/Danemark	18 082	14 499	17 912	12 674
Oil/Huile	16 408	12 419	9 231	6 310
France	6 307	4 185	4 633	3 145
Netherlands/Pays-Bas	4 469	3 101	1 691	1 248
Denmark/Danemark	2 082	1 492	1 883	1 050
Other/Autresg)	3 356	16 709	838	1 857

a) Provisional figures (Jan.-Nov.)/Chiffres provisoires (janv.-nov.).
b) Including fillets, liver, eggs, etc./Y compris les filets, le foie, les oeufs, etc.
c) Including tuna, liver, eggs, etc./Y compris le thon, le foie, les oeufs, etc.
d) Including meal, liver, eggs, etc./Y compris la farine, le foie, les oeufs, etc.
e) Including crustaceans and molluscs (also canned)/Y compris crustacés et mollusques (également en conserve).
f) Including waste and fooder/Y compris les déchets et les aliments pour animaux.
g) Coral and sponges/Coraux et éponges.

ITALY/ITALIE

Table III(b)/Tableau III(b)

EXPORTS BY MAJOR PRODUCTS AND BY COUNTRY/EXPORTATIONS PAR PRINCIPAUX PRODUITS ET PAR PAYS
1985 & 1986[a]

Quant. : tons/tonnes
Val. : Million L

	1985		1986	
	Quant.	Val.	Quant.	Val.
Total fish and fish products/Total poisson et produits de la pêche	156 665	270 560	93 530	224 510
Fresh, chilled/Frais, sur glace[b]	32 056	70 445	20 629	58 716
Tuna/Thon	780	3 515	697	3 217
France	682	2 917	594	2 112
Sardines	6 869	6 282	5 058	5 222
France	6 793	6 196	4 930	5 006
Anchovies/Anchois	13 778	21 242	5 970	12 969
Spain/Espagne	11 483	16 598	5 414	11 904
France	2 234	4 538	545	1 035
Other/Autres	4 635	9 509	3 927	10 192
France	2 437	3 241	1 985	2 874
Germany/Allemagne	1 393	3 793	981	3 272
Fillets/Filets	1	7	3	10
Liver, eggs/Foie, oeufs, etc.	3	122	1	20
Frozen (excl. fillets)/Congelés (non compris les filets)[c]	21 360	34 856	18 323	32 632
Tuna/Thon	201	1 451	232	1 927
Sardines	12 328	8 899	10 160	8 313
France	11 088	7 856	9 130	7 576
Yugoslavia/Yougoslavie	684	377	533	308
Anchovies/Anchois	1 571	1 806	317	625
Spain/Espagne	612	509	62	108
Germany/Allemagne	223	376	40	72
Bel./Lux.	496	630	185	169
France	153	472	9	17
Other/Autres	5 290	17 767	6 290	17 884
Greece/Grèce	2 077	5 507	2 209	4 956
Switzerland/Suisse	859	4 280	670	3 752
France	1 198	2 994	1 061	2 874
Liver, eggs/Foie, oeufs, etc.	6	18	54	199
Frozen fillets/Filets congelés	337	2 491	416	1 875
Salted, dried, smoked/Salés, séché, fumés[d]	1 180	5 261	1 101	5 037
Anchovies/Anchois	546	2 597	749	3 054
France	100	499	122	603
Australia/Australie	117	616	216	830
Cod and saithe/Morue et lieu noir	61	392	173	661
Other/Autres	273	993	-	-
Czechoslovakia/Tchécoslovaquie	80	180	-	-
United States/Etats-Unis	60	232	-	-
Cod and saithe fillets/Filets de morue et lieu noir	20	90	19	126
Denmark/Danemark	12	70	19	126
Other fillets/Autres filets	117	1 209	-	-
Liver, eggs/Foie, oeufs, etc.	1	83	-	-
Meal/Farine	150	108	-	-

ITALY/ITALIE

Table III(b)(cont'd)/Tableau III(b)(suite)

EXPORTS BY MAJOR PRODUCTS AND BY COUNTRY/EXPORTATIONS PAR PRINCIPAUX PRODUITS ET PAR PAYS
1985 & 1986[a]

Quant. : tons/tonnes
Val. : Million L

	1985		1986	
	Quant.	Val.	Quant.	Val.
Canned/En conserve	12 592	63 277	9 705	54 355
Sardines	8 949	37 283	6 117	25 405
United Kingdom/Royaume-Uni	1 092	3 104	1 354	3 865
Germany/Allemagne	660	2 340	225	1 003
France	1 521	4 948	1 393	4 731
Greece/Grèce	1 452	5 966	1 127	5 002
Tuna/Thon	2 043	14 800	1 789	13 567
Bel./Lux.	538	3 981	456	3 591
Greece/Grèce	913	6 617	812	6 232
Anchovies/Anchois	731	5 937	1 175	10 829
France	228	2 069	542	5 172
Shellfish/Coquillages[e]	40 401	58 529	21 296	58 846
Canned/En conserve	1 058	6 190	1 037	6 636
Oysters/Huîtres	1 361	1 763	968	1 485
Spain/Espagne	1 305	1 583	945	1 340
Mussels/Moules	4 074	3 461	1 775	3 332
Spain/Espagne	3 781	3 020	1 529	2 821
Switzerland/Suisse	219	316	111	347
Other molluscs, not frozen/Autres mollusques non congelés	25 216	21 444	10 889	22 625
Spain/Espagne	24 431	19 051	10 561	21 603
Meal/Farine[f]	45 014	33 989	22 008	13 052
Yugoslavia/Yougoslavie	38 301	29 275	14 903	8 793
Oil/Huile	26	28	44	19
Other/Autres[g]	3 699	1 684	8	378

a) Provisional figures (Jan.-Nov.)/Chiffres provisoires (janv.-nov.).

b) Including fillets, liver, eggs, etc./Y compris les filets, le foie, les oeufs, etc.

c) Including tuna, liver, eggs, etc./Y compris le thon, le foie, les oeufs, etc.

d) Including meal, liver, eggs, etc./Y compris la farine, le foie, les oeufs, etc.

e) Including crustaceans and molluscs (also canned)/Y compris crustacés et mollusques (également en conserve).

f) Including waste and fooder/Y compris les déchets et les aliments pour animaux.

g) Coral and sponges/Coraux et éponges.

JAPAN/JAPON

Table I/Tableau I

LANDINGS AND VALUES/DEBARQUEMENTS ET VALEURS
1984-1986

Quant. : '000 tons/tonnes
Val. : billion Y

	1984		1985		1986a)	
	Quant.	Val.	Quant.	Val.	Quant.	Val.
TOTAL	12 816	2 947	12 197	2 902	12 677	
Marine fisheries/Pêches maritimes	11 501	2 232	10 877	2 192	11 287	
Tuna/Thon	366	378	391	370		
Bigeye/Thon obèse	131	166	149	169		
Yellowfin/Albacore	115	79	134	78		
Skipjack/Frigate mackerel/Listao/Auxide	468	85	339	87		
Salmon/Trout/Saumon/Truite	157	113	203	175		
Jack mackerel/Scads/Carangue	234	77	225	64		
Sea bream/Dorade	27	47	26	45		
Pacific mackerel/Maquereau du Pacifique	814	79	773	71		
Sardines	4 179	85	3 866	72		
Anchovy/Anchois	224	18	206	18		
Saury/Balaou	210	29	246	17		
Yellowtail/Sériole	41	31	33	27		
Flounder/Flet	264	114	214	106		
Cod/Morue	114	29	118	25		
Alaska pollack/Morue du Pacifique occidental	1 621	122	1 532	126		
Prawns/Shrimps/Crevettes	62	87	53	82		
Crab/Crabe	99	48	100	54		
Squid, common/Calmar commun	174	94	133	94		
Squid, other/Autres calmars	352	150	398	171		
Octopus/Poulpe	43	20	40	19		
Seaweeds/Algues	184	35	184	27		
Marine culture/Aquaculture	1 111	517	1 088	522	1 190	
Yellowtail/Sériole	152	143	151	143		
Oysters/Huîtres	257	31	251	30		
Seaweeds/Algues	578	155	523	147		
Inland water fisheries/Pêches d'eau douce	107	61	110	61	106	
Inland water culture/Cultures d'eau douce	97	123	96	115	94	

292

JAPAN/JAPON

Table II/Tableau II

EXTERNAL TRADE IN FISH AND FISH PRODUCTS/ECHANGES INTERNATIONAUX DE POISSON ET PRODUITS DE LA PECHE
1985 & 1986

Quant. : '000 tons/tonnes
Val. : Y billion/milliard

	IMPORTS/IMPORTATIONS				EXPORTS/EXPORTATIONS			
	1985		1986		1985		1986	
	Quant.	Val.	Quant.	Val.	Quant.	Val.	Quant.	Val.
Total fish and fish products/ Total poisson et produits de la pêche	1 577.3	1 176.0	1 868.5	1 137.8	786.4	287.6	760.4	317.5
Fresh, chilled/Frais, sur glace	58.7	66.9	81.5	90.9	19.5	2.6	18.1	2.0
Frozen/Congelés	695.8	339.5	804.1	313.1	95.8	20.1	135.5	19.0
Frozen fillets/Fillets congelés	16.9	8.8	20.0	8.3	19.7	12.9	14.9	10.0
Cured/Salés, séchés, fumés	27.3	60.6	23.8	39.9	2.0	3.5	1.6	2.6
Canned/En conserve	0.5	0.6	0.9	0.8	163.4	55.1	124.9	35.5
Shellfish/(live, fresh, chilled, salted, dried, canned)/Crustacés et coquillages (vivants, frais, sur glace, salés, séchés, en conserve)	544.2	562.1	624.8	560.3	18.4	24.1	18.1	23.1
Meal/Farine	80.2	8.1	161.4	12.7	157.4	15.0	167.2	13.7
Oil/Huile	7.7	3.5	7.6	4.5	250.0	14.6	225.2	7.0
Other/Autres	146.0	125.9	144.4	108.2	160.2	147.9	54.9	114.6

JAPAN/JAPON

Table III/Tableau III

SELECTED JAPANESE IMPORTS AND EXPORTS OF FISH AND FISH PRODUCTS/
IMPORTATIONS ET EXPORTATIONS JAPONAISES SELECTIONNEES DE POISSON ET PRODUITS DE LA PECHE
1983-1986

Quant.: '000 tons/tonnes
Val.: million Y

	1983		1984		1985		1986	
	Quant.	Val.	Quant.	Val.	Quant.	Val.	Quant.	Val.
IMPORTS/IMPORTATIONS								
Fresh/Frais								
Yellowfin tuna/Albacore	11.7	13 396	12.7	13 433	13.0	12 986	14.5	12 659
Bluefin tuna/Thon rouge	0.9	2 154	1.0	2 479	1.0	2 778	2.2	4 975
Spanish mackerel/Maquereau espagnol	6.6	3 338	4.4	2 016	2.5	1 593	3.2	1 899
Sea bream/Dorade	2.1	2 382	0.2	46	2.5	2 960	3.5	3 906
Frozen/Congelés								
Skipjack tuna/Listao	9.8	1 623	0.8	104	6.1	1 991	2.9	390
Yellowfin tuna/Albacore	46.0	19 608	36.1	19 252	62.4	23 396	52.3	17 602
Bigeye tuna/Thon obèse	51.6	28 571	44.8	35 955	41.9	29 477	57.3	28 257
Billfish/Voilier	13.2	5 819	11.3	5 000	11.2	4 977	15.2	5 722
Herring/Hareng	53.2	19 408	55.1	16 403	71.6	22 276	62.7	15 914
Cod/Morue	68.1	17 503	93.7	22 052	112.2	26 514	-	-
Horse mackerel/Chinchard	14.8	3 194	20.4	1 822	29.3	7 085	34.5	4 890
Cod roe/Rogue de morue	13.0	9 959	16.1	10 291	18.5	14 469	-	-
Salmon/Saumon	97.8	83 809	93.0	82 981	115.5	115 773	133.4	89 504
Sea bream/Dorade	19.0	6 653	20.6	7 054	21.3	7 261	16.0	4 453
Surimi							130.7	29 546
Frozen shellfish/Coquillages congelés								
Spiny lobster/Langouste	7.2	16 859	7.5	16 897	8.7	21 046	9.2	18 513
Shrimp/Crevettes	148.6	300 978	169.1	301 628	182.9	314 511	212.8	306 722
Crab/Crabe	17.7	21 353	22.7	23 063	33.9	33 531	44.4	37 736
Squid/Calmar	101.7	55 739	102.6	57 118	112.9	64 932	43.5	11 920
Octopus/Poulpe	95.3	45 569	108.2	49 730	98.6	55 187	107.0	59 005
Abalone/Ormeau	1.0	1 730	1.1	2 337	0.9	2 426	0.8	2 194
Clams	17.4	3 790	18.5	3 922	20.5	4 059	21.9	4 130
Cuttlefish/Seiche							81.7	47 682
Salted, dried/Salés, séchés								
Salmon roe/Rogue de saumon	8.8	18 765	9.2	18 722	10.2	19 035	9.5	13 362
Herring roe/Rogue de hareng	8.6	31 282	9.3	24 217	7.7	31 234	6.8	19 410
EXPORTS/EXPORTATIONS								
Fresh/Frais								
Skipjack tuna/Listao	11.3	1 301	14.6	1 557	16.8	1 895	16.9	1 684
Frozen/Congelés								
Skipjack tuna/Listao	35.6	5 623	53.9	8 689	16.4	2 384	40.4	4 427
Albacore/Germon	3.0	1 007	12.1	5 397	7.5	3 127	6.5	1 586
Yellowfin tuna/Albacore	5.2	1 301	3.6	755	4.8	972	4.8	756
Saury/Balaou	12.9	2 564	13.8	2 804	19.2	3 292	18.9	2 411
Cod/Morue	4.5	1 174	10.6	1 790	5.3	2 203	6.2	2 841
Frozen fillets/Filets congelés								
Flatfish/Poissons plats	6.8	3 711	4.3	2 136	6.6	3 762	3.1	1 540
Cod/Morue	6.4	2 244	6.3	1 923	4.2	1 283	2.1	720
Fresh, frozen shellfish/Coquillages frais, congelés								
Shrimp/Crevettes	2.2	4 366	2.2	4 544	2.0	4 041	1.7	2 524
Squid/Calmar	9.7	4 486	5.8	2 608	4.4	1 942	2.7	1 103
Canned/En conserve								
Albacore/Germon	9.5	8 110	9.4	8 577	11.3	9 740	10.3	6 503
Skipjack/Listao	27.1	17 938	36.0	20 205	21.1	12 960	17.9	8 517
Mackerel/Maquereau	68.6	20 899	63.7	16 461	59.1	15 728	38.6	8 768
Sardines	68.7	16 385	75.0	16 752	69.2	14 632	55.9	10 544
Oysters/Huîtres	1.1	989	1.0	682	1.6	1 123	0.9	466
Surimi-based products/Produits à base de surimi							40.9	25 127

JAPAN/JAPON

Table IV(a)/Tableau IV(a)

IMPORTS BY MAJOR PRODUCTS AND BY COUNTRY/IMPORTATIONS PAR PRINCIPAUX PRODUITS ET PAR PAYS
1985 & 1986

Quant. : '000 tons/tonnes
Val. : million yen

	1985		1986	
	Quant.	Val.	Quant.	Val.
Total fish and fish products/Total poisson et produits de la pêche	1 577.3	1 176.0	1 868.5	1 137.8
Fresh, chilled/Frais, sur glace	58.7	66.9	81.5	90.0
Yellowfin tuna/Albacore	13.0	12 986.0	14.5	12.6
Bluefin tuna/Thon rouge	1.0	2 778.0	2.1	4 975.0
Seabream/Dorade	2.5	2 960.0	3.5	3 406.0
New Zealand/Nouvelle Zélande	4.9	4 186.0	3.9	3 684.0
Panama	4.8	1 879.0	4.5	1 420.0
Rep. of Korea/Rép de Corée	4.6	1 198.0	4.3	989.0
Frozen, excl. fillets/Congelés, non compris les filets	695.8	339.5	804.1	313.1
Salmon/Saumon	115.5	115 773.0	113.4	89 504.0
United States/Etats-Unis	102.5	102 547.0	96.0	74 421.0
Canada	9.9	9 857.0	15.1	14 306.0
Norway/Norvège	0.4	438.0	0.8	1 222.0
Frozen fillets/Filets congelés				
Tuna/Thon	16.9	8.8	20.0	8.3
Spain/Espagne	3.3	17 545.0	2.5	3 094.0
United States/Etats-Unis	0.9	8 065.0	0.4	12 000.0
Italy/Italie	0.1	1 412.0	0.1	525.0
Shellfish (live, fresh, chilled, salted, dried and canned)/ Crustacés et coquillages (vivants, frais, sur glace, salés, séchés et en conserve)	544.2	562.1	624.8	560.3
Shrimp/Crevettes	182.9	314 511.0	212.8	306 722.0
Meal/Farine	80.2	8.1	161.4	12.7
United States/Etats-Unis	29.9	6 868.0	36.4	4 622.0
Ecuador/Equateur	22.5	4 629.0	67.3	3 997.0
Chile/Chili	14.3	7 663.0	40.6	2 397.0
Oil/Huile	7.7	3.5	7.6	4.5
Indonesia/Indonesie	0.2	0.2	1.5	1 360.0
United States/Etats-Unis	2.7	0.3	2.1	238.0
Spain/Espagne	0.8	0.8	0.8	0.9
Other/Autres	146.0	125.9	144.4	108.2
Rep. of Korea/Rép de Corée	26.7	9 500	25.9	7 002.0
China/Chine	2.5	0.6	2.9	0.5

JAPAN/JAPON

Table IV(b)/Tableau IV(b)

EXPORTS BY MAJOR PRODUCT AND BY COUNTRY/EXPORTATIONS PAR PRINCIPAUX PRODUITS ET PAR PAYS
1985 & 1986

Quant. : '000 tons/tonnes
Val. : million Y

	1985		1986	
	Quant.	Val.	Quant.	Val.
Total fish and fish products/Total poisson et produits de la pêche	786.4	287.6	760.4	317.0
Fresh, chilled/Frais, sur glace				
Skipjack tuna/Listao	16.3	1.9	16.9	1.7
Yellowfin tuna/Listao	2.2	0.3	0.1	0.0
Frozen, excl. fillets/Congelés, non compris les filets	95.8	20.1	135.5	19.0
Tuna/Listao	48.7	9.6	70.0	9.5
Thailand/Thaïlande	5.2	1.0	26.2	3.0
Frozen fillets/Filets congelés	19.7	12.9	14.9	10.0
Shark/Requin	2.0	0.8	3.0	1.4
United States/Etats-Unis	0.9	0.4	1.2	0.6
Shellfish (live, fresh, chilled, salted, dried and canned)/Coquillages (vivants, frais, sur glace, salés, séchés et en conserve)	18.4	24.1	18.1	23.1
Cuttlefish/Seiche	4.4	2.0	29.0	1.1
United States/Etats-Unis	0.8	0.2	0.8	0.8
Meal/Farine	157.4	15.0	167.2	13.7
Taiwan	148.3	13.9	157.8	12.7
Oil/Huile	250.0	14.6	225.2	7.0
	134.5	7.6	90.0	2.6
Other/Autres	2.8	2.1	2.6	2.1
Taiwan	2.6	2.0	2.6	1.9

NETHERLANDS/PAYS-BAS

Table I/Tableau I

LANDINGS SOLD FOR HUMAN CONSUMPTION/DEBARQUEMENTS DESTINES A LA CONSOMMATION HUMAINE
1985 & 1986[a]

Quant. : tons/tonnes landed weight/Poids débarqué)
Val. : '000 Gld/fl

SPECIES/ESPECES	1985		1986	
	Quant.	Val.	Quant.	Val.
Roundfish/Poissons ronds				
Cod/Morue	21 973	70 973	26 427	83 244
Whiting/Merlan	13 538	23 285	6 147	11 905
Haddock/Eglefin	1 224	3 378	2 890	6 733
Flatfish/Poissons plats				
Plaice/Plie	65 378	176 520	89 976	238 436
Sole	9 964	200 874	15 975	259 434
Herring/Hareng, etc.				
Herring/Hareng	97 646	65 423	81 274	59 330
Horse mackerel/Chinchard) Mackerel/Maquereau)	120 542	86 195	134 159	111 352
Lake IJssel/Lac Ijssel				
Eel/Anguille	681	7 918	677	7 854
Perch/Perche	598	2 621	596	2 612
Shellfish/Coquillages et crustacés				
Mussels/Moules	74 692	67 677	101 089	49 310
Shrimp/Crevettes	7 078	34 882	6 268	26 568
Oysters/Huîtres	1 054	16 371	887	13 894

a) Provisional figures/Chiffres provisoires.

NETHERLANDS/PAYS-BAS

Table II/Tableau II

EXTERNAL TRADE IN FISH AND FISH PRODUCTS/ECHANGES INTERNATIONAUX DE POISSON ET PRODUITS DE LA PECHE
1985 & 1986a)

Quant. : tons/tonnes
Val. : '000 Gld/Fl

	IMPORTS/IMPORTATIONS				EXPORTS/EXPORTATIONS			
	1985		1986		1985		1986	
	Quant.	Val.	Quant.	Val.	Quant.	Val.	Quant.	Val.
Total fish and fish products/Total poisson et produits de la pêche	525 774	1 015 973	476 328	946 687	511 938	1 790 971	559 982	1 870 012
Fresh, chilled/Frais, sur glace	53 867	232 270	53 935	272 890	70 132	515 082	71 411	519 801
Frozen/Congelés	57 326	74 742	52 607	70 429	242 618	386 854	288 014	418 604
Frozen fillets/Filets congelés	5 858	37 278	6 454	42 833	33 443	267 068	33 415	258 478
Salted, dried, smoked/Salés, séchés, fumés	7 920	25 617	9 847	30 545	22 447	87 110	21 351	81 660
Canned/En conserve	16 434	122 205	19 158	126 477	22 088	140 588	24 172	146 719
Shellfish (live, fresh, chilled, salted, dried and canned)/Crustacés et coquillages (vivants, frais, sur glace, salés, séchés et en conserve)	54 144	196 403	51 586	201 851	83 327	343 354	76 837	395 379
Meal/Farine	95 874	93 446	115 882	101 203	21 836	27 047	26 731	27 945
Oil/Huile	234 351	234 012	166 859	100 459	16 047	23 838	18 051	21 426
Other/Autres								

a) Provisional figures and estimates/Chiffres provisoires et estimations.

NETHERLANDS/PAYS-BAS

Table III(a)/Tableau III(a)

IMPORTS BY MAJOR PRODUCTS AND BY COUNTRY/IMPORTATIONS PAR PRINCIPAUX PRODUITS ET PAR PAYS
1985 & 1986a)

Quant. : tons/tonnes (product weight/poids du produit)
Val. : '000 Gld/fl

	1985		1986	
	Quant.	Val.	Quant.	Val.
Total fish and fish products/Total poisson et produits de la pêche				
Fresh, chilled/Frais, sur glace	525 774	1 015 973	476 328	946 687
Herring/Hareng	53 867	232 270	53 935	272 890
Denmark/Danemark	12 461	15 725	12 006	15 536
(France)	6 694	10 985	7 907	12 439
Mackerel/Maquereau	6 487	5 288	3 994	2 946
United Kingdom/Royaume-Uni	4 663	3 917	2 978	2 236
Cod/Morue	8 509	28 367	9 637	34 756
Denmark/Danemark	6 180	22 720	6 472	26 708
Frozen, excl. fillets/Congelés, non compris les filets	57 326	74 742	52 607	70 429
Herring/Hareng	8 617	12 437	11 728	14 499
Ireland/Irlande	403	620	283	425
Denmark/Danemark	2 971	5 814	2 493	4 650
United Kingdom/Royaume-Uni	3 403	3 206	4 053	3 666
Mackerel/Maquereau	39 401	21 694	34 605	17 971
United Kingdom/Royaume-Uni	23 955	13 587	26 925	13 672
Frozen fillets/Filets congelés	5 858	37 278	6 454	42 833
Cod fillets/Filets de morue	2 299	17 016	2 313	18 069
Germany/Allemagne	1 119	7 910	1 208	8 984
Saithe fillets/Filets de lieu noir	2 339	11 659	2 244	11 900
Germany/Allemagne	1 245	6 016	1 311	6 580
Cured/Salés, séchés, fumés	7 920	25 617	9 847	30 545
Herring/Hareng	6 768	14 863	8 290	16 419
Ireland/Irlande	971	1 465	993	1 355
Denmark/Danemark	5 028	10 767	4 928	10 706
Canned/En conserve	16 434	122 205	19 158	126 477
Shellfish (live, fresh, chilled, salted, dried and canned)/Crustacés et coquillages (vivants, frais, sur glace, salés, séchés et en conserve)	54 144	196 403	51 586	201 851
Shrimp/Crevettes	12 594	130 273	12 722	130 045
Germany/Allemagne	5 672	41 065	4 820	35 394
Denmark/Danemark	2 122	12 112	2 231	12 176
Malaysia/Malaysie	1 209	14 363	1 210	11 549
Meal/Farine	95 874	93 446	115 882	101 203
Chile/Chili	63 622	55 849	85 693	72 038
Oil/Huile	234 351	234 012	166 859	100 459
Japan/Japon	102 001	98 647	65 449	38 176

a) Provisional figures and estimates/Chiffres provisoires et estimations.

NETHERLANDS/PAYS-BAS

Table III(b)/Tableau III(b)

EXPORTS BY MAJOR PRODUCT AND BY COUNTRY/EXPORTATIONS PAR PRINCIPAUX PRODUITS ET PAR PAYS
1985 & 1986[a]

Quant. : tons/tonnes (product weight/poids du produit)
Val. : '000 Gld/fl

	1985		1986	
	Quant.	Val.	Quant.	Val.
Total fish and fish products/Total poisson et produits de la pêche	511 938	1 790 971	559 982	1 870 012
Fresh, chilled/Frais, sur glace	70 132	515 082	71 411	519 801
Herring/Hareng	7 295	9 711	6 822	7 909
Germany/Allemagne	5 769	7 403	5 346	5 674
Denmark/Danemark	295	222	41	31
Cod/Morue	15 783	55 869	14 812	55 625
France	8 308	32 766	3 799	21 107
United Kingdom/Royaume-Uni	4 180	10 422	3 239	8 800
Plaice/Plie	12 868	40 799	14 934	47 802
United Kingdom/Royaume-Uni	7 008	24 747	7 110	23 845
Denmark/Danemark	4 107	9 062	5 279	14 110
Sole	10 950	187 966	7 194	156 212
France	3 628	57 077	1 473	30 637
Italy/Italie	3 523	57 292	2 721	58 720
Frozen, excl. fillets/Congelés, non compris les filets	242 618	386 854	288 014	418 604
Herring/Hareng	57 754	53 303	93 326	76 458
Germany/Allemagne	10 852	8 456	13 662	12 274
Poland/Pologne	11 303	9 550	16 253	12 447
Japan/Japon	10 218	13 370	8 680	9 425
Mackerel/Maquereau	117 481	129 218	122 998	105 699
Nigeria	63 489	71 674	64 729	55 408
Egypt/Egypte	17 282	17 665	11 193	8 244
Sole	7 581	105 807	7 878	137 616
Italy/Italie	5 042	64 759	5 476	95 128
Frozen fillets/Filets congelés	33 443	267 068	33 415	258 478
Cured/Salés, séchés, fumés	22 447	87 110	21 351	81 660
Herring/Hareng	17 577	54 146	16 931	50 028
Germany/Allemagne	11 439	31 737	10 123	29 194
Greece/Grèce	848	4 895	724	3 658
Canned/En conserve	22 088	140 588	24 172	146 719
Shellfish (live, fresh, chilled, salted, dried and canned)/Coquillages (vivants, frais, sur glace, salés, séchés et en conserve)	83 327	343 354	76 837	395 379
Shrimp/Crevettes	10 868	135 266	12 840	153 984
France	3 668	29 017	3 902	31 126
BLEU/UEBL	4 120	69 674	5 067	81 635
Mussels/Moules	54 370	85 433	46 283	98 392
France	22 079	29 839	19 635	30 851
BLEU/UEBL	30 776	51 183	24 109	60 680
Meal/Farine	21 836	27 047	26 731	27 945
BLEU/UEBL	18 985	23 623	21 922	23 395
Oil/Huile	16 047	23 868	19 051	21 426
Germany/Allemagne	3 348	4 536	4 197	3 874

a) Provisional figures and estimates/Chiffres provisoires et estimations.

NEW ZEALAND/NOUVELLE ZELANDE

Table I/Tableau I

FISHING FLEET AND FISHERMEN/FLOTTE DE PECHE ET PECHEURS
1985 & 1986

	1985	1986
	Number/Nombre	Number/Nombre
Total vessels - Classification by overall vessel length/ Total des navires - Classification par longueur hors-tout	2 281	1 844
less than 6m/moins de 6m	619	499
6m and under/et moins de 9m	563	398
9m " " " " " 12m	549	450
12m " " " " " 15m	281	224
15m " " " " " 18m	114	99
18m " " " " " 21m	60	53
21m " " " " " 24m	38	35
24m " " " " " 27m	5	7
27m " " " " " 30m	6	9
30m " " " " " 33m	6	7
more than 33m/plus 33m	40	63
Total commercial fishermen/Total des pêcheurs commerciaux	n.a	n.a

Source : New Zealand Ministry of Agriculture and Fisheries/Ministère de l'Agriculture et des Pêches de Nouvelle-Zélande.

NEW ZEALAND/NOUVELLE-ZELANDE

Table II/Tableau II

LANDINGS/DEBARQUEMENTS
1984 & 1985

Quant. : tons (live weight)/tonnes (poids vif)

	1984	1985
1. National landings in domestic ports by domestic vessels/Débarquements nationaux dans les ports nationaux[a]		
For direct human consumption/Pour la consommation humaine directe :		
Finfish/Poisson : Orange roughy/Mérou	21 068	26 649
Snapper/Vivaneau	9 244	9 090
Red cod/Morue	11 280	13 926
Shellfish/Coquillages: Rock lobster/Langouste	5 473	5 489
Scallops/Coquilles St. Jacques	4 660	3 204
Oysters (wild)/Huîtres	9 354	8 755
For other purposes/Pour d'autres buts)	[b]	[b]
TOTAL 1	166 049	160 090
2. Catch by foreign chartered vessels/Prises de navires affrétés par des étrangers		
Finfish/Poisson : Hoki	29 710	27 931
Orange roughy/Mérou	16 203	13 349
Oreo dories	14 487	17 808
Shellfish/Coquillages: Squid/Calmar	54 871	43 084
TOTAL 2	156 250	145 251
3. Catch by foreign vessels/Prises par des navires étrangers		
Finfish/Poisson : Hoki	11 380	10 743
Southern blue whiting/Merlan bleu du Sud	9 978	7 176
Barracouta	7 125	4 674
Shellfish/Coquillages: Squid/Calmar	60 373	42 718
TOTAL 3	118 403	88 531
TOTAL 1 + 2	322 299	305 341
TOTAL 1 + 2 + 3	440 702	393 872

a) Landed values are no longer estimated. However, rankings are on the basis of approximate landed value./Les valeurs débarquées ne sont plus estimées. Cependant, les données sont classées sur la base de la valeur débarquée approximative.
b) Negligible/Négligeable.

Note : There are only small quantities of tuna caught by domestic vessels landed in foreign ports. The landing of fish from foreign vessels into domestic ports has only recently been approved./Il n'y a que de petites quantités de thon prises par des navires domestiques débarqués dans les ports étrangers. Le débarquement de poisson à partir de navires étrangers dans des ports nationaux n'a été approuvé que récemment.

NEW ZEALAND/NOUVELLE-ZÉALANDE

Table III/Tableau III

EXTERNAL TRADE IN FISH AND FISH PRODUCTS/ECHANGES INTERNATIONAUX DE POISSON ET PRODUITS DE LA PECHE
1985 & 1986

Quant. : tons/tonnes
Val. : Imp. NZ$'000 CIF
Exp. NZ$'000 FOB

	IMPORTS/IMPORTATIONS				EXPORTS/EXPORTATIONS			
	1985		1986		1985		1986	
	Quant.	Val.	Quant.	Val.	Quant.	Val.	Quant.	Val.
Total fish and fish products/Total poisson et produits de la pêche	6 859.3	36 377.5	8 076.3	42 717.0	145 170	545 112	158 183	657 342
Fresh, chilled/Frais, sur glace	38.1	221.4	5.1	69.1	5 882	41 843	8 335	63 663
Frozen/Congelés	361.9	302.1	544.0	695.0	57 903	106 961	67 331	147 682
Frozen fillets/Filets congelés	-	-	4.7	50.5	26 153	159 576	35 138	219 353
Salted, dried, smoked/Salés, séchés, fumés	27.2	182.7	36.2	271.3	1 470	2 449	2 835	3 966
Processed (incl. canned)/Transformés (y compris en conserve)	5 213.6	22 372.5	6 397.7	29 510.3	3 762	13 506	3 301	11 464
Shellfish (live, fresh, chilled, salted, dried, canned)/Coquillages (vivants, frais, sur glace, salés, séchés, en conserve)a)	1 218.5	13 298.8	1 088.6	12 120.8	50 000	220 777	41 242	211 214

a) Includes squid and small quantities of fishmeal and oil./Y compris le calmar et de petites quantités de farine et d'huile de poisson.

NEW ZEALAND/NOUVELLE-ZELANDE

Table IV(a)/Tableau IV(a)

IMPORTS OF FISHERIES PRODUCTS/IMPORTATIONS DES PRODUITS DE LA PECHE
1985 & 1986

Quant. : tons/tonnes
Val. : NZ$ CIF

	1985		1986		% change 1986/1985	
	Quant.	Val.	Quant.	Val.	Quant.	Val.
Total fish and fish products/Total poisson et produits de la pêche	6 859.3	36 377 498	8 076.3	42 717 362	17.7	17.4
FISH						
Live/Vivants	-	20 711	-	47 455	-	129.1
Fresh, chilled/Frais, sur glace	0.1	544	1.1	7 890	2 525.0	1 350.3
Frozen/Congelés	361.9	302 109	544.1	695 288	50.3	130.1
Fillets, fresh or chilled/Filets, frais ou sur glace	38.1	200 193	4.0	13 787	-89.5	-93.1
Frozen fillets/Filets congelés	-	-	4.7	50 503	-	-
Salted, dried, smoked/Salés, séchés, fumés						
Meal/Farine	-	-	0.2	2 727	-	-
Cod (not in fillets)/Morue (pas des filets)	0.1	380	0.2	2 383	900	527.1
Fish, dried, salted or in brine/Poisson, séchés, salés ou en saumure	9.6	100 936	15.8	146 349	64.6	45.0
Smoked/Fumés	17.7	81 367	20.0	119 891	13.0	47.3
Prepared or preserved/Préparés ou préservés						
Pastes and similar/Pâtes et similaires	18.0	126 488	10.3	90 204	-42.8	28.7
Herrings, pilchards/Harengs	364.3	952 422	478.2	1 453 001	31.3	52.6
Sardines	2 552.0	6 622 669	2 646.1	7 153 471	3.7	8.0
Salmon/Saumon	1 177.6	9 766 151	1 923.6	13 594 381	63.3	39.2
Other/Autres	1 096.7	4 828 006	987.5	5 703 180	-9.9	18.1
Otherwise packed/Emballés différemment	0.8	7 082	343.6	1 351 960	42 850.0	18 990.0
Calivar	4.0	69 690	7.3	164 109	82.5	135.5
CRUSTACEANS & MOLLUSCS (live)/ CRUSTACES ET MOLLUSQUES (vivants)			1.1	35 008	-	-
Fresh, chilled, frozen/Fraîches, sur glace, congelées						
Shrimps, prawns/Crevettes	142.4	2 265 839	111.7	1 918 303	-21.6	-15.3
Other/Autres	85.3	1 015 398	41.5	482 743	-51.3	-52.4
Salted in brine or dried/Salés en saumure ou séchés	2.6	59 550	1.9	36 808	-26.9	-38.2
Simply boiled in water/Bouillis à l'eau	13.7	203 869	3.0	46 318	-78.1	-77.3
Prepared or preserved/Préparés ou préservés						
Pastes and similar/Pâtes et similaires	138.0	1 153 542	3.1	28 055	-97.8	-97.6
Canned (includes airtight containers)/En conserve (y compris les récipients étanches)	200.5	1 790 469	209.6	1 706 441	4.5	-4.7
Otherwise packed/Emballés différemment	636.0	6 810 083	717.8	7 867 107	12.9	15.5

NEW ZEALAND/NOUVELLE-ZELANDE

Table IV(b)/Tableau IV(b)

EXPORTS TO MAJOR MARKETS/EXPORTATIONS AUX PRINCIPAUX MARCHES
1985 & 1986

Quant. : tons/tonnes
Val. : NZ$'000 FOB

	1985		1986	
	Quant.	Val.	Quant.	Val.
UNITED STATES/ETATS-UNIS				
Total fish and fish products/Total poisson et produits de la pêche	15 324	194 158	21 629	262 799
Finfish, fresh, chilled (excl. fillets)/Poisson maigre, frais, sur glace (non compris les filets)	621	4 264	817	6 166
Finfish, frozen (excl. fillets)/Poisson maigre, congelé, (non-compris les filets)	1 061	4 084	2 128	4 932
Finfish fillets, fresh or chilled/Filets de poisson maigre, frais ou sur glace	871	8 589	1 183	12 101
Finfish fillets, frozen/Filets de poisson maigre, congelés	9 897	103 354	14 057	156 008
Finfish, dried, salted, in brine/Poisson maigre, séché, salé, en saumure	22	157	42	315
Finfish, processed/Poisson maigre, transformé	119	914	18 230	179 537
Rock lobster, fresh, chilled, frozen/Langouste, fraîche, sur glace, congelée	1 547	64 830	1 860	75 445
Squid, fresh, chilled, frozen/Calmar, frais, sur glace, congelé	347	1 789	492	1 912
Other shellfish, fresh, chilled, frozen/Autres crustacés et coquillages, frais, sur glace, congelés	844	6 011	1 026	5 597
Other shellfish, processed/Autres crustacés et coquillages, transformés	12	137	22	308
JAPAN/JAPON				
Total fish and fish products/Total poisson et produits de la pêche	70 746	172 857	76 135	232 240
Finfish, fresh, chilled (excl. fillets)/Poisson maigre, frais, sur glace (non compris les filets)	2 466	18 266	3 728	32 664
Finfish, frozen (excl. fillets)/Poisson maigre, congelé (non compris les filets)	38 978	74 796	39 990	106 817
Finfish fillets, fresh or chilled/Filets de poisson maigre, frais ou sur glace	9 862	26	47	238
Finfish fillets, frozen/Filets de poisson maigre, congelés	2 320	4 950	9 287	22 212
Finfish, dried, salted, in brine/Poisson maigre, séché, salé, en saumure	180	176	1 128	1 562
Finfish, processed/Poisson maigre, transformé	408	1 050	505	1 663
Rock lobster, fresh, chilled, frozen/Langouste, fraîche, sur glace, congelée	583	13 111	935	22 018
Squid, fresh, chilled, frozen/Calmar, frais, sur glace, congelé	24 286	56 052	18 888	39 242
Other shellfish, fresh, chilled, frozen/Autres crustacés et coquillages, frais, sur glace, congelés	646	3 018	1 081	4 785
Other shellfish, processed/Autres crustacés et coquillages transformés	18	124	22 547	150
Other/Autres	851	1 187	525	889
AUSTRALIA/AUSTRALIE				
Total fish and fish products/Total poisson et produits de la pêche	20 060	78 742	21 083	73 723
Finfish, fresh, chilled (excl. fillets)/Poisson maigre, frais, sur glace (non compris les filets)	1 286	6 594	1 555	7 702
Finfish, frozen (excl. fillets)/Poisson maigre, congelé (non compris les filets)	4 868	6 986	6 801	9 818
Finfish fillets, fresh or chilled/Filets de poisson maigre, frais ou sur glace	372	2 064	428	2 241
Finfish fillets, frozen/Filets de poisson maigre, congelés	7 515	36 356	5 673	27 150
Finfish, dried, salted, in brine/Poisson maigre, séché, salé, en saumure	594	1 569	319	707
Finfish, processed/Poisson maigre, transformé	2 274	7 243	2 153	7 323
Rock lobster, fresh, chilled, frozen/Langouste, fraîche, sur glace, congelée	331	6 324	203	3 726
Squid, fresh, chilled, frozen/Calmar, frais, sur glace, congelé	1 418	5 819	1 270	6 342
Other shellfish, fresh, chilled, frozen/Autres crustacés et coquillages, frais, sur glace, congelés	1 118	4 880	1 280	7 284
Other shellfish, processed/Autres crustacés et coquillages transformés	25	688	88	566
Other/Autres	270	221	1 312	864
TOTAL EXPORTS : ALL COUNTRIES/ TOTAL EXPORTATIONS : TOUS LES PAYS	145 170	545 112	158 183	657 342

NEW ZEALAND/NOUVELLE-ZELANDE

Table V/Tableau V

FOREIGN LICENSED VESSELS - ALLOCATIONS AND CATCHES/
NAVIRES ETRANGERS BENEFICIANT DE PERMIS - QUANTITES ALLOUEES ET PRISES[a]

(tons/tonnes : live weight/poids vif)

	Allocations		Catch/Prises	
	1.10.84-30.9.85	1.10.85-30.9.86	1.10.84-30.9.85	1.10.85-30.9.86
Finfish trawl/Pêche au chalut (poisson maigre)				
Japan/Japon	55 100	56 100	22 546	24 514
USSR/URSS	11 500	25 050	10 905	13 526
Republic of Korea/République de Corée	18 050	6 150	12 710	4 715
Total trawl catch/Total pêche au chalut			46 161	42 659
Bottom longline/Pêche aux lignes de fond				
Japan/Japon	-	-	-	-
Republic of Korea/République de Corée	-	-	-	-
Total longline/Total lignes de fond	-	-	-	-
Total finfish catch/Total des prises de poisson maigre			46 161	42 659
Squid trawl/Pêche au chalut (calmar)				
Japan/Japon	7 920	7 920	6 719	6 524
USSR/URSS	8 500	8 700	8 406	8 365
Republic of Korea/République de Corée	2 490	2 240	2 289	1 382
Total trawl catch/Total des prises au chalut			17 414	
Jig/Turlutte				
Japan/Japon	85[b]	24 000	23 897	23 170
Republic of Korea/République de Corée	6[b]	1 500	583	1 081
Total jig catch/Total des prises à la turlutte	-		24 480	24 251
Total squid catch/Total des prises de calmar	-		41 894	
Southern bluefin tuna/Thon rouge				
Japan/Japon		33[b]	1 718	1 337

a) Prov.

b) Vessels recruited/Navires recrutés.

NEW ZEALAND/NOUVELLE-ZELANDE

Table VI/Tableau VI

FOREIGN LICENSED FISHING FEES/NAVIRES ETRANGERS BENEFICIANT DE PERMIS
1986 & 1987

Species/Espèces	Management/Gestion	Quant. allocated/allouée (tonnes)	Licence fee/ Permis ($/tonnes)	Revenue/ Revenu
Alfonsino	Central (West), Challenger	10	173	1 733
	Chatham Rise, Southland, Sub-Antarctic	165	173	28 595
Barracouta/ Thyrsite	Central (West), Challenger	765	70	53 189
	Chatham Rise	360	70	25 030
	Southland, Sub-Antarctic	1 980	70	137 666
Blue cod/Morue charbonnière	Central (West), Challenger	10	70	700
	Chatham Rise	10	53	525
	Southland, Sub-Antarctic	35	60	2 100
Blue moki	Central (West), Challenger	10	200	2 000
	Chatham Rise	10	30	300
	Southland, Sub-Antarctic	15	30	450
Bluenose	Central (West), Challenger	10	90	900
	Chatham Rise, Southland, Sub-Antarctic	15	83	1 238
Bleu warehou	Central (West), Challenger	30	161	4 838
	Chatham Rise, Southland, Sub-Antarctic	300	107	32 200
Elephant fish	Central (West), Challenger	10	200	2 000
	Chatham Rise	10	200	2 000
	Southland, Sub-Antarctic	5	200	1 000
Flatfish/ Poisson plat	Central (West), Challenger	10	200	2 000
	Chatham Rise, Southland, Sub-Antarctic	30	200	6 000
Gemfish	Central (West), Challenger	10	115	1 150
	Chatham Rise, Southland, Sub-Antarctic	500	70	35 000
Grey mullet/ Mulet labeon	Central (West), Challenger	10	60	600
	Chatham Rise, Southland, Sub-Antarctic	15	60	900
Gurnard/Grondin	Central (West), Challenger	25	187	4 675
	Chatham Rise, Southland, Sub-Antarctic	80	200	16 000
Hake/Merlu	Central (West), Challenger, Southland	610	166	101 283
	Chatham Rise	100	166	16 604
Hapuku/Bass	Central (West), Challenger	10	200	2 000
	Chatham Rise	10	200	2 000
	Southland, Sub-Antarctic	65	200	13 000
Hoki	All areas/Toutes zones	15 750	62	980 660
Jack mackerel/ Chinchard	Central (West), Challenger	4 790	70	333 041
John Dory/ Saint-Pierre	Central (West), Challenger	25	113	2 813
	Chatham Rise, Sub-Antarctic	15	113	1 688
Ling/Lingue	Central (West), Challenger	10	200	2 000
	Chatham Rise	260	110	28 600
	Southland	880	110	96 600
	Sub-Antarctic	875	110	96 250
Orange roughy/ Mérou	Central (West), Challenger	20	200	4 000
	Chatham Rise, Southland, Sub-Antarctic	75	200	15 000
Oreo dory	Central (West), Challenger, Southland	50	53	2 646
	Chatham Rise	110	53	5 822
	Sub-Antarctic	480	53	25 404
Red cod	Central (West), Challenger	50	23	1 125
	Chatham Rise, Southland, Sub-Antarctic	1 850	18	32 375
Rig	Central (West), Challenger	10	200	2 000
	Chatham Rise, Southland, Sub-Antarctic	25	200	5 000
School shark	Central (West), Challenger	30	200	6 000
	Chatham Rise	20	200	4 000
	Southland, Sub-Antarctic	100	200	20 000
Silver warehou	Central (West), Challenger	20	138	2 760
	Chatham Rise, Southland, Sub-Antarctic	1 660	138	229 111
Snapper/Vivaneau	Central (West), Challenger	10	200	2 000
	Chatham Rise, Southland, Sub-Antarctic	15	200	3 000
Stargazer/ Uranoscope	Central (West), Challenger	10	35	350
	Chatham Rise	125	45	5 625
	Southland, Sub-Antarctic	195	55	10 725
Tarakihi	Central (West), Challenger	75	200	15 000
	Chatham Rise	40	200	8 000
	Southland, Sub-Antarctic	40	200	8 000
Trevally/Carangue australienne	Central (West), Challenger	10	168	1 680
	Chatham Rise	15	72	1 080
	Southland, Sub-Antarctic			
Mixed/Divers	Sub-Antarctic	No limit	40	-
	Elsewhere/Ailleurs	No limit	49	-
Estimated finfish total revenue/Revenu total estimé pour le poisson maigre				2 450 229
Squid/Calmar	Southern Islands	12 800	239	3 055 094
	Elsewhere Jig/Turlutte ailleurs	29 000	261	7 569 000
	Elsewhere all methods/Ailleurs, toutes méthodes	5 760	239	1 374 792
Estimated squid total revenue/Revenu total estimé pour le calmar				11 998 887
Tuna - albacore fishery/Thon - pêche au germon		na	$8 000 per vessel/navire	
Tuna - bluefin fishery/Thon - pêche au thon rouge		na	$138 000 per vessel/navire	
Fish carriers/Navires transporteurs		na	$5.00a)	
Support vessels/Navires de support		na	$2.50b)	

a) Per tonne of carrying capacity for each voyage./Par tonne de capacité de transport pour chaque voyage.

b) Per gross registered ton for each voyage./Par tonneau de jauge brute pour chaque voyage.

NORWAY/NORVEGE

Table I/Tableau I

LANDINGS AND VALUES/DEBARQUEMENTS ET VALEURS
1985[a] & 1986[a]

Quant. : '000 tons/tonnes
Val. : '000 NKr/KrN

	1985		1986	
	Quant.	Val.	Quant.	Val.
TOTAL LANDINGS/DEBARQUEMENTS TOTAUX[b]	2 071.3	4 480.2	1 842.2	4 989.4
For direct human consumption/Pour la consommation humaine directe	868.7	3 766.6	907.5	4 461.2
Finfish/Poisson				
Cod/Morue	243.8	1 215.3	263.0	1 632.9
Haddock/Eglefin	24.8	113.5	57.8	282.9
Saithe/Lieu noir	201.6	576.7	127.9	514.2
Whiting/Merlan	0.1	0.4	0.1	0.5
Plaice/Plie	0.8	3.2	0.6	3.0
Greenland halibut/Flétan noir	5.5	25.3	7.8	46.7
Redfish/Sébaste	21.8	83.9	24.2	96.9
Other demersal/Autres démersaux	72.7	420.5	77.0	505.2
Herring/Hareng	87.1	172.2	157.4	239.4
Mackerel/Maquereau	45.5	89.5	68.1	107.0
Sprat	10.3	38.7	4.7	20.3
Other pelagic/Autres pélagiques	40.9	101.6	40.2	51.2
Sharks/Requins	4.7	14.4	3.9	13.3
Heads/Têtes		0.4		0.4
Other species/Autres espèces	0.5	9.1	0.5	8.1
Unspecified/Non spécifiés[c]	1.5	9.5	15.8	109.8
Shellfish (crustaceans, molluscs and squid)/Coquillages (crustacés, mollusques et calmar)				
Shrimp/Crevettes	91.0	829.2	56.2	812.5
Other shellfish/Autres coquillages	2.4	18.8	2.2	16.7
Squid/Calmar	13.7	44.7	0.1	0.3
For other purposes/Pour d'autres buts	1 202.6	713.6	934.7	528.2
Capelin/Capelan	606.1	362.3	240.8	154.2
Herring/Hareng	152.0	102.6	167.6	88.2
Mackerel/Maquereau	70.3	59.4	88.3	68.4
Sprats	6.8	5.4	0.1	0.2
Norway pout/Tacaud norvégien	113.9	63.0	64.5	34.2
Sandeel/Lançon	14.7	8.0	85.8	49.1
Blue whiting/Merlan bleu	232.3	97.8	279.5	122.5
Other/Autres	2.7	3.5	5.3	4.4
Squid/Calmar	0.1	0.0	-	-
Sharks/Requins	3.1	7.5	2.5	3.9
Heads/Têtes	-	2.7	-	2.3
Unspecified/Non spécifiés	0.5	1.3	0.2	0.9

a) Preliminary figures/Chiffres provisoires.

b) Excluding seaweed and fish farming/Non compris les algues et l'aquaculture.

c) Including scallops/Y compris les coquilles St. Jacques.

NORWAY/NORVEGE

Table II/Tableau II

EXTERNAL TRADE IN FISH AND FISH PRODUCTS/ECHANGES INTERNATIONAUX DE POISSON ET PRODUITS DE LA PECHE
1985 & 1986

Quant. : '000 metric tons/tonnes métriques
Val. : Million Nkr/KrN

	IMPORTS/IMPORTATIONS				EXPORTS/EXPORTATIONS			
	1985		1986		1985		1986	
	Quant.	Val.	Quant.	Val.	Quant.	Val.	Quant.	Val.
Total fish and fish products/Total poisson et produits de la pêche	127.6	620.9	140.1	790.2	828.6	8 076.6	730.3	8 911.2
Fresh, chilled/Frais, sur glace	21.0	96.4	9.5	58.8	74.5	1 409.5	149.0	1 897.4
Frozen, whole/Congelés, entiers	15.6	38.7	8.5	44.2	85.1	650.8	100.2	602.0
Frozen fillets/Filets congelés	0.6	8.3	0.9	17.3	86.4	1 175.3	85.6	1 373.0
Salted, dried, smoked/Salés, séchés, fumés	1.6	17.4	4.6	89.8	88.4	1 626.0	84.2	2 183.1
Canned/En conserve[a]	2.0	33.5	2.3	45.6	11.1	287.8	11.3	287.8
Other prepared fish/Autres poissons préparés	1.3	27.5	1.4	34.0	22.1	355.3	22.3	408.3
Shellfish (live, fresh, chilled, salted, dried and canned)/Coquillages (vivants, frais, sur glace, salés, séchés et en conserve)	20.7	199.9	16.8	316.0	41.2	1 320.1	26.8	1 248.0
Meal/Farine	2.5	11.2	5.1	18.0	173.7	555.8	95.8	235.6
Oil/Huile[b]	51.1	156.7	71.8	115.2	114.3	300.6	35.4	97.5
Other/Autres	10.3	26.0	19.2	51.3	131.8	510.0	119.8	578.5

a) In airtight containers/En récipients étanches.

b) Oil of herring and other fish or marine mammals, fish liver oil, whale oil/Huile de hareng et autres poisson ou mammifères marins, huile de foie de morue, huile de baleine.

NORWAY/NORVEGE

Table III(a)/Tableau III(a)

IMPORTS BY MAJOR PRODUCTS AND BY COUNTRY/IMPORTATIONS PAR PRINCIPAUX PRODUITS ET PAR PAYS
1985 & 1986

Quant. : '000 metric tons/tonnes métriques
Val. : Million NKr/KrN

	1985		1986	
	Quant.	Val.	Quant.	Val.
Total fish and fish products/Total poisson et produits de la pêche	127.6	620.9	140.1	790.2
Fresh, chilled (incl. fillets)/Frais, sur glace (y compris les filets)	21.0	96.4	9.5	58.8
Salmonidae, live (i.e. fry, smolt)/Salmonidés, vivants (c.à.d. appâts, tacon)	0.4	48.3	0.3	31.7
Trout/Truite	0.5	19.0	0.2	8.2
Capelin/Capelan	17.8	9.8	8.3	4.9
Sweden	0.5	25.5	0.3	20.2
Finland/Finlande	0.4	21.4	0.2	8.6
Frozen (excl. fillets)/Congelés (non compris les filets)	15.6	38.7	8.5	44.2
Cod/Morue	0.0	0.4	2.1	19.5
Sprat, anchovy, sardine/Sprat, anchois, sardines	4.8	12.3	4.1	8.2
USSR/URSS	7.3	10.5	3.8	21.2
Sweden/Suède	1.6	5.2	2.2	6.3
Frozen fillets/Filets congelés	0.6	8.3	0.9	17.3
Other flatfish than Greenland halibut/Poisson plat autre que le flétan noir	0.2	4.5	0.3	8.3
Denmark/Danemark	0.2	3.9	0.3	7.6
Netherlands/Pays-Bas	0.1	1.6	0.2	3.8
Dried, salted/Séchés, salés	1.6	17.4	4.6	89.8
Cod/Morue	1.4	14.9	3.0	66.1
Saithe/Lieu noir	0.0	0.2	1.0	13.8
Canada	0.1	1.4	2.9	58.2
Denmark/Danemark	1.0	9.9	0.6	16.0
Crustaceans and molluscs/Crustacés et mollusques	20.7	199.1	16.8	316.0
Deep water prawns/Crevettes d'eaux profondes	13.6	120.9	11.4	216.0
Molluscs (other than oysters)/Mollusques (autres que les huîtres)	5.9	28.9	3.9	36.2
Lobster/Homard	0.2	10.2	0.2	14.0
Denmark/Danemark	0.9	18.3	2.6	63.4
USSR/URSS	12.0	79.9	4.6	54.0
Fish, prepared or canned/Poisson préparé ou en conserve	3.5	66.3	3.7	79.6
Herring/Hareng	1.5	23.7	1.9	37.6
Breaded fillets/Filets panés	0.3	6.2	0.4	9.9
Anchovy/Anchois	0.3	7.2	0.2	8.3
Sweden/Suède	1.7	29.0	1.4	31.8
Denmark/Danemark	1.0	17.9	1.2	23.3
Meal and powder of fish, crustaceans or molluscs/Farine et poudre de poisson, crustacés ou mollusques	2.5	11.2	5.1	18.0
Denmark/Danemark	0.0	0.1	4.0	15.0
Oil of herring and other fish/Huile de hareng et autres poissons	41.3	111.0	71.8	115.2
Iceland/Islande	27.5	69.4	30.3	56.5
Denmark/Danemark	8.1	20.8	13.9	18.8
Other/Autres	20.8	71.7	19.2	51.3
Fish liver oil/Huile de foie de poisson	10.5	45.7	7.4	23.3
Fish offal/Déchets de poisson	10.0	10.6	11.4	11.2
Aquarium fish/Poissons d'aquarium	0.0	7.9	0.0	9.0
Faroe Islands/Iles Féroé	5.2	5.4	10.9	10.9
Sweden/Suède	0.0	4.2	0.0	4.8

NORWAY/NORVEGE

Table III(b)/Tableau III(b)

EXPORTS BY MAJOR PRODUCTS AND BY COUNTRY/EXPORTATIONS PAR PRINCIPAUX PRODUITS ET PAR PAYS
1985 & 1986

Quant. : '000 metric tons/tonnes métriques
Val. : Million NKr/KrN

	1985		1986	
	Quant.	Val.	Quant.	Val.
Total fish and fish products/Total poisson et produits de la pêche	837.0	8 191.1	730.3	8 911.2
Fresh, chilled (incl. fillets)/Frais, sur glace (y compris les filets)	74.5	1 409.5	149.0	1 897.4
Reared salmon/Saumon d'élevage	21.4	1 160.9	34.4	1 458.7
Saithe/Lieu noir	15.0	55.9	11.3	56.1
Herring/Hareng	23.0	61.3	69.3	145.4
United States/Etats-Unis	7.5	387.2	9.5	426.4
France	7.9	224.1	13.6	351.5
Denmark/Danemark	28.5	210.5	47.5	341.4
Germany/Allemagne	17.8	224.6	19.4	271.5
Frozen (excl. fillets)/Congelés (non compris les filets)	85.1	650.8	100.2	602.0
Reared salmon/Saumon d'élevage	2.5	147.7	4.5	205.1
Mackerel/Maquereau	17.8	54.6	50.9	128.2
Herring/Hareng	26.6	119.1	28.2	110.8
Japan/Japon	36.9	258.0	33.5	165.4
France	1.8	76.2	5.9	190.4
Germany/Allemagne	6.5	41.5	14.5	102.8
United Kingdom/Royaume-Uni	6.5	57.7	9.0	66.4
Frozen fillets/Filets congelés	86.4	1 175.3	85.6	1 373.0
Cod/Morue	33.5	634.9	38.7	806.1
Saithe/Lieu noir	45.4	421.0	23.8	282.8
Haddock/Eglefin	3.8	77.2	9.2	184.4
United Kingdom/Royaume-Uni	25.4	404.6	28.4	501.0
Germany/Allemagne	22.7	194.3	21.4	236.0
United States/Etats-Unis	10.4	186.5	9.6	185.0
Klipfish	58.2	994.9	50.0	1 188.9
Cod/Morue	19.5	473.6	18.7	566.7
Saithe/Lieu noir	23.5	269.0	14.2	238.7
Ling/Lingue	7.6	147.5	7.5	203.6
Brazil/Brésil	15.1	247.8	20.3	521.7
Portugal	5.8	117.1	6.7	162.1
France	3.9	88.9	3.5	99.9
Italy/Italie	4.4	102.1	3.0	79.4
Stockfish	6.3	348.6	12.9	585.9
Cod/Morue	5.0	304.8	8.1	472.0
Saithe/Lieu noir	0.8	19.6	3.1	60.2
Tusk/Brosme	0.3	11.1	1.3	33.8
Italy/Italie	3.9	261.9	3.0	322.5
Nigeria	-	-	6.4	154.5
Cameroon/Cameroun	0.7	21.8	1.0	30.2
Sweden/Suède	0.2	8.1	0.2	11.9
Other salted fish/Autres poissons salés	23.9	282.5	21.3	408.3
Cod/Morue	5.7	99.1	8.8	186.5
Herring/Hareng	11.8	43.7	5.7	37.9
Italy/Italie	6.2	92.7	6.1	160.5
Sweden/Suède	4.5	53.3	3.5	59.9
Portugal	1.0	14.9	1.7	33.4
Crustaceans and molluscs/Crustacés et mollusques	41.2	1 320.1	26.8	1 248.0
Deep water prawns/Crevettes d'eaux profondes	32.4	1 221.4	22.2	1 073.6
United Kingdom/Royaume-Uni	9.6	381.9	5.2	320.0
Sweden/Suède	7.3	212.6	5.6	252.6
United States/Etats-Unis	7.2	294.3	3.2	188.5
Fish, prepared or canned/Poisson préparé ou en conserve	33.3	643.1	33.5	696.1
Herring/Hareng	7.1	190.3	8.1	187.7
Breaded fillets/Filets panés	12.5	207.1	11.5	232.1
Sprat and sardines/Sprat et sardines	2.4	76.3	2.1	68.1
United States/Etats-Unis	4.8	172.7	5.9	162.6
Sweden/Suède	6.7	111.3	6.9	132.8
France	4.3	63.3	3.4	59.6
Switzerland/Suisse	2.6	47.4	2.8	59.6
Meal and powder of fish, crustaceans or molluscs/Farine et poudre de poisson, crustacés ou mollusques	173.7	555.8	95.8	235.6
Sweden	88.7	273.7	58.9	176.5
Switzerland/Suisse	22.1	67.3	9.8	27.3
Finland/Finlande	24.1	85.9	5.3	17.4
United Kingdom/Royaume-Uni	12.0	42.6	5.2	10.8
Oil of herring and other fish/Huile de hareng et autres poissons	114.3	300.6	35.4	97.5
United Kingdom/Royaume-Uni	73.3	186.5	17.8	26.6
Germany/Allemagne	22.1	55.5	13.5	24.6
Netherlands/Pays-Bas	9.6	25.1	0.8	1.2
Other/Autres	131.8	510.0	119.8	578.5
Hardened fat/Graisses solides	46.4	214.4	38.3	166.4
Roe, liver, milt/Rogue, foie, laitance	6.9	77.7	6.8	103.8
Other fish products/Autres produits de poisson	69.4	61.4	66.0	63.1

PORTUGAL

Table I/Tableau I

LANDINGS AND VALUES/DEBARQUEMENTS ET VALEURS
1985 & 1986

Quant. : tons/tonnes (live weight/poids vif)
Val. : '000 Esc

	1985		1986	
	Quant.	Val.	Quant.	Val.
Total landings in national ports/Débarquements totaux dans les ports nationaux	388 842	31 940 356	396 585	47 465 658
For direct human consumption/Pour la consommation humaine directe				
Fish/Poisson				
Species/Espèces: Sardine	111 887	3 017 791	103 558	3 323 316
Hake/Merlu	29 450	3 384 453	37 638	7 413 350
Horse mackerel/Chinchard	23 687	4 439 646	28 558	5 940 574
Shellfish/Crustacés et coquillages				
Species/Espèces: Crustaceans/Crustacés	1 833	1 329 202	2 935	2 340 956
Molluscs/Mollusques	9 985	1 753 022	10 863	2 522 359

PORTUGAL

Table II/Tableau II

EXTERNAL TRADE IN FISH AND FISH PRODUCTS/ECHANGES INTERNATIONAUX DE POISSON ET PRODUITS DE LA PECHE
1985 & 1986a)

Quant.: '000 tons/tonnes
Val.: '000 Esc

	IMPORTS/IMPORTATIONS				EXPORTS/EXPORTATIONS			
	1985		1986		1985		1986	
	Quant.	Val.	Quant.	Val.	Quant.	Val.	Quant.	Val.
Total fish and fish products/Total poisson et produits de la pêche	146	34 212 594	155	38 769 506	65	18 368 727	74	22 563 873
Fresh, chilled/Frais, sur glace	13	1 949 066	29	3 948 691	8	2 691 375	10	4 422 586
Frozen/Congelés	34	4 246 859	42	5 577 093	10	2 106 646	18	3 709 425
Frozen fillets/Filets congelés	0.1	29 046	0.1	30 040	1	350 878	2	618 448
Cured/Salés, séché, fumés	91	27 184 087	69	28 014 830	1	303 582	1	560 662
Canned/En conserve	0	22 676	0.3	93 762	36	9 224 629	35	10 898 133
Shellfish (live, fresh, chilled, salted, dried and canned)/Coquillages (vivants, frais, sur glace, salés, séchés et en conserve)	0.3	50 525	0.5	80 808	3	408 214	4	775 219
Meal/Farine	5	269 035	10	535 288	-	-	-	-
Oil/Huile	2	140 680	2	117 552	3	1 038 574	2	449 400

a) Prov.

SPAIN/ESPAGNE

Table I/Tableau I

EXTERNAL TRADE IN FISH AND FISH PRODUCTS/ECHANGES INTERNATIONAUX DE POISSON ET PRODUITS DE LA PECHE
1985 & 1986

Quant. : tons/tonnes
Val. : Ptas million

	IMPORTS/IMPORTATIONS				EXPORTS/EXPORTATIONS			
	1985		1986		1985		1986	
	Quant.	Val.	Quant.	Val.	Quant.	Val.	Quant.	Val.
Total fish and fish products/Total poisson et produits de la pêche	323 355	70 131	351 141	100 997	230 726	60 429	220 178	55 436
Fresh, chilled/Frais, sur glace	79 265	16 814	58 916	23 149	12 816	4 515	28 215	6 803
Frozen/Congelés	67 955	12 788	91 355	15 765	54 186	7 611	61 728	8 701
Frozen fillets/Filets congelés	-	-	-	-	-	-	-	-
Cured/Salés, séchés, fumés	21 198	6 932	28 710	10 783	20 105	7 116	14 592	6 365
Canned/En conserve	1 922	881	3 327	1 800	29 749	10 433	19 106	8 118
Shellfish (live, fresh, chilled, salted dried, and canned)/Coquillages (vivants, frais, sur glace, salés, séchés et en conserve)	50 926	8 804	65 674	33 025	30 219	4 383	29 156	5 370
Meal/Farine	14 195	806	9 434	475	1 453	77	5 897	278
Oil/Huile	7 338	792	10 485	725	9 961	1 197	3 838	699
Other (Cephalopods)/Autres (Cephalopodes)	80 556	22 314	83 240	15 274	72 237	25 097	57 646	19 102

SPAIN/ESPAGNE

Table II(a)/Tableau II(a)

IMPORTS BY MAJOR PRODUCTS AND BY COUNTRY/IMPORTATIONS PAR PRINCIPAUX PRODUITS ET PAR PAYS
1985 & 1986

Quant. : tons/tonnes
Val. : Ptas million

	1985		1986	
	Quant.	Val.	Quant.	Val.
Total fish and fish products/Total poisson et produits de la pêche	323 355	70 131	351 141	100 997
Fresh, chilled/Frais, sur glace	79 265	16 814	58 916	23 149
Cod/Morue	6 819	1 859	4 689	1 711
Hake/Merlu	9 167	4 071	11 691	6 073
Codfish/Espèces similaires	1 021	188	1 986	357
France	20 783	5 800	14 193	6 669
United Kingdom/Royaume-Uni	10 573	3 392	17 440	6 238
Denmark/Danemark	4 347	1 700	3 675	1 922
Frozen, incl. fillets/Congelés, y compris filets	67 955	12 788	91 355	15 765
Hake/Merlu	17 320	3 732	25 797	4 678
Tuna/Thon	13 458	1 873	21 886	3 136
Liver and roes/Foie et rogues	769	274	1 285	590
Chile/Chili	13 368	3 232	11 012	2 493
Morocco/Maroc	8 624	1 367	9 183	1 385
South Korea/Corée du Sud	5 230	909	3 503	650
Shellfish (live, fresh, chilled, salted, dried and canned)/Coquillages (vivants, frais, sur glace, salés, séchés et en conserve)	50 926	8 804	65 674	33 025
Grooved carpet shell and quahaug/Palourde et praire	40 852	5 108	35 380	11 218
Canned/Conserves	8 789	3 173	8 608	4 327
Shrimp/Crevette	311	170	244	249
Italy/Italie	28 337	2 261	15 781	3 406
Netherlands/Pays-Bas	6 834	2 173	7 392	3 561
France	3 416	1 022	3 171	1 873
Meal/Farine	14 195	806	9 434	475
Senegal	-	-	2 048	106
Denmark/Danemark	109	9	-	-
Oil/Huile	7 338	792	10 485	725
Portugal	3 010	524	1 832	315
Norway/Norvège	1 346	85	39	8
Iceland/Islande	1 491	84	5 870	298
Other/Autres	103 676	30 127	115 277	27 858
Frozen cephalopods/Céphalopodes congelés	61 347	12 931	57 956	13 712
Frozen shellfish/Coquillages congelés	13 849	8 476	26 632	20 962
Dried and salted fish/Poisson séché et salé	21 198	6 932	28 710	10 783
Poland/Pologne	16 887	2 903	11 269	1 444
Iceland/Islande	11 391	3 954	12 243	5 140
Morocco/Maroc	8 175	2 438	32 885	4 119

SPAIN/ESPAGNE

Table II(b)/Tableau II(b)

EXPORTS BY MAJOR PRODUCTS AND BY COUNTRY/EXPORTATIONS PAR PRINCIPAUX PRODUITS ET PAR PAYS
1985 & 1986

Quant. : tons/tonnes
Val. : Ptas million

	1985		1986	
	Quant.	Val.	Quant.	Val.
Total fish and fish products/Total poisson et produits de la pêche	230 726	60 429	220 178	55 436
Fresh, chilled/Frais, sur glace	12 816	4 515	28 215	6 803
Tuna/Thon	2 713	1 404	2 792	1 213
Italy/Italie	2 497	1 251	3 970	1 752
Japan/Japon	1 533	1 001	817	622
France	1 828	484	3 745	833
Frozen, incl. fillets/Congelés, y compris les filets	54 186	7 611	61 728	8 701
Tuna/Thon	26 749	4 444	42 071	5 957
Hake/Merlu	48 051	868	5 745	809
Cod/Morue	141	40	184	55
Italy/Italie	16 047	2 847	25 857	4 131
Portugal	13 379	1 448	12 439	1 367
France	4 453	642	4 480	655
Shellfish (live, fresh, chilled, salted, dried and canned)/Coquillages (vivants, frais, sur glace, salés, séchés et en conserve)	30 219	4 383	29 156	5 370
Canned/En conserve	7 431	2 634	6 650	2 656
Mussels/Moules	22 309	1 654	20 926	1 441
Shrimp/Crevette	14	29	8	23
France	11 366	1 389	11 971	1 455
Italy/Italie	12 430	892	10 854	1 411
United States/Etats-Unis	1 841	837	1 197	644
Meal/Farine	1 453	77	5 897	278
Portugal	1 400	72	3 555	171
Italy/Italie	5	2	1 048	49
Denmark/Danemark	24	1	-	-
Oil/Huile	9 961	1 197	3 838	699
Japan/Japon	836	699	659	530
Portugal	3 018	155	1 375	81
Norway/Norvège	47	42	1	1
Other/Autres	122 091	42 646	91 344	33 585
Frozen cephalopods/Céphalopodes congelés	66 780	22 944	52 338	18 647
Canned fish/Poisson en conserve	29 749	10 433	19 106	8 118
Dried and salted fish/Poisson séché et salé	20 105	7 116	14 592	6 365
Japan/Japon	44 151	17 551	31 498	14 269
Italy/Italie	14 648	6 415	9 114	4 077
United States/Etats-Unis	6 144	3 033	5 160	2 563

SWEDEN/SUEDE

Table I/Tableau I

LANDINGS AND VALUES/DEBARQUEMENTS ET VALEURS

1985 & 1986[a)]

Quant. : '000 tons/tonnes (landed weight)/(poids débarqué)
Val. : Million SKr/KrS

	1985		1986	
	Quant.	Val.	Quant.	Val.
1. National landings in domestic ports[b)]				
For direct consumption/Pour la consommation directe	156.7	632.3	136.6	626.5
Finfish/Poisson: Cod/Morue	51.3	250.2	43.4	275.2
Herring/Hareng	86.7	139.2	75.5	119.9
Shellfish/Coquillages: Norway lobster/Langoustine	1.1	42.8	1.2	53.6
Deepwater prawn/Crevette nordique	1.5	37.9	1.4	41.1
For other purposes/Pour d'autres buts[c)]	41.4	23.0	36.7	18.1
Herring/Hareng	32.0	d)		
Sprat	12.9	d)		
Blue whiting/Merlan bleu	3.6	d)		
TOTAL 1	198.1	655.2	173.2	644.7
2. National landings in foreign ports/Débarquements nationaux dans les ports étrangers				
For direct consumption/Pour la consommation directe	18.1	51.8	14.2	48.2
Herring/Hareng	16.4	39.0	12.1	31.2
Cod/Morue	0.7	4.2	1.0	7.2
For other purposes/Pour d'autres buts	9.7	5.8	13.1	6.5
TOTAL 2	27.8	57.6	27.3	54.8
3. Foreign landings in domestic ports/Débarquements étrangers dans les ports nationaux				
For direct consumption/Pour la consommation directe	0.1	1.2	0.1	1.5
Salmon/Saumon	0.1	1.2	0.1	1.5
For other purposes/Pour d'autres buts	-	-	-	-
TOTAL 3	0.1	1.2	0.1	1.5
TOTAL 1 + 3	198.2	656.4	173.3	646.2

a) Preliminary figures./Chiffres provisoires.

b) Quantities from freshwater fishing are not included. The price supplements are not included in the value./Non compris les quantités pêchées en eau douce. Les suppléments de prix ne sont pas compris dans la valeur.

c) Reduction to meal and oil/Réduction en farine et huile.

d) Breakdown of species not available/Ventilation en espèces non disponible.

SWEDEN/SUEDE

Table II/Tableau II

EXTERNAL TRADE IN FISH AND FISH PRODUCTS/ECHANGES INTERNATIONAUX DE POISSON ET PRODUITS DE LA PECHE
1985 & 1986

Quant. : '000 tons (product weight)/tonnes (poids du produit)
Val. : SKr/KrS million

	IMPORTS/IMPORTATIONS				EXPORTS/EXPORTATIONS			
	1985		1986		1985		1986	
	Quant.	Val.	Quant.	Val.	Quant.	Val.	Quant.	Val.
Total fish and fish products/Total poisson et produits de la pêche	240.3	2 125.8	229.6	2 393.7	130.8	715.2	113.7	709.9
Fresh, chilled/Frais, sur glace a) b)	9.1	201.7	11.2	246.0	86.5	362.4	69.9	329.1
Frozen, whole/Congelés, entiers	6.8	161.4	8.3	160.7	7.3	47.1	5.1	29.5
Frozen fillets/Filets congelés	14.0	266.1	13.7	312.4	1.8	33.0	1.0	20.4
Other prepared frozen fish products/Autres produits de poisson congelé, préparés c)	7.4	127.0	8.6	169.8	1.1	23.2	0.9	22.3
Salted, dried, smoked/Salés, séchés, fumés d)	11.8	147.0	13.0	195.5	1.3	22.0	1.5	33.6
Canned/En conserve e)	10.3	177.0	11.0	197.9	5.0	94.2	5.3	111.7
Shellfish (live, fresh, chilled, salted, dried and canned)/Coquillages (vivants, frais, sur glace, salés, séchés et en conserve)	20.7	566.9	20.1	707.7	1.9	75.0	2.4	103.9
Meal/Farine	112.7	371.0	108.6	331.7	0.0	0.1	2.9	8.0
Oil/Huile	20.1	57.5	10.5	23.0	2.4	6.0	2.4	3.6
Other/Autres f)	27.5	50.0	24.4	49.0	23.5	52.2	22.4	47.8

a) Including landings abroad/Y compris les débarquements dans les ports étrangers.
b) Including fresh fillets/Y compris les filets frais.
c) Breaded fish fillets, fishsticks, etc/Filets de poisson pané, bâtonnets de poisson, etc.
d) Including spiced fish/Y compris le poisson assaisonné.
e) Including semi-preserves/Y compris les semi-conserves.
f) Including fish-waste/Y compris les déchets.

SWEDEN/SUEDE

Table III(a)/Tableau III(a)

IMPORTS BY MAJOR PRODUCTS AND BY COUNTRY/IMPORTATIONS PAR PRINCIPAUX PRODUITS ET PAR PAYS
1985 & 1986

Quant. : '000 tons/tonnes
Val. : million SKr/KrS

	1985		1986	
	Quant.	Val.	Quant.	Val.
Total fish and fish products/Total poisson et produits de la pêche	240.3	2 125.7	229.6	2 393.7
Fresh, chilled/Frais, sur glace a)	9.1	201.7	11.2	246.0
Plaice/Plie	1.5	27.9	1.7	35.8
Salmon/Saumon	1.2	74.1	1.9	88.3
Haddock	1.1	7.0	1.0	7.8
Mackerel/Maquereau	1.0	6.7	1.0	7.1
Denmark/Danemark	4.9	82.4	5.3	105.6
Norway/Norvège	3.0	89.2	5.1	112.6
Frozen, excl. fillets/Congelés, non compris les filets	6.8	161.4	8.3	160.7
Salmon/Saumon	4.3	117.9	5.5	115.8
United States/Etats-Unis	2.6	70.5	3.1	59.0
Canada	2.2	55.4	2.9	62.9
Frozen fillets/Filets congelés	14.0	266.1	13.7	312.4
Cod/Morue	6.2	114.6	5.8	129.7
Plaice/Plie	2.6	62.1	3.1	82.7
Saithe/Lieu noir	2.6	33.9	1.5	24.2
Denmark/Danemark	6.0	114.0	5.9	140.3
Norway/Norvège	5.0	86.3	3.8	82.0
Other prepared frozen fish products/Autres produits de poisson congelé, préparés d)	7.4	127.0	8.6	169.8
Fishsticks/Bâtonnets de poisson	6.6	111.7	7.4	140.7
Norway/Norvège	4.4	67.7	5.1	88.6
Denmark/Danemark	2.6	49.3	2.8	60.2
Salted, dried, smoked/Salés, séchés, fumés b)	11.8	147.0	13.0	195.5
Herring/Hareng	7.0	57.9	8.0	62.5
Roe/Rogue	2.5	36.0	3.5	63.9
Sprat	1.2	6.0	0.4	1.8
Norway/Norvège	5.9	90.2	5.3	104.2
Iceland/Islande	2.8	27.2	4.1	45.3
Denmark/Danemark	1.3	15.0	1.9	19.1
Canned/En conserve c)	10.3	177.0	11.0	197.9
Tuna/Thon	3.3	57.4	4.3	64.2
Mackerel/Maquereau	3.5	42.4	3.1	41.2
Roe/Rogue	1.2	40.2	1.2	51.9
Denmark/Danemark	4.3	57.9	3.9	59.2
Norway/Norvège	1.8	40.0	1.8	44.5
Thailand/Thaïlande	1.7	28.0	2.8	40.0
Shellfish (live, fresh, chilled, salted, dried, frozen and canned)/Coquillages (vivants, frais, sur glace, salés, séchés, congelés et en conserve)	20.7	566.9	20.1	707.7
Prawn/Crevettes	13.8	388.0	14.2	523.1
Crayfish/Ecrevisses	3.5	98.9	2.7	101.6
Crab/Crabe	0.7	27.5	0.8	26.3
Norway/Norvège	6.2	197.3	5.7	259.6
Greenland/Groënland	3.2	75.6	4.7	117.4
Denmark/Danemark	2.9	61.7	2.8	67.2
Turkey/Turquie	2.5	73.2	1.8	76.8
Meal/Farine	112.7	371.0	108.6	331.7
Norway/Norvège	89.0	292.4	56.2	171.2
Denmark/Danemark	16.1	54.8	20.0	64.4
Iceland/Islande	6.7	21.1	28.4	82.4
Oil/Huile	20.1	57.5	10.5	23.0
United States/Etats-Unis	19.2	54.0	8.6	15.9
Other/Autres	27.5	50.0	24.4	49.0
Fish waste/Déchets de poisson	26.9	29.5	23.5	24.0
Denmark/Danemark	22.2	24.4	19.6	20.5
Norway/Norvège	2.0	2.0	1.7	1.4

a) Including fresh fillets/Y compris les filets frais.
b) Including spiced fish/Y compris le poisson assaisonné.
c) Including semi-preserves/Y compris les semi-conserves.
d) Breaded fish fillets, fishsticks, etc./Filets de poisson panés, bâtonnets de poisson, etc.

SWEDEN/SUEDE

Table III(b)/Tableau III(b)

EXPORTS BY MAJOR PRODUCTS AND BY COUNTRY/EXPORTATIONS PAR PRINCIPAUX PRODUITS ET PAR PAYS
1985 & 1986

Quant. : '000 tons/tonnes
Val. : Million Skr/KrS

	1985		1986	
	Quant.	Val.	Quant.	Val.
Total fish and fish products/Total poisson et produits de la pêche	130.8	715.2	113.7	709.9
Fresh, chilled/Frais, sur glace[a]	86.5	362.4	69.9	329.1
Herring/Hareng	49.2	117.5	36.0	94.8
Cod/Morue	22.0	140.4	15.4	127.6
Industrial fish/Poisson industriel	10.2	6.0	13.1	6.6
Denmark/Danemark	71.8	242.5	59.6	224.3
Germany/Allemagne	3.3	43.9	2.6	34.8
Finland/Finlande	2.5	8.2	2.4	11.1
Austria/Autriche	1.8	10.4	2.2	16.8
Frozen, excl. fillets/Congelés, non compris les filets	7.3	47.1	5.1	29.5
Herring/Hareng	4.3	23.3	1.1	4.6
Sprat	1.7	5.4	2.5	8.4
Norway/Norvège	1.6	5.3	1.9	6.9
Denmark/Danemark	0.8	9.5	0.6	7.1
Frozen fillets/Filets congelés	1.8	33.0	1.0	20.4
Cod/Morue	1.5	28.0	0.8	15.4
Other prepared frozen fish products/Autres produits de poisson congelé, préparés[d]	1.1	23.2	0.9	22.3
Salted, dried, smoked/Salés, séchés, fumés[b]	1.3	22.0	1.5	33.6
Canned/En conserve[c]	5.0	94.2	5.3	111.7
Anchovies/Anchois	1.6	31.2	1.7	35.2
Tidbits	2.1	31.9	2.2	39.4
Finland/Finlande	2.7	47.2	2.9	54.6
Norway/Norvège	1.6	27.4	1.6	29.4
Shellfish (live, fresh, chilled, salted, dried, frozen and canned)/ Coquillages (vivants, frais, sur glace, salés, séchés, congelés et en conserve)	1.9	75.0	2.4	103.9
Prawns/Crevettes	0.7	30.5	0.8	46.6
Crayfish/Ecrevisse	1.0	36.7	1.1	46.3
Denmark/Danemark	0.9	33.9	1.0	47.7
Meal/Farine	0.0	0.1	2.9	8.0
Denmark/Danemark	0.0	0.0	2.9	8.0
Oil/Huile	2.4	6.0	2.4	3.6
Denmark/Danemark	1.4	3.4	1.3	1.7
Norway/Norvège	1.0	2.4	1.1	1.9
Other/Autres	23.5	52.2	22.4	47.8
Fishwaste/Déchets de poisson	23.0	24.4		
Finland/Finlande	20.4	25.8	19.3	21.5
Denmark/Danemark	2.7	2.8	2.6	1.7

a) Including fresh fillets and landings abroad/Y compris les filets frais et les débarquements dans les ports étrangers.

b) Including spiced fish/Y compris le poisson assaisonné.

c) Including semi-preserves/Y compris les semi-conserves.

d) Breaded fish fillets, fishsticks, etc./Filets de poisson panés, bâtonnets de poisson, etc.

SWEDEN/SUEDE

Table IV/Tableau IV

SWEDISH FISHING IN FOREIGN ZONES/PECHE SUEDOISE DANS LES ZONES ETRANGERES

(tons/tonnes)

Area/Zone	Country/Pays Group of countries/ Groupe de pays	Species/Especes	1985	1986	1987
North Sea/ Mer du nord	Norway/Norvège	Herring/Hareng	1 000	1 000	1 000
		Cod/Morue	700	475	450
		Haddock/Eglefin	1 930	2 475	1 700
		Whiting/Merlan)			160*
		Saithe/Lieu noir)	700	925	
		Pollack/Lieu jaune)			750
		Mackerel/Maquereau	125	200	200
		Reduction/Réduction	800	800	800
		Shrimp/Crevette)	Traditional fishing		
		Other/Autres)	Pêche traditionnelle		
	EEC/CEE	Herring/Hareng	1 350	1 575	1 600
		Cod/Morue	150	150	150
		Haddock/Eglefin	400	400	400
		Whiting/Merlan	20	20	20
		Ling/Lingue	200	100	-
		Mackerel/Maquereau	-	-	300
		Other/Autres	-	-	150
Baltic Sea/ Mer Baltique	EEC/CEE	Herring/Hareng	1 600	1 810	2 130
		Cod/Morue	1 250	1 000	500
	Finland/Finlande	Cod/Morue)	Traditional fishing		
		Herring/Hareng)	Pêche traditionnelle		
	GDR/RDA	Herring/Hareng	400	300	300
		Cod/Morue	700	600	600
	Poland/Pologne	Cod/Morue	6 000a)	4 500a)	4 203
		Salmon/Saumon	70	60	60
	USSR/URSS	Cod/Morue	4 000	6 000a)	6 000
		Salmon/Saumon	115	115	115

FOREIGN FISHING IN SWEDISH ZONE/PECHE ETRANGERE DANS LA ZONE SUEDOISE

Area/Zone	Country/Pays	Species/Especes	1985	1986	1987
Baltic Sea/ Mer Baltique	EEC/CEE	Herring/Hareng	1 300	2 800b)	3 000b)
		Cod/Morue	3 100	5 600c)	5 160c)
		Salmon/Saumon	20	20	25
	Finland/Finlande	Herring/Hareng)	Traditional fishing		
		Cod/Morue)	Pêche traditionnelle		
	GDR/RDA	Herring/Hareng	2 000	2 000	2 500
	Poland/Pologne	Herring/Hareng	14 800a)	13 920a)	13 250
		Cod/Morue	200	150	-
		Salmon/Saumon	2	20	10
	USSR/URSS	Herring/Hareng	14 200	20 050a)	22 500

a) This includes additions agreed while the Agreement is in force./Ceci comprend les additions qui ont fait l'objet d'une entente au cours de la période couverte par l'Accord.

b) Including 1 500 tons according to the subsidiary protocol to the free trade agreement with the EEC./Comprend 1 500 tonnes en vertu du protocole subsidiaire à l'accord de libre-échange avec la CEE.

c) Including 2 500 tons according to the subsidiary protocol to the free trade agreement with the EEC./Comprend 2 500 tonnes en vertu du protocole subsidiaire à l'accord de libre-échange avec la CEE.

TURKEY/TURQUIE

Table I/Tableau I

LANDINGS AND VALUES/DEBARQUEMENTS ET VALEURS
1985 & 1986

Quant. : tons/tonnes
Val. : TL/LT million

	1985		1986	
	Quant.	Val.	Quant.	Val.
National landings in domestic ports/Débarquements nationaux dans les ports nationaux				
For direct human consumption/Pour la consommation humaine directe	463 929			
Finfish/Poisson				
Anchovy/Ancois	185 074			
Horse mackerel/Chinchard	105 184			
Sardine (Pilchard)	17 693			
Mackerel/Maquereau	22 270			
Whiting/Merlan	17 410			
Atlantic bonito/Bonite atlantique	12 281			
Blue whiting/Merlan bleu	8 383			
Other/Autres	82 943			
Shellfish/Coquillages				
Shrimp/Crevettes	7 517			
Mussels/Moules	2 077			
Other/Autres	3 097			
For other purposes/Pour d'autres buts	114 144			
Anchovy/Anchois	99 502			
Horse mackerel/Chinchard	11 400			
Other/Autres	3 242			
TOTAL	578 073			

TURKEY/TURQUIE

Table II/Tableau II

EXTERNAL TRADE IN FISH AND FISH PRODUCTS/ECHANGES INTERNATIONAUX DE POISSON ET PRODUITS DE LA PECHE
1985 & 1986a)

Quant. : '000 tons/tonnes
Val. : US$

	IMPORTS/IMPORTATIONS				EXPORTS/EXPORTATIONS			
	1985		1986		1985		1986	
	Quant.	Val.	Quant.	Val.	Quant.	Val.	Quant.	Val.
Total fish and fish products/Total poisson et produits de la pêche	357	705.6	1 799	3 455.1	11 444	20 041.2	8 397	22 336.9
Fresh, chilled/Frais, sur glace	194	85.0	446	229.4	7 891	11 669.9	5 755	13 271.3
Frozen/Congelés								
Frozen fillets/Filets congelés								
Salted, dried, smoked/Salés, séchés, fumés								
Canned/En conserve	5	87.0	35	85.3	489	763.6	389	527.2
Shellfish (live, fresh, chilled, salted, dried and canned)/Coquillages (vivants, frais, sur glace, salés, séchés et en conserve)	156	519.2	1 317	3 125.6	3 034	7 573.7	2 251	8 522.5
Meal/Farine								
Oil/Huile								
Other/Autres	1	14.5	1	14.8	29	34.1	3	15.3

a) February to October only/De février à octobre seulement.

TURKEY/TURQUIE

Table III(a)/Tableau III(a)

IMPORTS BY MAJOR PRODUCTS AND BY COUNTRY/IMPORTATIONS PAR PRINCIPAUX PRODUITS ET PAR PAYS
1985 & 1986

Quant. : tons/tonnes
Val. : US$

	1985		1986	
	Quant.	Val.	Quant.	Val.
Total fish and fish products/Total poisson et produits de la pêche	356.9	705.7	1 798.7	3 455.1
Fresh, chilled, frozen, frozen fillets/Frais, sur glace, congelés, filets congelés	194.3	85.1	446.3	229.4
Freshwater fish/Poisson d'eau douce	131.7	54.4	186.0	67.2
Mackerel/Maquereau	37.0	16.6	244.6	142.6
Other sea fish/Autres poissons de mer	25.6	13.9	15.6	19.5
Singapore/Singapour	88.6	30.5	150.8	53.2
Netherlands/Pays-Bas	80.6	40.9	186.8	97.2
Germany/Allemagne	19.8	11.1	3.0	2.8
Other/Autres	5.2	2.4	105.5	76.3
Salted, dried, smoked, canned/Salés, séchés, fumés, en conserve	5.4	87.1	34.6	85.3
Herring/Hareng	–	–	30.8	7.1
Other/Autres	5.4	87.1	3.8	78.2
N. Cyprus/Nord de Chypre	–	–	30.8	7.1
Denmark/Danemark	3.6	53.4	1.5	27.7
Other/Autres	1.8	33.6	2.3	50.5
Shellfish (live, fresh, chilled, salted, dried, canned)/Coquillages (vivants, frais, sur glace, salés, séchés, en conserve)	156.3	519.2	1 317.2	3 125.6
Sea snail/Escargots de mer	156.3	519.2	1 237.2	2 994.2
Shrimp/Crevettes	–	–	14.1	23.8
Other/Autres	–	–	65.8	107.5
Germany/Allemagne	118.9	376.3	967.9	2 374.1
Greece/Grèce	37.4	142.8	213.8	500.1
Other/Autres	–	–	55.5	120.2
Other/Autres				
France	0.7	14.5	0.5	14.8

TURKEY/TURQUIE

Table III(b)/Tableau III(b)

IMPORTS BY MAJOR PRODUCTS AND BY COUNTRY/IMPORTATIONS PAR PRINCIPAUX PRODUITS ET PAR PAYS
1985 & 1986

Quant. : tons/tonnes
Val. : US$

	1985		1986	
	Quant.	Val.	Quant.	Val.
Total fish and fish products/Total poisson et produits de la pêche	11 444.2	20 041.2	8 397.5	22 336.9
Fresh, chilled, frozen, frozen fillets/Frais, sur glace, congelés, filets congelés				
Carp/Carpe	7 891.1	11 669.9	5 755.0	13 271.8
Eel/Anguille	220.1	246.9	275.9	327.4
Atlantic bonito/Bonite atlantique	224.5	757.3	166.8	735.5
Tuna/Thon	33.0	65.6	16.9	45.7
Other/Autres	712.7	2 345.4	731.1	5 624.3
	6 700.7	8 254.6	4 564.2	6 538.9
Greece/Grèce	3 089.7	4 099.2	1 903.9	3 303.6
Italy/Italie	498.4	1 467.3	435.8	1 269.8
Denmark/Danemark	46.2	134.8	27.3	106.1
Germany/Allemagne	353.5	578.7	388.5	590.9
Other/Autres	3 903.4	5 389.8	2 999.5	8 001.4
Salted, dried, smoked, canned/Salés, séchés, fumés, en conserve	489.4	763.6	388.8	527.2
Fresh fish/Poisson frais	1.1	5.1	0.1	0.5
Sea fish/Poisson de mer	488.3	758.5	388.7	526.7
Italy/Italie	210.4	108.4	110.2	65.9
France	151.2	149.4	182.2	143.3
Germany/Allemagne	20.2	35.7	39.0	59.4
Other/Autres	107.6	470.0	57.4	258.5
Shellfish (live, fresh, chilled, salted, dried, canned)/Coquillages (vivants, frais, sur glace, salés, séchés, en conserve)	3 034.6	7 573.7	2 250.9	8 522.5
Sea snail/Escargots de mer	315.9	1 170.5	216.5	2 762.7
Shrimp/Crevettes	342.9	1 356.6	238.2	1 338.4
Crayfish/Ecrevisses	1 541.6	3 844.3	431.6	2 112.8
Other/Autres	834.1	1 202.4	1 364.5	2 308.3
Other/Autres	29.0	34.1	2.7	15.3

UNITED KINGDOM/ROYAUME-UNI

Table I(a)/Tableau I(a)

LANDINGS INTO THE U.K. BY U.K. VESSELS/DEBARQUEMENTS AU ROYAUME-UNI PAR DES NAVIRES DU ROYAUME-UNI
1985 & 1986

Quant. : tons/tonnes (landed weight/poids débarqué)
Val. : £'000

	1985		1986	
	Quant.	Val.	Quant.	Val.
TOTAL LANDINGS/DEBARQUEMENTS TOTAUX	762 053	323 824	716 583	361 368
For direct human consumption/Pour la consommation humaine directe				
Cod/Morue	598 882	255 692	568 331	282 164
Haddock/Eglefin	89 968	69 943	76 409	69 126
Plaice/Plie	123 923	67 517	130 509	79 035
Saithe/Lieu noir	20 292	13 816	21 158	16 040
Whiting/Merlan	13 745	3 863	17 520	6 358
Other demersal/Autres démersaux	49 762	18 988	40 984	18 460
Total demersal/Total démersaux	58 179	51 663	55 451	66 526
	355 869	225 790	342 031	255 532
Herring/Hareng	74 075	10 624	98 620	11 585
Mackerel/Maquereau	159 048	17 967	132 053	14 344
Sprat	7 735	895	3 777	445
Other pelagic/Autres pélagiques	2 155	416	1 971	258
Total pelagic/Total pélagiques	243 013	29 902	226 300	26 632
SHELLFISH/CRUSTACES ET COQUILLAGES	74 805	64 916	87 270	77 210
For other purposes/Pour d'autres buts				
Demersal/Démersaux	88 366	3 216	60 982	1 994
Herring/Hareng	46 385	1 384	41 199	1 184
Mackerel/Maquereau	21 349	869	7 509	296
Sprat	15 175	728	10 094	405
Other pelagic/Autres pélagiques	2 728	105	2 061	101
Total pelagic/Total pélagiques	2 703	126	102	5
Shellfish/Crustacés et coquillages	41 955	1 828	19 766	807
	26	4	17	3

UNITED KINGDOM/ROYAUME-UNI

Table I(b)/Tableau I(b)

UK LANDINGS BY FOREIGN VESSELS/DEBARQUEMENTS DU ROYAUME-UNI PAR DES NAVIRES ETRANGERS
1985 & 1986a)

Quant. : tons/tonnes (landed weight/poids débarqué)
Val. : £'000

	1985		1986	
	Quant.	Val.	Quant.	Val.
TOTAL LANDINGS/DEBARQUEMENTS TOTAUX	52 772	29 466	53 119	26 264
For direct human consumption/Pour la consommation humaine directe	40 198	27 507	37 232	24 777
Cod/Morue	22 397	17 464	17 850	16 180
Haddock/Eglefin	2 866	2 321	2 572	2 254
Plaice/Plie	3 342	2 440	1 246	932
Saithe/Lieu noir	3 048	1 252	2 234	900
Whiting/Merlan	78	29	117	46
Other demersal/Autres démersaux	4 523	3 481	3 692	3 494
Total demersal/Total démersaux	36 254	26 987	27 711	23 806
Herring/Hareng	1 333	199	8 644	877
Mackerel/Maquereau	2 611	321	811	88
Sprat	-	-	-	-
Other pelagic/Autres pélagiques	-	-	65	6
Total pelagic/Total pélagiques	3 944	520	9 521	971
SHELLFISH/CRUSTACES ET COQUILLAGES	1 454	1 428	861	885
For other purposes/Pour d'autres buts	11 120	531	15 026	602
Demersal/Démersaux	175	5	4 654	139
Herring/Hareng	1 088	43	1 054	48
Mackerel/Maquereau	2 618	122	2 166	86
Sprat	8	b)	-	-
Other pelagic/Autres pélagiques	7 231	361	7 153	329
Total pelagic/Total pélagiques	10 945	526	10 373	463
Shellfish/Crustacés et coquillages	-	-	-	-

a) Includes provisional figures for foreign landings into Northern Ireland/Y compris les chiffres provisoires concernant les débarquements étrangers en Irlande du Nord.

b) Less than half a unit./Moins d'une demie unité.

Table I(c)/Tableau I(c)

LANDINGS BY BRITISH VESSELS IN FOREIGN PORTS/
DEBARQUEMENTS PAR LES NAVIRES BRITANNIQUES DANS LES PORTS ETRANGERS
1985 & 1986a)

tons/tonnes (live weight/poids vif)

	1985	1986
Total landings/Débarquements totaux	46 489	35 684
Cod/Morue	1 164	1 111
Haddock/Eglefin	716	602
Hake/Merlu	2 548	1 683
Mackerel/Maquereau	24 559	17 011
Horse mackerel/Chinchard	177	72
Herring/Hareng	4 468	5 896
Blue whiting/merlan bleu	3	95
Saithe/Lieu noir	6 599	6 331
Other/Autres	6 255	2 883

a) Prov.

UNITED KINGDOM/ROYAUME-UNI

Table II/Tableau II

EXTERNAL TRADE IN FISH AND FISH PRODUCTS/ECHANGES INTERNATIONAUX DE POISSON ET PRODUITS DE LA PECHE
1985 & 1986

Quant. : '000 tons/tonnes
Val. : £ million

	IMPORTS/IMPORTATIONS				EXPORTS/EXPORTATIONS			
	1985		1986		1985		1986	
	Quant.	Val.	Quant.	Val.	Quant.	Val.	Quant.	Val.
Total fish and fish products/Total poisson et produits de la pêche	886	727	820	834	304	263	350	329
Fresh, chilled/Frais, sur glace	112	83	113	99	212	85	241	105
Frozen/Congelés	51	51	47	51	16	11	19	14
Frozen fillets/Filets congelés	87	134	96	174	10	17	13	23
Salted, dried, smoked/Salés, séchés, fumés	3	5	3	5	8	19	11	23
Canned/En conserve	86	183	105	220	13	27	11	25
Shellfish (live, fresh, chilled, salted, dried and canned)/Crustacés et coquillages (vivants, frais, sur glace, salés, séchés et en conserve)	46	145	50	200	38	101	44	134
Meal/Farine	236	61	236	57	3	1	6	2
Oil/Huile	265	65	170	28	4	2	5	3
Other/Autres	-	-	-	-	-	-	-	-

UNITED KINGDOM/ROYAUME-UNI

Table III(a)/Tableau III(a)

IMPORTS BY MAJOR PRODUCTS AND BY COUNTRY/IMPORTATIONS PAR PRINCIPAUX PRODUITS ET PAR PAYS
1985 & 1986

Quant. : '000 tons/tonnes
Val. : £ million

	1985			1986	
	Quant.	Val.		Quant.	Val.
Total fish and fish products/ Total poisson et produits de la pêche	886	727	Total fish and fish products Total poisson et produits de la pêche	820	834
Fresh and chilled/ Frais et sur glace	112	83	Fresh and chilled/ Frais et sur glace	113	99
Cod/Morue	46	29	Cod/Morue	50	42
Plaice/Plie	14	9	Plaice/Plie	16	13
Haddock/Eglefin	16	11	Haddock/Eglefin	13	12
Iceland/Islande	44	24	Iceland/Islande	57	47
Irish Rep./Rép. d'Irlande	26	11	Irish Rep./Rép. d'Irlande	22	9
Denmark/Danemark	16	11	Netherlands/Pays-Bas	12	13
Frozen (excl. fillets)/ Congelés (non compris les filets)	51	51	Frozen (excl. fillets)/ Congelés (non compris les filets)	47	51
Cod/Morue	18	11	Cod/Morue	14	12
Salmon/Saumon	5	15	Salmon/Saumon	5	12
Herring/Hareng	4	2	Herring/Hareng	3	1
Canada	8	7	Norway/Norvège	5	6
Norway/Norvège	7	6	Canada	6	5
Netherlands/Pays-Bas	6	4	Iceland/Islande	4	8
Frozen fillets/Filets congelés	87	134	Frozen Fillets/Filets congelés	96	174
Cod/Morue	65	106	Cod/Morue	74	142
Haddock/Eglefin	6	12	Haddock/Eglefin	7	14
Saithe/Lieu noir	5	4	Saithe/Lieu noir	5	5
Norway/Norvège	26	39	Norway/Norvège	28	49
Iceland/Islande	21	31	Iceland/Islande	25	43
Denmark/Danemark	13	22	Denmark/Danemark	19	39
Salted, dried, smoked/ Salés, séchés, fumés	3	5	Salted, dried, smoked/ Salés, séché, fumés	3	5
Cod/Morue	a)	1	Cod/Morue	a)	1
Herring/Hareng	a)	a)	Herring/Hareng	1	a)
Iceland/Islande	1	1	Iceland/Islande	1	1
Canned/En conserve	86	183	Canned/En conserve	105	220
Tuna/Thon	24	46	Tuna/Thon	37	61
Salmon/Saumon	19	72	Salmon/Saumon	28	96
Pilchards	11	8	Pilchards	11	7
Sardines	9	10	Sardines	9	11
Denmark/Danemark	14	24	United States/Etats-Unis	15	44
United States/Etats-Unis	12	43	Canada	11	43
Japan/Japon	11	17	Thailand/Thailande	18	26
			Denmark/Danemark	12	26

UNITED KINGDOM/ROYAUME-UNI

Table III(a) (cont'd)/Tableau III(a) (suite)

IMPORTS BY MAJOR PRODUCTS AND BY COUNTRY/IMPORTATIONS PAR PRINCIPAUX PRODUITS ET PAR PAYS
1985 & 1986

Quant. : '000 tons/tonnes
Val. : £ million

	1985			1986	
	Quant.	Val.		Quant.	Val.
Shellfish (all presentations)/ Coquillages (toutes présentations)	46	145	Shellfish (all presentations)/ Coquillages (toutes présentations)	50	200
Shrimps & prawns/Crevettes	20	61	Shrimps & prawns/Crevettes	26	90
Norway/Norvège	10	36	Denmark/Danemark	8	33
Denmark/Danemark	8	20	Norway/Norvège	5	30
India/Inde	5	15	Iceland/Islande	4	27
			India/Inde	6	19
Meal/Farine	236	61	Meal/Farine	236	57
Other (inc. crustaceans & molluscs)/Autres (y compris crustacés et mollusques)	201	50	Other (inc. crustaceans & molluscs)/Autres (y compris crustacés & mollusques)	218	52
Herring/Hareng	36	11	Herring/Hareng	18	5
Germany/Allemagne	86	20	Germany/Allemagne	117	27
Chile/Chili	52	12	Iceland/Islande	37	10
Iceland/Islande	38	11	Chile/Chili	33	8
Oil/Huile	265	65	Oil/Huile	170	28
Liver/Foie	3	2	Liver/Foie	2	1
Norway/Norvège	72	18	Iceland/Islande	39	8
Iceland/Islande	66	17	Japan/Japon	44	6
United States/Etats-Unis	41	9	Norway/Norvège	23	4
Japan/Japon	31	7	Chile/Chili	21	3

a) Less than 0.5 tonnes or £500/Moins de 0.5 tonne ou £500.

UNITED KINGDOM/ROYAUME-UNI

Table III(b)/Tableau III(b)

EXPORTS BY MAJOR PRODUCTS AND BY COUNTRY/EXPORTATIONS PAR PRINCIPAUX PRODUITS ET PAR PAYS
1985 & 1986

Quant. : '000 tons/tonnes
Val. : £ million

	1985			1986	
	Quant.	Val.		Quant.	Val.
Total fish & fish products/ Total poisson et produits de la pêche	304	263	Total fish & fish products/ Total poisson et produits de la pêche	350	329
Fresh & chilled/ Frais et sur glace	212	85	Fresh & chilled/ Frais et sur glace	241	105
Mackerel/Maquereau	124	15	Mackerel/Maquereau	128	16
Herring/Hareng	55	8	Herring/Hareng	78	9
USSR/URSS	73	9	USSR/URSS	112	12
France	28	36	France	30	47
Netherlands/Pays-Bas	22	10	GDR/RDA	40	4
GDR/RDA	32	4	Netherlands/Pays-Bas	11	13
Frozen (excl. fillets)/ Congelés (non compris les filets)	16	11	Frozen (excl. fillets)/ Congelés (non compris les filets)	19	14
Mackerel/Maquereau	7	2	Mackerel/Maquereau	9	3
France	6	3	France	7	5
Denmark/Danemark	2	1	Germany/Allemagne	2	2
Germany/Allemagne	2	1	Netherlands/Pays-Bas	2	2
Frozen fillets/Filets congelés	10	17	Frozen fillets/Filets congelés	13	23
Haddock/Eglefin	3	8	Haddock/Eglefin	4	9
Whiting/Merlan	2	3	Whiting/Merlan	2	3
Cod/Morue	1	2	Cod/Morue	2	5
United States/Etats-Unis	3	8	United States/Etats-Unis	4	9
Australia/Australie	2	3	France	3	3
France	2	2	Germany/Allemagne	2	2
Salted, dried, smoked/ Salés, séchés, fumés	8	19	Salted, dried, smoked/ Salés, séchés, fumés	11	23
Herring/Hareng	3	3	Herring/Hareng	5	4
France	1	3	France	2	4
Irish Rep./Rép. d'Irlande	1	2	Netherlands/Pays-Bas	2	2
Italy/Italie	1	2	Italy/Italie	1	3
Canned/En conserve	13	27	Canned/En conserve	11	25
Salmon/Saumon	1	4	Salmon/Saumon	1	3
Sardines	1	2	Sardines	1	2
Irish Rep./Rép. d'Irlande	6	13	Irish Rep./Rép. d'Irlande	6	13
Netherlands/Pays-Bas	1	2	Germany/Allemagne	1	1
Australia/Australie	1	1	Netherlands/Pays-Bas	1	1

UNITED KINGDOM/ROYAUME-UNI

Table III(b) (cont'd)/Tableau III(b) (suite)

EXPORTS BY MAJOR PRODUCTS AND BY COUNTRY/EXPORTATIONS PAR PRINCIPAUX PRODUITS ET PAR PAYS
1985 & 1986

Quant. : '000 tons/tonnes
Val. : £ million

	1985			1986	
	Quant.	Val.		Quant.	Val.
Shellfish (all presentations)/ Coquillages (toutes présentations)	38	101	Shellfish (all presentations)/ Coquillages (toutes présentations)	44	134
Shrimps & prawns/Crevettes	9	31	Shrimps & prawns/Crevettes	12	46
Crabs & crayfish/Crabe et écrevisses	9	14	Crabs & crayfish/Crabe et écrevisses	7	11
France	19	40	France	21	51
Spain/Espagne	7	12	Spain/Espagne	9	21
United States/Etats-Unis	2	11	United States/Etats-Unis	3	9
Italy/Italie	2	7	Italy/Italie	3	9
Meal/Farine	3	1	Meal/Farine	6	2
Herring/Hareng	1	a)	Herring/Hareng	2	1
Irish Rep./Rép. d'Irlande	1	a)	Irish Rep./Rép. d'Irlande	2	a)
Oil/Huile	4	2	Oil/Huile	5	3
Liver/Foie	1	1	Liver/Foie	1	1
			Irish Rep./Rép. d'Irlande	3	1

a)   Less than 0.5 tonnes or £500/Moins de 0.5 tonne ou £500.

UNITED STATES/ETATS-UNIS

Table I/Tableau I

LANDINGS AND VALUES/DEBARQUEMENTS ET VALEURS
1985 & 1986

Quant. : tons/tonnes
Val. : '000$

	1985		1986	
	Quant.	Val.	Quant.	Val.
TOTAL LANDINGS/DEBARQUEMENTS TOTAUX	2 838 500	2 326 200	2 735 500	2 762 800
For direct human consumption/Pour la consommation humaine directe :	1 517 500	2 182 800	1 565 700	2 616 700
Finfish/Poisson	1 054 900	1 062 500	1 055 000	1 191 300
Pacific salmon/Saumon du Pacifique	329 500	439 800	298 700	493 900
Tuna/Thon	37 700	52 500	24 800	44 600
Groundfish/Poissons de fond[a]	245 400	122 100	256 100	131 000
Flounders/Flets	88 800	129 100	76 7000	124 600
Others/Autres	353 500	319 000	398 700	387 200
Shellfish, etc./Crustacés et coquillages, etc.	462 600	1 120 300	510 700	1 425 400
Shrimp/Crevettes	151 400	472 900	181 500	662 700
Crab/Crabe	153 100	203 000	161 300	270 100
Scallop meat/Chair de coquille St. Jacques	13 600	93 000	10 100	107 000
Lobster/Homard	23 300	129 200	24 000	139 800
Clam meat/Chair de clam	68 300	128 300	65 900	134 900
Oyster meat/Chair d'huître	20 000	70 100	18 400	78 100
Squid/Calmar	22 200	11 300	34 200	14 600
Other/Autres	10 700	12 500	15 300	18 200
For other purposes/Pour d'autres buts (e.g. reduction to fish meal and oil/réduction en farine et huîle)	1 321 000	143 400	1 169 800	146 100
Menhaden	1 242 600	100 700	1 084 700	93 800
California anchovy/Anchois de Californie	6 600	2 700	6 100	2 500
Other/Autres	71 800	40 000	79 0000	49 800

a) Includes cod, cusk, haddock, hake, ocean perch, pollock, rockfish, whiting and wolffish, excludes joint venture harvests (1985 were 911 000 tons worth $104 million and 1986 harvests were 1.3 million tons worth $155 million)/Y compris morue, brosme, églefin, merlu, sébaste, lieu noir, rascasse, merlan et loup, non compris les prises des entreprises conjointes (1985 : 911 000 tonnes d'une valeur de $104 million et 1986 : 1.3 million tons d'une valeur de $155 million).

Source : U.S. Commercial landings by species, Fisheries of the United States, 1984./Débarquements commerciaux des Etats-Unis par espèces.

UNITED STATES/ETATS-UNIS

Table II(a)/Tableau II(a)

UNITED STATES IMPORTS OF MAJOR FISH PRODUCTS BY MAJOR COUNTRY AND PRODUCT FORM/
IMPORTATIONS AMERICAINES DE PRINCIPAUX PRODUITS DE LA PECHE PAR PRINCIPAUX PAYS ET FORME DE PRODUIT

1985 & 1986

Quant. : '000 tons/tonnes
Val. : US$ million

	1985		1986	
	Quant.	Val.	Quant.	Val.
Albacore/Germon (fresh, frozen/frais, congelé)	64.2	114.1	73.3	117.2
Taiwan	15.4	24.9	37.2	58.9
Japan/Japon	23.0	41.4	12.3	18.8
South Africa/Afrique du sud	7.0	12.9	5.5	9.2
Yellowfin/Albacore (fresh, frozen, whole/frais, congelé, entier)	36.2	35.7	42.8	39.2
Panama	10.8	10.3	14.6	12.0
Venezuela	12.5	12.7	13.0	12.1
Skipjack/Listao (fresh, frozen/frais, congelé)	69.9	54.5	83.4	70.2
France	12.4	10.9	25.4	22.4
Ghana			39.6	15.8
Brazil/Brésil	12.9	10.2	11.3	9.3
Cod/Morue (fresh/fraiche)	14.1	11.7	13.2	14.5
Canada	14.1	11.6	12.9	14.2
Cusk, haddock, hake, pollock/Brosme, eglefin, merlu, lieu noir (fresh/frais)	20.3	16.5	20.3	20.5
Canada	20.3	16.3	20.1	20.2
Halibut/Flétan (fresh/frais)	4.9	18.8	4.1	19.8
Canada	4.8	18.3	4.0	19.3
Salmon/Saumon (fresh/frais)	8.7	56.4	12.9	77.8
Norway/Norvège	6.3	46.3	8.9	62.1
Salmon/Saumon (frozen/congelé)	3.5	19.2	5.5	22.8
Canada	3.4	18.2	5.1	20.4
Swordfish/Espadon (fresh/frais)	3.4	14.9	5.0	28.8
Spain/Espagne	1.3	6.5	1.6	11.3
Taiwan	1.2	4.1	1.4	5.4
Flounder and other flatfish, except halibut/Flet et autres poissons plats, sauf flétan (fresh/frais)	9.0	11.8	10.1	14.5
Canada	8.2	7.5	9.5	10.5
Cod blocks over 10 lb./Blocs de morue de plus de 10 livres	74.6	162.7	78.3	215.9
Canada	33.4	75.0	40.6	112.6
Denmark/Danemark	19.3	41.1	13.9	38.9
Iceland/Islande	15.7	34.4	12.7	36.4
Haddock blocks over 10 lb./Blocs d'eglefin de plus de 10 livres	7.4	17.5	11.1	31.4
Norway/Norvège	1.1	1.9	3.2	8.7
Iceland/Islande	3.3	8.2	2.6	7.8
Denmark/Danemark	1.3	3.1	2.6	7.7
Pollock blocks over 10 lb./Blocs de lieu noir de plus de 10 livres	36.0	43.9	31.6	44.3
Rep. of Korea/Rép. de Corée	18.1	21.7	16.6	23.7
Atlantic ocean perch, fillets/Perche atlantique, filets (fresh, frozen, over quota/fraiche, congelée en plus du contingent)	24.1	56.9	22.2	65.0
Canada	16.3	41.2	16.8	51.2
Iceland/Islande	7.4	14.7	7.9	12.2
Cod fillets/Filets de morue (fresh, over quota/frais, en plus du contingent)	8.5	22.2	9.4	29.8
Canada	8.0	20.8	9.1	29.0
Cod fillets/Filets de morue (frozen, over quota/congelés, en plus du contingent)	76.4	221.4	72.2	237.2
Canada	41.4	114.6	40.5	129.8
Iceland/Islande	22.2	69.2	22.0	74.9
Denmark/Danemark	5.7	18.7	3.8	13.4
Norway/Norvège	3.8	10.4	2.3	7.9
Cusk, haddock, hake, pollock/Brosme, églefin, merlu, lieu noir (frozen, over quota/congelés, en plus du contingent)	26.2	69.3	22.9	67.6
Iceland/Islande	9.2	25.1	7.9	22.6
Canada	6.3	13.9	6.2	16.6
Denmark/Danemark	5.5	17.6	4.1	16.0
Pike pickerel perch, fillets/Perche, filets (frozen/congelée)	2.1	14.3	2.1	15.0
Canada	2.1	14.1	2.0	14.8
Turbot fillets/Filets de turbot (frozen/congelés)	9.7	26.3	4.7	14.8
Canada	3.5	10.9	2.8	9.3
Cod, cusk, haddock, hake, pollock/Morue, brosme, églefin, merlu, lieu noir (smoked, pickled, whole/fumés, au vinaigre, entiers)	11.5	27.4	12.6	34.4
Canada	10.4	24.3	11.9	31.5
Cod, cusk, haddock, hake, pollock/Morue, brosme, églefin, merlu, lieu noir (smoked, pickled, processed/fumés, au vinaigre, transformés)	4.3	13.9	4.1	15.1
Canada	4.2	13.6	3.9	14.6
Pollock/Lieu noir (canned, not in oil/en conserve, pas à l'huile)	6.5	18.4	4.3	13.5
Japan/Japon	6.2	17.6	3.7	11.8
Sardines (canned, not in oil, not over 15 lb./en conserve, pas à l'huile, pas au-delà de 15 livres)	5.8	11.5	6.6	12.7
Japan/Japon	2.3	2.1	2.4	2.6
Norway/Norvège	1.8	6.0	1.7	5.6
Albacore/Germon (canned, not in oil, not over 15 lb./en conserve, pas à l'huile, pas au-delà de 15 livres)	4.8	15.6	4.9	15.4
Japan/Japon	2.3	7.5	1.9	5.9
Taiwan	1.9	6.3	1.8	5.8

UNITED STATES/ETATS-UNIS

Table II(a)(cont'd)/Tableau II(a)(suite)

UNITED STATES IMPORTS OF MAJOR FISH PRODUCTS BY MAJOR COUNTRY AND PRODUCT FORM/
IMPORTATIONS AMERICAINES DE PRINCIPAUX PRODUITS DE LA PECHE PAR PRINCIPAUX PAYS ET FORME DE PRODUIT

1985 & 1986

Quant. : '000 tons/tonnes
Val. : US$ million

	1985		1986	
	Quant.	Val.	Quant.	Val.
Other tuna/Autres thons (canned, not in oil, not over 15 lb./En conserve, pas à l'huile, pas au-delà de 15 livres)	37.8	76.5	34.6	70.5
Thailand/Thaïlande	23.7	48.3	24.2	49.5
Philippines	5.5	10.2	4.5	8.4
Taiwan	7.0	19.0	3.0	6.6
Other tuna/Autres thons (canned, not in oil/En conserve, pas à l'huile)	54.3	116.5	67.6	142.0
Thailand/Thaïlande	31.5	62.7	44.1	87.8
Taiwan	7.0	19.0	8.2	22.1
Philippines	8.4	15.5	8.1	14.5
Japan/Japon	8.9	11.5	2.5	7.8
Anchovy/Anchois (canned, not in oil/en conserve, pas à l'huile)	2.3	11.3	2.6	16.5
Spain/Espagne	1.3	6.8	1.5	10.1
Sardines (canned, smoked, counts/lb/en conserve, fumées)	3.9	13.7	4.0	15.1
Norway/Norvège	3.2	11.8	3.5	13.5
Balls, cakes, pudding/Boulettes, bâtonnets (canned, not in oil, not over 15lb./en conserve, pas à l'huile, pas au-delà de 15 livres)	5.4	15.3	3.3	10.8
Japan/Japon	4.8	13.9	2.4	8.5
Balls, cakes, pudding/Boulettes, bâtonnets (canned, not in oil/en conserve, pas à l'huile) [Surimi based products/Produits à base de surimi]	na	na	2.4	8.0
Japan/Japon	na	na	2.2	7.1
Clams, except razor (fresh, frozen/Frais, congelés)	4.4	12.6	4.6	15.1
Canada	4.1	11.9	4.2	14.3
Crabmeat/Chair de crabe (fresh, frozen/fraiche, congelée)	5.9	48.3	6.6	71.7
Canada	4.2	37.9	4.0	49.8
Crabmeat, except snow crab/Chair de crabe, sauf crabe des neiges	2.8	18.1	3.5	20.2
Thailand/Thaïlande	1.2	6.8	1.2	6.5
Crab, except crabmeat/Crabe, sauf chair de crabe (fresh, frozen/frais, congelé)	4.5	22.7	4.9	22.3
Rep. of Korea/Rép. de Corée	0.2	0.7	1.8	10.2
Canada	2.9	10.1	3.4	7.4
USSR/URSS	0.8	9.2	0.5	0.7
Oysters/Huîtres (canned, smoked/en conserve, fumées)	3.2	13.3	3.6	14.3
Rep. of Korea/Rép. de Corée	2.8	11.4	3.3	13.3
Oysters/Huîtres (canned/en conserve)	9.7	16.5	10.9	18.0
Hong Kong	5.0	5.7	5.4	6.1
Abalone/Ormeau (fresh, frozen, canned/frais, congelé, en conserve)	1.3	14.7	1.2	15.5
Chile/Chili	0.4	2.5	0.4	2.4
Mexico/Mexique	0.3	6.6	0.4	7.6
Lobster/Homard (live, fresh/vivant, frais)	12.5	88.9	13.4	98.3
Canada	12.4	87.1	13.3	97.4
Lobster tails (rock)/Queues de homard (fresh, frozen/frais, congelées)	13.4	260.4	12.5	232.4
Australia/Australie	4.2	100.6	3.5	78.4
Brazil/Brésil	2.3	38.8	1.6	28.9
New Zealand/Nouvelle-Zélande	1.3	28.8	1.3	28.1
South Africa/Afrique du Sud	1.1	29.5	1.0	24.4
Honduras	0.7	10.4	0.9	13.1
Lobster/Homard (fresh, frozen/frais, congelé)	9.2	115.6	10.1	134.2
Canada	3.0	37.5	3.6	51.1
Honduras	1.5	20.0	1.6	21.3
Bahamas	0.7	10.0	1.6	8.8
Scallops/Coquilles St. Jacques (fresh, frozen/fraîches, congelées)	19.1	147.1	21.7	192.6
Canada	5.4	58.6	6.1	67.6
Japan/Japon	4.9	43.6	5.3	50.4
Panama			3.6	22.3
Iceland/Islande	1.8	10.6	2.1	16.5
Shrimp with shell/Crevettes non décortiquées (fresh, frozen/fraîches, congelées)	105.5	866.6	118.9	1 080.1
Ecuador/Equateur	1.9	16.3	27.5	273.6
Mexico/Mexique	26.2	255.4	27.1	270.5
Panama	8.5	66.3	9.3	74.4
China/Chine	2.6	17.9	6.4	48.2
Brazil/Brésil	7.4	52.0	5.9	48.5
Taiwan	3.0	23.4	4.8	42.8
Venezuela	3.0	27.3	4.0	38.1
El Salvador	2.8	17.3	3.4	20.0
Costa Rica	3.3	18.4	3.1	19.3
Thailand/Thaïlande	2.9	22.1	2.5	22.4
Honduras	2.4	18.0	2.5	24.5
Guyana/Guyane	1.0	8.9	2.3	21.6
Pakistan	1.6	9.7	1.8	13.5
Columbia/Colombie	1.4	12.9	1.8	16.9
Philippines	1.8	18.0	1.7	17.9
Peru/Pérou	1.9	16.3	1.6	15.3
French Guyana/Guyane française	1.0	10.1	1.5	17.0
Guatemala	1.7	12.3	1.3	9.8

UNITED STATES/ETATS-UNIS

Table II(a)(cont'd)/Tableau II(a)(suite)

UNITED STATES IMPORTS OF MAJOR FISH PRODUCTS BY MAJOR COUNTRY AND PRODUCT FORM/
IMPORTATIONS AMERICAINES DE PRINCIPAUX PRODUITS DE LA PECHE PAR PRINCIPAUX PAYS ET FORME DE PRODUIT

1985 & 1986

Quant. : '000 tons/tonnes
Val. : US$ million

	1985		1986	
	Quant.	Val.	Quant.	Val.
Shrimp/Crevettes (canned/en conserve)	7.8	32.2	7.1	29.4
Thailand/Thaïlande	4.7	26.3	3.9	16.1
Shrimp peeled Ran/Crevettes décortiquées (fresh, frozen/fraîches, congelées)	35.2	173.3	41.7	221.3
Taiwan	8.5	37.5	7.8	38.3
India/Inde	7.1	25.1	7.2	27.1
Mexico/Mexique	4.3	40.9	6.6	60.3
Thailand/Thaïlande	3.2	15.3	4.0	18.6
Pakistan	2.7	8.2	3.5	12.0
Shrimp peeled other/Autres crevettes décortiquées (fresh, frozen/fraîches, congelées)	14.5	79.3	13.7	102.7
Taiwan	1.5	13.1	2.8	31.3
India/Inde	2.1	7.6	2.5	9.1
Analogue products/Produits analogues	15.3	48.2	15.6	58.5
Japan/Japon	15.0	47.3	14.4	54.9
Total edible fishery products/Total des produits de la pêche comestibles	1 249.2	4 064.3	1 351.2	4 813.5
Total non-edible fishery products/Total des produits de la pêche non comestibles		2 614.3		2 812.8
Grand Total/Total Général		6 678.6		7 626.3

UNITED STATES/ETATS-UNIS

Table II(b)/Tableau II(b)

UNITED STATES EXPORTS OF MAJOR FISH PRODUCTS BY MAJOR COUNTRY AND PRODUCT FORM/
EXPORTATIONS AMERICAINES DE PRINCIPAUX PRODUITS DE LA PECHE PAR PRINCIPAUX PAYS ET FORME DE PRODUIT

1985 & 1986

Quant. : '000 tons/tonnes
Val. : US$ million

	1985		1986	
	Quant.	Val.	Quant.	Val.
Herring/Hareng (fresh, chilled, frozen, whole, eviscerated/frais, sur glace, congelé, entier, éviscéré)	43.4	70.6	38.5	66.2
Japan/Japon	33.8	54.7	28.9	46.8
Rep. of Korea/Rép. de Corée	5.8	10.6	5.2	11.8
Sablefish/Morue charbonnière (fresh, chilled, frozen, whole, eviscerated/fraîche, sur glace, congelée, entière, éviscérée)	7.2	22.1	12.1	36.1
Japan/Japon	7.2	22.1	11.9	35.5
Salmon, chinook/Saumon royale (fresh, chilled, frozen/frais, sur glace, congelé)	2.0	10.2	2.3	11.5
Japan/Japon	1.4	7.5	1.6	8.3
Salmon, chum/Saumon keta (fresh, chilled, frozen/frais, sur glace, congelé)	17.3	47.4	21.3	59.3
Japan/Japon	8.1	20.6	10.9	30.6
France	1.9	5.5	3.0	8.2
United Kingdom/Royaume-Uni			1.6	4.9
Salmon, pink/Saumon rose (fresh, chilled, frozen/frais, sur glace congelé)	24.1	37.3	23.8	43.4
Japan/Japon	14.7	24.6	13.3	25.8
Rep. of Korea/Rép. de Corée	0.6	0.7	4.0	5.5
Canada	6.3	6.5	2.5	4.3
Salmon, silver/Saumon argenté (fresh, chilled, frozen/frais, sur glace congelé)	na	na	11.6	46.5
Japan/Japon	na	na	5.8	21.2
France	na	na	3.6	16.3
Canada	na	na	1.0	3.6
Salmon, sockeye/Saumon rouge (fresh, chilled, frozen/frais, sur glace congelé)	70.1	302.5	65.0	351.4
Japan/Japon	69.1	297.9	63.0	343.1
Salmon, other/Saumon, autres (fresh, chilled, frozen/frais, sur glace congelé)	17.4	64.5	8.4	34.2
Japan/Japon	9.7	34.3	3.8	16.5
France	3.1	14.0	2.2	9.2
Canada	2.4	7.7	1.0	3.2
Halibut/Flétan (fresh, chilled, frozen, whole, eviscerated/frais, sur glace, congelé, entier, éviscéré)	na	na	5.0	18.2
Japan/Japon	na	na	3.1	10.1
Canada	na	na	1.5	6.3
Cod, fillets, steaks, portions/Morue, filets, steaks, morceaux (fresh, frozen/frais, congelés)	na	na	4.3	12.7
Japan/Japon	na	na	3.9	11.3

UNITED STATES/ETATS-UNIS

Table II(b)(cont'd)/Tableau II(b)(suite)

UNITED STATES EXPORTS OF MAJOR FISH PRODUCTS BY MAJOR COUNTRY AND PRODUCT FORM/
EXPORTATIONS AMERICAINES DE PRINCIPAUX PRODUITS DE LA PECHE PAR PRINCIPAUX PAYS ET FORME DE PRODUIT

1985 & 1986

Quant. : '000 tons/tonnes
Val. : US$ million

	1985		1986	
	Quant.	Val.	Quant.	Val.
Fillets, steaks, portions, other than salmon or herring/Filets, steaks, morceaux autres que saumon ou hareng (fresh, chilled, frozen/Frais, sur glace, congelés)	7.9	24.6	9.6	32.1
Japan/Japon	2.3	8.8	3.7	14.5
Canada	1.9	6.0	2.9	9.2
Fish/Poisson (salted, dried, smoked/salé, seché, fumé)	3.3	9.8	0.9	2.9
Japan/Japon	1.1	4.9	0.4	1.5
Salmon, pink/Saumon, rose (canned/en conserve)	9.0	27.5	17.5	52.6
United Kingdom/Royaume-Uni	3.1	9.8	10.2	31.7
Australia/Australie	2.6	7.4	3.0	8.4
Canada	1.6	5.0	1.4	4.7
Salmon, sockeye/Saumon, rouge (canned/en conserve)	10.9	49.4	6.5	38.2
United Kingdom/Royaume-Uni	5.1	24.7	4.4	25.8
Herring roe/Rogue de hareng	6.9	44.6	4.8	19.1
Japan/Japon	3.6	33.4	2.8	13.8
Canada			1.2	2.9
Salmon roe/Rogue de saumon	9.1	66.7	8.7	72.0
Japan/Japon	9.0	65.1	8.5	68.5
Roe, other than salmon, pollock, sea urchin, herring/Rogue, autres que saumon, lieu noir, oursin, hareng	0.7	6.6	1.0	8.6
Taiwan	0.4	3.7	0.5	4.2
Shrimp/Crevettes (fresh, chilled/fraiches, sur glace)	0.9	5.8	1.9	12.8
Canada	0.4	2.1	0.8	4.7
Japan/Japon	0.2	1.9	0.5	4.2
Mexico/Mexique	0.3	1.5	0.4	2.6
Sea urchin/Oursin (fresh, chilled/frais, sur glace)	0.3	5.3	0.6	11.5
Japan/Japon	0.3	5.3	0.5	11.0
King crab/Crabe royale (frozen/congelé)	1.2	10.9	1.8	24.5
Japan/Japon	0.7	6.7	1.3	19.3
Canada	0.3	3.5	0.5	4.7
Shrimp/Crevettes (frozen/congelées)	5.9	42.4	7.2	56.3
Mexico/Mexique	2.4	14.2	2.9	21.5
Canada	2.8	21.5	2.3	18.5
Japan/Japon	0.5	4.9	1.6	12.8
Snow crab (opilio)/Crabe des neiges (frozen/congelé)	na	na	9.7	41.0
Japan/Japon	na	na	7.6	34.2
Snow crab, other/Crabe des neiges, autres (frozen/congelé)	na	na	4.6	32.6
Japan/Japon	na	na	3.4	28.1
Squid (loligo)/Calmar (frozen/congelé)	na	na	1.6	3.0
Spain/Espagne	na	na	0.7	1.5
Squid (other)/Calmar (autres) (frozen/congelé)	na	na	4.9	8.8
Japan/Japon	na	na	1.5	2.6
Norway/Norvège	na	na	0.6	0.8
Menhaden oils/Huile de menhaden	126.2	35.9	85.7	19.8
Netherlands/Pays-Bas	89.2	25.5	57.4	13.5
Belg./Lux.	8.3	2.1	6.5	1.7
United Kingdom/Royaume-Uni	16.4	4.8	6.1	1.4
Fishmeal/Farine de poisson	31.4	7.0	34.9	10.6
Egypt/Egypte	6.2	1.7	19.3	6.9
Germany/Allemagne	15.2	2.9	1.5	0.2
Canada	3.1	0.4	3.4	0.6
Marine shells/Coquillages	7.4	10.2	6.4	10.5
Japan/Japon	6.4	9.2	5.3	9.3
Seaweed/Algues	1.4	11.5	1.9	14.7
Japan/Japon	0.2	1.3	0.3	2.1
Netherlands/Pays-Bas	0.1	0.4	0.2	1.0
Total edible fishery products/Total des produits de la pêche comestibles	294.0	1 013.3	333.4	1 289.8
Total non-edible fishery products/Total des produits de la pêche non-comestibles		73.8		66.3
Grand Total/Total Général		1 084.1		1 356.1

Note: Does not include U.S. flag vessel catches transferred onto foreign vessels in the United States EEZ joint venture operations. 1985 joint venture harvests were 911 000 tons worth $104 million, 1986 joint venture harvests were 1.3 million tons worth $155 million./Non compris les prises des navires américains sur les navires étrangers dans les opérations conjointes de la ZEE des Etats-Unis. Les prises conjointes pour 1985 étaient de 911 000 tonnes d'une valeur de $104 million et celles de 1986 étaien de 1.3 million de tonnes d'une valeur de $155 million.

EEC/CEE

Table I/Tableau I

COMMON ACTION FOR RESTRUCTURING, MODERNISING AND DEVELOPING THE FISHING INDUSTRY AND FOR DEVELOPING AQUACULTURE
ACTION COMMUNE DE RESTRUCTURATION, DE MODERNISATION ET DE DEVELOPPEMENT DU SECTEUR DE LA PECHE ET DU DEVELOPPEMENT DU SECTEUR DE L'AQUACULTURE
1985a) & 1986b)

(Million d'ECUS/ECU)

MEMBER STATES/ETATS MEMBRES	VESSEL BUILDING/CONSTRUCTION DE BATEAUX		AQUACULTURE		ARTIFICIAL STRUCTURES/STRUCTURES ARTIFICIELLES	
	N° of projects/N° de projets	Amount granted/Montant accordé	N° of projects/N° de projets	Amount granted/Montant accordé	N° of projects/N° de projets	Amount granted/Montant accordé
Germany/Allemagne	100	7.7	2	0.3	-	-
Belgium/Belgique	15	0.8	-	-	-	-
Denmark/Danemark	203	5.3	22	1.9	-	-
Spain/Espagne	103	14.7	34	7.3	-	-
France	130	14.1	51	2.3	-	-
Greece/Grèce	136	8.8	-	-	-	-
Ireland/Irlande	77	4.5	12	2.8	-	-
Italy/Italie	294	13.3	11	3.0	6	3.2
Netherlands/Pays-Bas	28	1.0	20	3.4	-	-
Portugal	28	1.0	20	0.4	-	-
United Kingdom/Royaume-Uni	72	7.6	23	2.9	-	-
TOTAL	1 362	90.0	201	24.3	6	3.2

a) Budget year. The actual decisions were taken in 1986./Année d'imputation budgétaire. Les décisions d'octroi sont intervenues en 1986.

b) Budget year and actual decision./Année d'imputation budgétaire et de décision d'octroi.

EEC/CEE

Table II/Tableau II

COMMON ACTION TO IMPROVE THE CONDITIONS UNDER WHICH FISHERY PRODUCTS ARE PROCESSED AND MARKETED/
ACTION COMMUNE D'AMELIORATION DES CONDITIONS DE TRANSFORMATION ET DE COMMERCIALISATION DES PRODUITS DE LA PECHE
(1986)

MEMBER STATES/ ETATS MEMBRES	NUMBER OF PROJECTS/ NOMBRE DE PROJETS	CONTRIBUTION GIVEN IN MILLION OF ECU/ CONCOURS OCTROYE EN MILLIONS D'ECUS
Germany/Allemagne	9	0.9
Belgium/Belgique	2	0.1
Denmark/Danemark	28	3.0
Spain/Espagne	18	4.7
France	12	4.3
Greece/Grèce	7	0.8
Irelande/Irlande	5	1.5
Italy/Italie	11	7.2
Netherlands/Pays-Bas	7	1.2
Portugal	9	3.1
United Kingdom/Royaume-Uni	21	3.4
TOTAL	129	30.2

EEC/CEE

Table III/Tableau III

TOTAL QUOTAS BY MEMBER STATE/CONTINGENTS TOTAUX PAR ETAT MEMBRE[a)]
1986

(tons/tonnes)

SPECIES/ESPECES	BELGIUM/BELGIQUE	DENMARK/DANEMARK	GERMANY/ALLEMAGNE	SPAIN/ESPAGNE	FRANCE	IRELAND/IRLANDE	NETHERLANDS/PAYS-BAS	PORTUGAL	U.K./R.U.	(b)	TOTAL
Cod/Morue	7 010	157 930	56 750	-	28 370	11 720	19 020	-	96 670	-	377 470
	(9 330)	(183 380)	(69 310)	-	(39 820)	(11 520)	(26 820)	-	(128 240)	-	(468 420)
Haddock/Eglefin	2 270	24 980	9 710	-	23 540	4 050	1 560	-	179 520	-	245 630
	(1 960)	(21 450)	(8 430)	-	(21 130)	(4 050)	(1 340)	-	(159 070)	-	(217 430)
Saithe/Lieu noir	120	10 510	28 570	-	87 690	3 200	260	-	26 650	-	157 000
	(90)	(8 390)	(23 230)	-	(74 840)	(3 060)	(210)	-	(22 480)	-	(132 300)
Pollock/Lieu jaune	300	-	-	1 575	10 800	60	-	200	1 550	-	14 485
Whiting/Merlan	3 410	33 000	3 590	2 000	38 720	17 800	7 950	3 000	74 730	-	184 200
	(4 090)	(35 980)	(4 370)	-	(40 190)	(17 800)	(9 670)	-	(86 600)	-	(198 700)
Plaice/Plie	12 885	51 650	10 340	-	7 170	3 295	70 680	-	56 670	-	212 690
	(13 325)	(51 460)	(10 830)	-	(8 020)	(3 295)	(73 620)	-	(59 000)	-	(219 550)
Common sole/Sole commune	4 510	1 280	1 365	495	5 955	650	15 585	1 200	3 190	-	34 230
	(4 035)	(1 360)	(1 500)	-	(5 730)	(555)	(16 980)	-	(3 025)	-	(33 185)
Mackerel/Maquereau	450	10 500	21 830	19 000	15 650	71 250	32 370	5 500	197 150	-	373 700
	(330)	(8 350)	(24 330)	-	(16 330)	(80 000)	(35 330)	-	(220 330)	-	(385 700)
Sprat	30	60 460	1 790	-	350	-	350	-	2 620	87 000	152 600
	(1 400)	(139 270)	(4 730)	-	(2 700)	-	(3 350)	-	(43 000)	-	(194 450)
Horse mackerel/Chinchard	-	-	-	39 000	500	-	-	33 000	-	112 250c)	184 750
	-	-	-	-	-	-	-	-	-	(175 000)	(175 000)
Hake/Merlu	290	2 490	110	32 250	27 310	1 550	220	8 750	5 190	-	78 160
	(250)	(2 340)	(100)	-	(21 480)	(1 300)	(200)	-	(4 430)	-	(30 100)
Anchovies/Anchois	-	-	-	29 800	3 200	-	-	2 000	-	-	35 000
	-	-	-	-	(3 000)	-	-	-	-	-	(3 000)
Norway pout/Tacaud norvégien	-	-	-	-	-	-	-	-	-	300 000c)	300 000
	-	-	-	-	-	-	-	-	-	(340 000)	(340 000)
Blue whiting/Poutassou	-	-	-	30 000	-	-	-	10 000	-	315 000c)	355 000
	-	-	-	-	-	-	-	-	-	(277 000)	(277 000)
Monkfish/Baudroie	3 060	-	630	12 780	28 940	3 060	630	1 990	7 810	-	58 900
	(3 060)	-	(630)	-	(28 930)	(3 060)	(630)	-	(7 810)	-	(44 120)
Megrim/Cardine	390	-	-	17 950	8 710	2 960	-	400	3 450	-	33 860
	(390)	-	-	-	(8 110)	(2 960)	-	-	(3 450)	-	(14 910)
Norway lobster/Langoustine	-	-	-	3 360	13 290	9 300	-	3 000	22 550	-	51 500
Herring/Hareng	9 625	177 595	79 660	-	41 670	38 940	97 990	-	113 150	-	558 630
	(9 120)	(145 980)	(70 770)	-	(37 270)	(32 880)	(87 720)	-	(105 610)	-	(489 350)
Atlantic salmon/Saumon atlantique	-	807	63	-	-	-	-	-	-	-	870
	-	(853)	(67)	-	-	-	-	-	-	-	(920)

a) Quotas for Greece, Italy and Luxembourg are not given./Les quotas pour la Grèce, l'Italie et le Luxembourg n'ont pas été attribués.
b) Available for Member States./Disponibles pour les Etats Membres.
c) Available for Member States with the exception of Spain and Portugal./Disponibles pour les Etats Membres à l'exception de l'Espagne et du Portugal.

Notes: Figures in brackets refer to 1985./Les chiffres entre parenthèses se réfèrent à 1985.

TACs and quotas for 1987 were approved on 22.12.1986 by the Council of Ministers and published in OJ No. L 376 of 31.12.1986./Les TACs et contingents pour 1987 on été approuvés le 22.12.1986 par le Conseil des Ministres et publiés au JO No. L 376 du 31.12.1986.

WHERE TO OBTAIN OECD PUBLICATIONS
OÙ OBTENIR LES PUBLICATIONS DE L'OCDE

ARGENTINA - ARGENTINE
Carlos Hirsch S.R.L.,
Florida 165, 4° Piso,
(Galeria Guemes) 1333 Buenos Aires
Tel. 33.1787.2391 y 30.7122

AUSTRALIA - AUSTRALIE
D.A. Book (Aust.) Pty. Ltd.
11-13 Station Street (P.O. Box 163)
Mitcham, Vic. 3132   Tel. (03) 873 4411

AUSTRIA - AUTRICHE
OECD Publications and Information Centre,
4 Simrockstrasse,
5300 Bonn (Germany)   Tel. (0228) 21.60.45
Gerold & Co., Graben 31, Wien 1 Tel. 52.22.35

BELGIUM - BELGIQUE
Jean de Lannoy,
avenue du Roi 202
B-1060 Bruxelles   Tel. (02) 538.51.69

CANADA
Renouf Publishing Company Ltd/
Éditions Renouf Ltée,
1294 Algoma Road, Ottawa, Ont. K1B 3W8
Tel: (613) 741-4333
Toll Free/Sans Frais:
Ontario, Quebec, Maritimes:
1-800-267-1805
Western Canada, Newfoundland:
1-800-267-1826
Stores/Magasins:
61 rue Sparks St., Ottawa, Ont. K1P 5A6
Tel: (613) 238-8985
211 rue Yonge St., Toronto, Ont. M5B 1M4
Tel: (416) 363-3171

DENMARK - DANEMARK
Munksgaard Export and Subscription Service
35, Nørre Søgade, DK-1370 København K
Tel. +45.1.12.85.70

FINLAND - FINLANDE
Akateeminen Kirjakauppa,
Keskuskatu 1, 00100 Helsinki 10 Tel. 0.12141

FRANCE
OCDE/OECD
Mail Orders/Commandes par correspondance :
2, rue André-Pascal,
75775 Paris Cedex 16
Tel. (1) 45.24.82.00
Bookshop/Librairie : 33, rue Octave-Feuillet
75016 Paris
Tel. (1) 45.24.81.67 or/ou (1) 45.24.81.81
Librairie de l'Université,
12a, rue Nazareth,
13602 Aix-en-Provence   Tel. 42.26.18.08

GERMANY - ALLEMAGNE
OECD Publications and Information Centre,
4 Simrockstrasse,
5300 Bonn   Tel. (0228) 21.60.45

GREECE - GRÈCE
Librairie Kauffmann,
28, rue du Stade, 105 64 Athens Tel. 322.21.60

HONG KONG
Government Information Services,
Publications (Sales) Office,
Information Services Department
No. 1, Battery Path, Central

ICELAND - ISLANDE
Snæbjörn Jónsson & Co., h.f.,
Hafnarstræti 4 & 9,
P.O.B. 1131 - Reykjavik
Tel. 13133/14281/11936

INDIA - INDE
Oxford Book and Stationery Co.,
Scindia House, New Delhi 1 Tel. 331.5896/5308
17 Park St., Calcutta 700016   Tel. 240832

INDONESIA - INDONÉSIE
Pdii-Lipi, P.O. Box 3065/JKT.Jakarta
Tel. 583467

IRELAND - IRLANDE
TDC Publishers - Library Suppliers,
12 North Frederick Street, Dublin 1
Tel. 744835-749677

ITALY - ITALIE
Libreria Commissionaria Sansoni,
Via Lamarmora 45, 50121 Firenze
Tel. 579751/584468
Via Bartolini 29, 20155 Milano Tel. 365083
Editrice e Libreria Herder,
Piazza Montecitorio 120, 00186 Roma
Tel. 6794628
Libreria Hœpli,
Via Hœpli 5, 20121 Milano   Tel. 865446
Libreria Scientifica
Dott. Lucio de Biasio "Aeiou"
Via Meravigli 16, 20123 Milano Tel. 807679
Libreria Lattes,
Via Garibaldi 3, 10122 Torino   Tel. 519274
La diffusione delle edizioni OCSE è inoltre assicurata dalle migliori librerie nelle città più importanti.

JAPAN - JAPON
OECD Publications and Information Centre,
Landic Akasaka Bldg., 2-3-4 Akasaka,
Minato-ku, Tokyo 107   Tel. 586.2016

KOREA - CORÉE
Kyobo Book Centre Co. Ltd.
P.O.Box: Kwang Hwa Moon 1658,
Seoul   Tel. (REP) 730.78.91

LEBANON - LIBAN
Documenta Scientifica/Redico,
Edison Building, Bliss St.,
P.O.B. 5641, Beirut   Tel. 354429-344425

MALAYSIA - MALAISIE
University of Malaya Co-operative Bookshop Ltd.,
P.O.Box 1127, Jalan Pantai Baru,
Kuala Lumpur   Tel. 577701/577072

NETHERLANDS - PAYS-BAS
Staatsuitgeverij
Chr. Plantijnstraat, 2 Postbus 20014
2500 EA S-Gravenhage   Tel. 070-789911
Voor bestellingen: Tel. 070-789880

NEW ZEALAND - NOUVELLE-ZÉLANDE
Government Printing Office Bookshops:
Auckland: Retail Bookshop, 25 Rutland Stseet,
Mail Orders, 85 Beach Road
Private Bag C.P.O.
Hamilton: Retail: Ward Street,
Mail Orders, P.O. Box 857
Wellington: Retail, Mulgrave Street, (Head Office)
Cubacade World Trade Centre,
Mail Orders, Private Bag
Christchurch: Retail, 159 Hereford Street,
Mail Orders, Private Bag
Dunedin: Retail, Princes Street,
Mail Orders, P.O. Box 1104

NORWAY - NORVÈGE
Tanum-Karl Johan
Karl Johans gate 43, Oslo 1
PB 1177 Sentrum, 0107 Oslo 1 Tel. (02) 42.93.10

PAKISTAN
Mirza Book Agency
65 Shahrah Quaid-E-Azam, Lahore 3 Tel. 66839

PORTUGAL
Livraria Portugal,
Rua do Carmo 70-74, 1117 Lisboa Codex
Tel. 360582/3

SINGAPORE - SINGAPOUR
Information Publications Pte Ltd
Pei-Fu Industrial Building,
24 New Industrial Road No. 02-06
Singapore 1953   Tel. 2831786, 2831798

SPAIN - ESPAGNE
Mundi-Prensa Libros, S.A.,
Castelló 37, Apartado 1223, Madrid-28001
Tel. 431.33.99
Libreria Bosch, Ronda Universidad 11,
Barcelona 7   Tel. 317.53.08/317.53.58

SWEDEN - SUÈDE
AB CE Fritzes Kungl. Hovbokhandel,
Box 16356, S 103 27 STH,
Regeringsgatan 12,
DS Stockholm   Tel. (08) 23.89.00
Subscription Agency/Abonnements:
Wennergren-Williams AB,
Box 30004, S104 25 Stockholm Tel. (08)54.12.00

SWITZERLAND - SUISSE
OECD Publications and Information Centre,
4 Simrockstrasse,
5300 Bonn (Germany)   Tel. (0228) 21.60.45
Librairie Payot,
6 rue Grenus, 1211 Genève 11
Tel. (022) 31.89.50
United Nations Bookshop/
Librairie des Nations-Unies
Palais des Nations,
1211 - Geneva 10
Tel. 022-34-60-11 (ext. 48 72)

TAIWAN - FORMOSE
Good Faith Worldwide Int'l Co., Ltd.
9th floor, No. 118, Sec.2
Chung Hsiao E. Road
Taipei   Tel. 391.7396/391.7397

THAILAND - THAILANDE
Suksit Siam Co., Ltd.,
1715 Rama IV Rd.,
Samyam Bangkok 5   Tel. 2511630

TURKEY - TURQUIE
Kültur Yayinlari Is-Türk Ltd. Sti.
Atatürk Bulvari No: 191/Kat. 21
Kavaklidere/Ankara   Tel. 25.07.60
Dolmabahce Cad. No: 29
Besiktas/Istanbul   Tel. 160.71.88

UNITED KINGDOM - ROYAUME-UNI
H.M. Stationery Office,
Postal orders only: (01)211-5656
P.O.B. 276, London SW8 5DT
Telephone orders: (01) 622.3316, or
Personal callers:
49 High Holborn, London WC1V 6HB
Branches at: Belfast, Birmingham,
Bristol, Edinburgh, Manchester

UNITED STATES - ÉTATS-UNIS
OECD Publications and Information Centre,
2001 L Street, N.W., Suite 700,
Washington, D.C. 20036 - 4095
Tel. (202) 785.6323

VENEZUELA
Libreria del Este,
Avda F. Miranda 52, Aptdo. 60337,
Edificio Galipan, Caracas 106
Tel. 32.23.01/33.26.04/31.58.38

YUGOSLAVIA - YOUGOSLAVIE
Jugoslovenska Knjiga, Knez Mihajlova 2,
P.O.B. 36, Beograd   Tel. 621.992

Orders and inquiries from countries where Distributors have not yet been appointed should be sent to:
OECD, Publications Service, Sales and Distribution Division, 2, rue André-Pascal, 75775 PARIS CEDEX 16.

Les commandes provenant de pays où l'OCDE n'a pas encore désigné de distributeur peuvent être adressées à :
OCDE, Service des Publications. Division des Ventes et Distribution. 2. rue André-Pascal. 75775 PARIS CEDEX 16.

71055-09-1987

OECD PUBLICATIONS, 2, rue André-Pascal, 75775 PARIS CEDEX 16 - No. 44159 1987
PRINTED IN FRANCE
(53 87 01 1) ISBN 92-64-13026-8